LONDON MATHEMATICAL SOCIETY LECTURE NOTE SERIES

Managing Editor: Professor J.W.S. Cassels, Department of Pure Mathematics and Mathematical Statistics, University of Cambridge, 16 Mill Lane, Cambridge CB2 1SB, England

The titles below are available from booksellers, or, in case of difficulty, from Cambridge University Press.

46 p-adic analysis: a short course on recent work, N. KOBLITZ
50 Commutator calculus and groups of homotopy classes, H.J. BAUES
59 Applicable differential geometry, M. CRAMPIN & F.A.E. PIRANI
66 Several complex variables and complex manifolds II, M.J. FIELD
69 Representation theory, I.M. GELFAND *et al*
86 Topological topics, I.M. JAMES (ed)
87 Surveys in set theory, A.R.D. MATHIAS (ed)
88 FPF ring theory, C. FAITH & S. PAGE
89 An F-space sampler, N.J. KALTON, N.T. PECK & J.W. ROBERTS
90 Polytopes and symmetry, S.A. ROBERTSON
92 Representation of rings over skew fields, A.H. SCHOFIELD
93 Aspects of topology, I.M. JAMES & E.H. KRONHEIMER (eds)
94 Representations of general linear groups, G.D. JAMES
96 Diophantine equations over function fields, R.C. MASON
97 Varieties of constructive mathematics, D.S. BRIDGES & F. RICHMAN
98 Localization in Noetherian rings, A.V. JATEGAONKAR
99 Methods of differential geometry in algebraic topology, M. KAROUBI & C. LERUSTE
100 Stopping time techniques for analysts and probabilists, L. EGGHE
104 Elliptic structures on 3-manifolds, C.B. THOMAS
105 A local spectral theory for closed operators, I. ERDELYI & WANG SHENGWANG
107 Compactification of Siegel moduli schemes, C.-L. CHAI
109 Diophantine analysis, J. LOXTON & A. VAN DER POORTEN (eds)
110 An introduction to surreal numbers, H. GONSHOR
113 Lectures on the asymptotic theory of ideals, D. REES
114 Lectures on Bochner-Riesz means, K.M. DAVIS & Y.-C. CHANG
116 Representations of algebras, P.J. WEBB (ed)
118 Skew linear groups, M. SHIRVANI & B. WEHRFRITZ
119 Triangulated categories in the representation theory of finite-dimensional algebras, D. HAPPEL
121 Proceedings of *Groups - St Andrews 1985*, E. ROBERTSON & C. CAMPBELL (eds)
122 Non-classical continuum mechanics, R.J. KNOPS & A.A. LACEY (eds)
128 Descriptive set theory and the structure of sets of uniqueness, A.S. KECHRIS & A. LOUVEAU
129 The subgroup structure of the finite classical groups, P.B. KLEIDMAN & M.W. LIEBECK
130 Model theory and modules, M. PREST
131 Algebraic, extremal & metric combinatorics, M.-M. DEZA, P. FRANKL & I.G. ROSENBERG (eds)
132 Whitehead groups of finite groups, ROBERT OLIVER
133 Linear algebraic monoids, MOHAN S. PUTCHA
134 Number theory and dynamical systems, M. DODSON & J. VICKERS (eds)
135 Operator algebras and applications, 1, D. EVANS & M. TAKESAKI (eds)
137 Analysis at Urbana, I, E. BERKSON, T. PECK, & J. UHL (eds)
138 Analysis at Urbana, II, E. BERKSON, T. PECK, & J. UHL (eds)
139 Advances in homotopy theory, S. SALAMON, B. STEER & W. SUTHERLAND (eds)
140 Geometric aspects of Banach spaces, E.M. PEINADOR & A. RODES (eds)
141 Surveys in combinatorics 1989, J. SIEMONS (ed)
144 Introduction to uniform spaces, I.M. JAMES
145 Homological questions in local algebra, JAN R. STROOKER
146 Cohen-Macaulay modules over Cohen-Macaulay rings, Y. YOSHINO
148 Helices and vector bundles, A.N. RUDAKOV *et al*
149 Solitons, nonlinear evolution equations and inverse scattering, M. ABLOWITZ & P. CLARKSON
150 Geometry of low-dimensional manifolds 1, S. DONALDSON & C.B. THOMAS (eds)
151 Geometry of low-dimensional manifolds 2, S. DONALDSON & C.B. THOMAS (eds)
152 Oligomorphic permutation groups, P. CAMERON
153 L-functions and arithmetic, J. COATES & M.J. TAYLOR (eds)
155 Classification theories of polarized varieties, TAKAO FUJITA
156 Twistors in mathematics and physics, T.N. BAILEY & R.J. BASTON (eds)
158 Geometry of Banach spaces, P.F.X. MÜLLER & W. SCHACHERMAYER (eds)
159 Groups St Andrews 1989 volume 1, C.M. CAMPBELL & E.F. ROBERTSON (eds)
160 Groups St Andrews 1989 volume 2, C.M. CAMPBELL & E.F. ROBERTSON (eds)
161 Lectures on block theory, BURKHARD KÜLSHAMMER
162 Harmonic analysis and representation theory, A. FIGA-TALAMANCA & C. NEBBIA
163 Topics in varieties of group representations, S.M. VOVSI
164 Quasi-symmetric designs, M.S. SHRIKANDE & S.S. SANE
166 Surveys in combinatorics, 1991, A.D. KEEDWELL (ed)
168 Representations of algebras, H. TACHIKAWA & S. BRENNER (eds)

169 Boolean function complexity, M.S. PATERSON (ed)
170 Manifolds with singularities and the Adams-Novikov spectral sequence, B. BOTVINNIK
171 Squares, A.R. RAJWADE
172 Algebraic varieties, GEORGE R. KEMPF
173 Discrete groups and geometry, W.J. HARVEY & C. MACLACHLAN (eds)
174 Lectures on mechanics, J.E. MARSDEN
175 Adams memorial symposium on algebraic topology 1, N. RAY & G. WALKER (eds)
176 Adams memorial symposium on algebraic topology 2, N. RAY & G. WALKER (eds)
177 Applications of categories in computer science, M. FOURMAN, P. JOHNSTONE & A. PITTS (eds)
178 Lower K- and L-theory, A. RANICKI
179 Complex projective geometry, G. ELLINGSRUD *et al*
180 Lectures on ergodic theory and Pesin theory on compact manifolds, M. POLLICOTT
181 Geometric group theory I, G.A. NIBLO & M.A. ROLLER (eds)
182 Geometric group theory II, G.A. NIBLO & M.A. ROLLER (eds)
183 Shintani zeta functions, A. YUKIE
184 Arithmetical functions, W. SCHWARZ & J. SPILKER
185 Representations of solvable groups, O. MANZ & T.R. WOLF
186 Complexity: knots, colourings and counting, D.J.A. WELSH
187 Surveys in combinatorics, 1993, K. WALKER (ed)
188 Local analysis for the odd order theorem, H. BENDER & G. GLAUBERMAN
189 Locally presentable and accessible categories, J. ADAMEK & J. ROSICKY
190 Polynomial invariants of finite groups, D.J. BENSON
191 Finite geometry and combinatorics, F. DE CLERCK *et al*
192 Symplectic geometry, D. SALAMON (ed)
193 Computer algebra and differential equations, E. TOURNIER (ed)
194 Independent random variables and rearrangement invariant spaces, M. BRAVERMAN
195 Arithmetic of blowup algebras, WOLMER VASCONCELOS
196 Microlocal analysis for differential operators, A. GRIGIS & J. SJÖSTRAND
197 Two-dimensional homotopy and combinatorial group theory, C. HOG-ANGELONI, W. METZLER & A.J. SIERADSKI (eds)
198 The algebraic characterization of geometric 4-manifolds, J.A. HILLMAN
199 Invariant potential theory in the unit ball of $\mathbf{C}^n$, MANFRED STOLL
200 The Grothendieck theory of dessins d'enfant, L. SCHNEPS (ed)
201 Singularities, JEAN-PAUL BRASSELET (ed)
202 The technique of pseudodifferential operators, H.O. CORDES
203 Hochschild cohomology of von Neumann algebras, A. SINCLAIR & R. SMITH
204 Combinatorial and geometric group theory, A.J. DUNCAN, N.D. GILBERT & J. HOWIE (eds)
205 Ergodic theory and its connections with harmonic analysis, K. PETERSEN & I. SALAMA (eds)
206 An introduction to noncommutative differential geometry and its physical applications, J. MADORE
207 Groups of Lie type and their geometries, W.M. KANTOR & L. DI MARTINO (eds)
208 Vector bundles in algebraic geometry, N.J. HITCHIN, P. NEWSTEAD & W.M. OXBURY (eds)
209 Arithmetic of diagonal hypersurfaces over finite fields, F.Q. GOUVÊA & N. YUI
210 Hilbert C*-modules, E.C. LANCE
211 Groups 93 Galway / St Andrews I, C.M. CAMPBELL *et al*
212 Groups 93 Galway / St Andrews II, C.M. CAMPBELL *et al*
214 Generalised Euler-Jacobi inversion formula and asymptotics beyond all orders, V. KOWALENKO, N.E. FRANKEL, M.L. GLASSER & T. TAUCHER
215 Number theory 1992–93, S. DAVID (ed)
216 Stochastic partial differential equations, A. ETHERIDGE (ed)
217 Quadratic forms with applications to algebraic geometry and topology, A. PFISTER
218 Surveys in combinatorics, 1995, PETER ROWLINSON (ed)
220 Algebraic set theory, A. JOYAL & I. MOERDIJK
221 Harmonic approximation, S.J. GARDINER
222 Advances in linear logic, J.-Y. GIRARD, Y. LAFONT & L. REGNIER (eds)
223 Analytic semigroups and semilinear initial boundary value problems, KAZUAKI TAIRA
224 Computability, enumerability, unsolvability, S.B. COOPER, T.A. SLAMAN & S.S. WAINER (eds)
225 A mathematical introdcution to string theory, S. ALBEVERIO, J. JOST, S. PAYCHA, S. SCARLATTI
226 Novikov conjectures, index theorems and rigidity I, S. FERRY, A. RANICKI & J. ROSENBERG (eds)
227 Novikov conjectures, index theorems and rigidity II, S. FERRY, A. RANICKI & J. ROSENBERG (eds)
228 Ergodic theory of $\mathbf{Z}^d$ actions, M. POLLICOTT & K. SCHMIDT (eds)
229 Ergodicity for infinite dimensional systems, G. DA PRATO & J. ZABCZYK
230 Prolegomena to a middlebrow arithmetic of curves of genus 2, J.W.S. CASSELS & E.V. FLYNN
231 Semigroup theory and its applications, K.H. HOFMANN & M.W. MISLOVE (eds)
232 The descriptive set theory of Polish group actions, H. BECKER & A.S. KECHRIS
233 Finite fields and applications, S. COHEN & H. NIEDERREITER (eds)
234 Introduction to subfactors, V. JONES & V.S. SUNDER
235 Number theory 1993–94, S. DAVID (ed)
236 The James forest, H. FETTER & B. GAMBOA DE BUEN
237 Sieve methods, exponential sums, and their applications in number theory, G.R.H. GREAVES, G. HARMAN & M.N. HUXLEY (eds)
238 Representation theory and algebraic geometry, A. MARTSINKOVSKY & G. TODOROV (eds)
239 Clifford algebras and spinors, P. LOUNESTO
240 Stable groups, FRANK O. WAGNER

London Mathematical Society Lecture Note Series. 236

The James Forest

Helga Fetter
CIMAT, Mexico

Berta Gamboa de Buen
CIMAT, Mexico

CAMBRIDGE
UNIVERSITY PRESS

PUBLISHED BY THE PRESS SYNDICATE OF THE UNIVERSITY OF CAMBRIDGE
The Pitt Building, Trumpington Street, Cambridge CB2 1RP, United Kingdom

CAMBRIDGE UNIVERSITY PRESS
The Edinburgh Building, Cambridge, CB2 2RU, United Kingdom
40 West 20th Street, New York, NY 10011-4211, USA
10 Stamford Road, Oakleigh, Melbourne 3166, Australia

First published 1997

Printed in the United Kingdom at the University Press, Cambridge

A catalogue record for this book is available from the British Library

ISBN 0 521 58760 3 paperback

Im Wald und auf der Heide,
da such ich meine Freude,
ich bin ein Jaegersmann.

Wilhelm Bornemann

CONTENTS

Preface 1

Chapter 1. Preliminaries 4

Chapter 2. The James space J 11

2.a Definition and fundamental properties of J 12

2.b Finite representability 17

2.c Complemented subspaces of J and the space $\mathfrak{J}$ 29

2.d Basic sequences and the primarity of J 44

2.e Isometries of the space J 61

2.f J as a conjugate space 72

2.g The dual of the James space 83

2.h The Banach-Saks properties and the spreading models of J and J^* 92

2.i J^* has cotype 2 102

2.j Appendix: π_λ-spaces, bounded approximation property and finite dimensional decompositions 116

2.k Appendix: $\mathrm{Lip}_\alpha([0,1],X)$ and $L_p(X)$ 126

(i) An isomorphism between $\ell_\infty(x)$ and $\mathrm{Lip}_\alpha([0,1],X)$ 126

(ii) About the space $L_p(X)$ of Bochner integrable functions 129

Chapter 3. The James tree space JT 134

3.a The space JT 134

3.b The fixed point property 161

3.c The conjugates of JT 167

3.d The norms of JT and JT^* have the Kadec-Klee property . . . 182

Chapter 4. What else is there about J and JT? 214

4.a More about J 214

4.b More about JT 217

4.c Generalizations of J 220

4.d Generalizations of JT 227

Chapter 5. Other pathological spaces 231

References 237

Index of citations 245

List of special symbols 249

Subject index 252

FOREWORD

As I believe is well known, I did not anticipate the wealth of mathematics that has resulted from the introduction of the spaces J and JT. In fact, the discovery of J was somewhat accidental. I had proved that a Banach space X is reflexive if each linear functional attains its supremum on the unit ball for any equivalent norm and if X has a basis with certain properties. This theorem lost interest when Victor Klee proved it without any assumption about a basis (Klee's theorem lost interest when I proved it with only the assumption that each linear functional attains its supremum on the given unit ball of X, but this theorem did not come easily!). However, the use of properties that had been assumed for the basis led to the realization that X^{**} could be described explicitly if X has a basis with a certain property (later called *shrinking* by M.M. Day). The definition of the space J isomorphic to J^{**} then came very easily. At a research conference about 24 years later, Charles Stegall asked what I thought about the conjecture that X has a subspace isomorphic with ℓ_1 if X is separable and X^{**} is not separable. He had asked this question the year before, but I had no ideas at that time. But this time I had been working on some other things that made the idea for JT come rather easily.

One always feels great pleasure when others discover applications of something one has done. Thus I feel deep gratitude for the work done by Helga Fetter Nathansky and Berta Gamboa de Buen in preparing this account of the mathematics that has developed from J and JT.

Robert C. James

PROLOGUE

When Stefan Banach introduced in the 30's the spaces which now carry his name, his aim was to provide a convenient framework for the solution of equations in infinitely many variables. Few examples of such spaces were known at that time: sequence spaces, function spaces.

The structure of Banach spaces was not as rich as that of Hilbert spaces (the inner product was missing), but it was general enough to handle a large variety of situations. The distinction was made, at an early stage, between reflexive and non-reflexive Banach spaces. The former enjoyed weak compactness properties, a tool which in many cases could replace inner products.

So Banach spaces developed smoothly, and many general theorems were proved, first by Banach himself, then by many others, for instance Steinhaus, Saks, and later Grothendieck (1950), Dvoretzky (1963).

The last open question, at this stage of the theory, was the existence of a basis, for separable spaces: it would have been nice to have some replacement tool for the so-convenient hilbertian basis, and the easy expansions it allows. The question was not too embarrassing, however: all known spaces had bases, and, despite the lack of success of Grothendieck on this question, one thought that some young and talented guy would soon come to settle the matter.

Unfortunately, before anyone could do it, and before the logicians could prove it undecidable, Per Enflo, in 1972, constructed an example of a separable Banach space with no basis. The opening of Pandora's box had awful consequences, and a lot of unexpected devils flew away: even the most ordinary spaces showed signs of disease, with pathological topologies and strange subspaces. Then the disease started spreading, and strange spaces started to show up: spaces with too few subspaces, or conversely too many, or just not the right ones, those which any civilized person would have expected.

Among the most horrible constructions, we cite those of R.C. James - the topic of the present book - who built a space isomorphic to its second dual, without being reflexive, and of B.S. Tsirelson, who created a space with only strange subspaces: none of them contained ℓ_p or c_0. The present author contributed to the general hysteria, by creating a space which had all the bad properties of both James's and Tsirelson's spaces, without enjoying any of the good ones.

Is the box going to close, and shall we see - as the legend wants - Hope leaving last? We don't know: we have not seen it yet.

Such a considerable flourish of examples had at least one consequence: everyone got lost. Nobody knew any longer what to expect, and even the most impetuous newcomers could hardly make any conjecture, which, for a mathematician, is a sad situation. The only general structure theorem which has been proved since then was Rosenthal's, dealing with ℓ_1 and weak Cauchy subsequences.

So, in order to describe all these strange things, have a look at the past and a guess at the future, a book was needed. Here it is. It has a major quality: around a single example, James' space and its variations, it presents almost all the deep tools introduced by the Geometry of Banach spaces. It will, moreover, have another benefit: to help the diffusion of the results.

This tremendous activity was confined to a small circle of specialists, and had very little impact on other branches of mathematics. This is unfortunate: the powerful tools which have been created over the last fifty years should have more applications to other fields, such as, for instance, Operator Theory, Harmonic Analysis, Numerical Analysis, Economics; some applications are already presented by Pelczynski [1], and, more recently, by Wojtaszczyk [1]. The topics of the present book, pathological subspaces, should find general applications to Harmonic Analysis (see Varopoulos [1] for a first step in this direction) and to Approximation Theory.

If such a confinement were to last too long, the net effect would be harmful. All the patiently developed material might be forgotten by the next generation, unless it had to use it. As Thomas Gray said in 1742:

Full many a gem of purest ray serene,
The dark unfathomed caves of ocean bear;
Full many a flower is born to blush unseen,
And waste its sweetness in the desert air.

Bernard Beauzamy

PREFACE

The James space J and the James tree space JT were constructed by Robert C. James in 1950 and 1974 respectively, to answer negatively several long standing conjectures in Banach space theory regarding the reflexivity of Banach spaces with enough good properties, such as for example having a basis or a separable dual. Since then these spaces have proved to be counterexamples to many other conjectures and have been the cornerstone for constructing other spaces which have enriched the wealth of existing Banach spaces.

On the other hand, the study of their inherent properties has created new branches in the geometry of Banach spaces, leading to the development of diverse topics such as the theory of quasi-reflexive spaces and the Banach spaces based on binary and other trees; the list of references in the bibliography, exceeding 100 titles, gives an indication of the vast amount of work devoted to the study of the subject, which nonetheless is far from exhausted.

Yet, to the best of our knowledge, a unified account of the theory of James spaces is still lacking. Therefore we think that a monograph on these spaces may prove to be useful for the students of these matters.

Given the size of the subject, a completely self-contained and exhaustive exposition seems impossible; hence a selection of the material was unavoidable. We chose to concentrate on the most classical papers dealing with James spaces; however, for the sake of completeness, we give a brief account of most of the new results in the last two chapters. Also, we intend this work to be accessible to graduate students, and it is for this reason that the proofs we do not give here are to be found in well known books. On the other hand, most of the proofs that are given here come from the original papers, because we feel that this may help the reader to go back to the original sources, although in some instances we make use of later works to simplify them. In every case it is indicated where the proofs can be found, and to spare the reader unnecessary work, a serious effort was made to give enough details so that the arguments can be followed easily.

The book is organized as follows:

In chapter one we specify the prerequisites for reading the book and mention some theorems needed later on. Most of the material of this chapter is now classical and can be found for instance in treatises by Beauzamy [1], Day [1], Lindenstrauss and Tzafriri [1] and Singer [1], cited in the references.

Chapters two and three are the core of the book.

Chapter two is devoted to the study of the James space J and its dual J^*. Here we discuss their most basic characteristics, such as the quasi-reflexivity or their complemented subspaces, giving complete proofs of almost every statement. Among the subjects covered in this chapter we mention the Banach-Saks property, the spreading models of J and the type and cotype of both spaces. We also introduce the important space $\mathfrak{J} = (\sum_n J_n)_{\ell_2}$ and its properties. The chapter includes two appendices with results not directly related to the James spaces, but necessary to complement Sections 2.c and 2.i.

Chapter three is dedicated to a similar study of the James tree space JT. The topics covered in this chapter include the somewhat reflexivity, the primarity and the fixed point property of the norm in JT, as well as the Kadec-Klee property of the norm of JT^*.

The last two chapters are included mainly for completeness, but also as a reference for further study.

In chapter four we state some other results about J and JT that did not fit into the body of chapters two and three, either because of their complexity or because they required much additional elaboration. Also we give a summary of some generalizations of J and JT which have appeared in the literature through the years.

In chapter five we talk about other pathological spaces and their properties. These spaces are not directly related to those of James, but have also been created to solve some of the many questions that have arisen in the geometry of Banach spaces, and are included here so that the reader can get an overview of how things stand today.

The results in these two chapters are included mostly without proofs, because as mentioned, they would have required too much additional material or else they are out of the scope of this monograph; however, for the interested reader, full references for this material are included in the bibliography.

Although we did our best to include all of the relevant results, we are well aware that this monograph only gives a part of the story, determined by our own preferences, knowledge and understanding of the subject, but we hope that the material included in this work gives a good idea of the many and important applications of J and JT in the geometry of Banach spaces.

We would like to thank Professor Robert C. James for his encouraging support and advice, Professor Bernard Beauzamy for his valuable criticism of this monograph and our colleague Fausto Ongay for his worthwhile suggestions.

Finally, we also would like to thank the Facultad de Matemáticas of the Universidad de Zaragoza and the Universidad Complutense de Madrid, where part of this work was written, for their warm hospitality.

CHAPTER 1. PRELIMINARIES

Anfang, bedenk' das Ende!

Kurfürst Georg Wilhelm von Brandenburg

The object of this chapter is to mention the prerequisites necessary for understanding this monograph. All the material in this chapter is classical and is included here for the sake of easy reference, but more complete expositions can be found in the books *Classical Banach Spaces I* by Lindenstrauss and Tzafriri [1], *Introduction to Banach Spaces and their Geometry* by Beauzamy [1] and in *Bases in Banach Spaces I* by Singer [1], to which the reader is referred for more details and proofs.

Besides a course on functional analysis, as constituted for instance by the first five chapters of Rudin [1], the reader will need some basic knowledge on classical Banach spaces and on Schauder bases; specifically he needs to know the properties of shrinking, boundedly complete and unconditional bases, as well as of block basic sequences. To fix some notations and terminology we now recall these definitions:

Definition 1.1. A sequence $\{x_n\}_{n=1}^{\infty}$ in a Banach space X is called a Schauder basis or simply a basis of X if for every $x \in X$ there is a unique sequence of scalars $\{a_n\}_{n=1}^{\infty}$ such that $x = \sum_{n=1}^{\infty} a_n x_n$.

A sequence $\{x_n\}_{n=1}^{\infty}$ which is the Schauder basis of its closed linear span is called a basic sequence.

A basis $\{x_n\}_{n=1}^{\infty}$ of a Banach space X is called unconditional if for every $x \in X$ its expansion $x = \sum_{n=1}^{\infty} a_n x_n$ in terms of the basis converges unconditionally.

The projections $\mathcal{P}_n : X \longrightarrow X$ defined by $\mathcal{P}_n(\sum_{i=1}^{\infty} a_i x_i) = \sum_{i=1}^{n} a_i x_i$ are called the natural projections associated to $\{x_n\}_{n=1}^{\infty}$ and the number $\sup_n \|\mathcal{P}_n\|$, which is finite, is called the basis constant of $\{x_n\}_{n=1}^{\infty}$.

A basis $\{x_n\}_{n=1}^{\infty}$ with basis constant one is called monotone.

Let $\{x_n\}_{n=1}^{\infty}$ be a basis of X. The functionals $x_n^* : X \longrightarrow \mathbb{R}$ given by

$$x_n^*(x_m) = \begin{cases} 1 \text{ if } m = n \\ 0 \text{ if } m \neq n \end{cases}$$

for every $m,n \in \mathbb{N}$ are called the biorthogonal functionals associated to the basis $\{x_n\}_{n=1}^{\infty}$.

Definition 1.2. A basis $\{x_n\}_{n=1}^{\infty}$ of a Banach space X is called boundedly complete, if for every sequence of scalars $\{a_n\}_{n=1}^{\infty}$ such that

$$\sup_k \|\textstyle\sum_{n=1}^{k} a_n x_n\| < \infty,$$

the series $\sum_{n=1}^{\infty} a_n x_n$ converges.

Definition 1.3. Let $\{x_n\}_{n=1}^{\infty}$ be a basis of a Banach space X. If for every $x^* \in X^*$, the norm of $x^*|_{[x_i]_{i=n}^{\infty}}$ tends to zero as n tends to infinity, the basis is called shrinking. Here X^* denotes the dual space of X and $x^*|_{[x_i]_{i=n}^{\infty}}$ the restriction of x^* to the closed linear span $[x_i]_{i=n}^{\infty}$ of $\{x_i\}_{i=n}^{\infty}$.

It can be shown that a basis $\{x_n\}_{n=1}^{\infty}$ is shrinking if and only if the biorthogonal functionals $\{x_n^*\}_{n=1}^{\infty}$ form a basis of X^*.

Definition 1.4. Let $\{x_n\}_{n=1}^{\infty}$ be a basic sequence in a Banach space X. A sequence of non-zero vectors $\{u_j\}_{j=1}^{\infty}$ in X of the form

$$u_j = \textstyle\sum_{n=p_j+1}^{p_{j+1}} a_n x_n$$

with $\{a_n\}_{n=1}^{\infty}$ scalars and $0 \le p_1 < p_2 <\ldots$ an increasing sequence of integers, is called a block basis of $\{x_n\}_{n=1}^{\infty}$.

It is easy to see that a block basis of $\{x_n\}_{n=1}^{\infty}$ is indeed a basic sequence in X and that its basis constant is less than or equal to the basis constant of $\{x_n\}_{n=1}^{\infty}$.

We will now state the basic results that will be used most often in the text. These are enunciated mostly without proof, but for a few exceptions, notably Corollary 1.9 which is one of the key tools we will use in the sequel.

We start with a result about spaces with shrinking bases. In this case the following important theorem gives a nice and useful way to represent their double dual.

Theorem 1.5. Let $\{x_n\}_{n=1}^{\infty}$ be a shrinking monotone basis of a Banach space X and $\{x_n^*\}_{n=1}^{\infty}$ the biorthogonal functionals associated to $\{x_n\}_{n=1}^{\infty}$. Then X^{**} can be identified with the space of all sequences of scalars $\{a_n\}_{n=1}^{\infty}$ such that

$$\sup_n \|\sum_{i=1}^n a_i x_i\| < \infty.$$

This correspondence for every $x^{**} \in X^{**}$ is given by

$$x^{**} \longleftrightarrow (x^{**}(x_1^*), x^{**}(x_2^*), \ldots).$$

The norm of x^{**} is equal to

$$\|x^{**}\| = \sup_n \|\sum_{i=1}^n x^{**}(x_i^*) x_i\|.$$

If $\{x_n\}_{n=1}^{\infty}$ is a shrinking basis then it is easily seen that the set of biorthogonal functionals associated to $\{x_n\}_{n=1}^{\infty}$ is a boundedly complete basis in X^*. The converse of this is also true:

Theorem 1.6. A Banach space X with a (monotone) boundedly complete basis $\{x_n\}_{n=1}^{\infty}$ is (isometrically) isomorphic to the dual of the subspace $Z = [x_n^*]_{n=1}^{\infty}$ of X^*, where $\{x_n^*\}_{n=1}^{\infty}$ is the set of biorthogonal functionals associated to $\{x_n\}_{n=1}^{\infty}$, via the isomorphism $T : X \longrightarrow Z^*$ given by $Tx(z) = z(x)$.

The following theorem classifying reflexive spaces with an unconditional basis is due to James [1].

Theorem 1.7. Let Y be a closed subspace of a Banach space X with an unconditional basis. Then Y is reflexive if and only if Y contains no subspace isomorphic to c_0 or ℓ_1.

Since the James space J neither is reflexive nor contains any subspace isomorphic to c_0 or ℓ_1 (Theorem 2.a.2), this shows that the condition of the existence of an unconditional basis is essential for the conclusion of the previous theorem.

Proposition 1.8, due to Krein, Milman and Rutman, shows that basic sequences are stable in the following sense: if a sequence is close enough to a given basic sequence in a Banach space, the perturbed sequence is also basic and equivalent to the original one, and if the space spanned by the first sequence is complemented, so is the second.

Proposition 1.8. Let $\{x_n\}_{n=1}^{\infty}$ be a basic sequence in a Banach space X with basis constant K such that for every n = 1, 2,...

$$\frac{1}{M} \le \|x_n\| \le M.$$

(i) Let $\{y_n\}_{n=1}^{\infty}$ be a sequence in X with

$$\sum_{n=1}^{\infty}\|x_n - y_n\| < \frac{1}{2KM}.$$

Then $\{y_n\}_{n=1}^{\infty}$ is also a basic sequence in X which is equivalent to $\{x_n\}_{n=1}^{\infty}$ via an isomorphism T with $\|T\| < 2$.

(ii) Assume that there is a projection P from X onto the closed linear span of $\{x_n\}_{n=1}^{\infty}$ which will be denoted by $[x_n]_{n=1}^{\infty}$. Let $\{y_n\}_{n=1}^{\infty}$ be a sequence in X with

$$\sum_{n=1}^{\infty}\|x_n - y_n\| < \frac{1}{2KM\|P\|}.$$

Then $Y = [y_n]_{n=1}^{\infty}$ is complemented in X.

We will prove the next corollary, which is an important application of the previous theorem, since it will be used several times in the course of the text.

Recall that a sequence $\{x_n\}_{n=1}^{\infty}$ is seminormalized if there exists a constant M such that $\frac{1}{M} \le \|x_n\| \le M$ and normalized if $\|x_n\| = 1$.

Corollary 1.9. Let X be a Banach space with a normalized basis $\{x_n\}_{n=1}^{\infty}$ and let $\{y_n\}_{n=1}^{\infty}$ be a seminormalized sequence in X such that for every $m \in \mathbb{N}$ the sequence $\{x_m^*(y_n)\}_{n=1}^{\infty}$ converges to zero. Then there exists a block basic sequence $\{y'_n\}_{n=1}^{\infty}$ of $\{x_n\}_{n=1}^{\infty}$ which is equivalent to a subsequence of $\{y_n\}_{n=1}^{\infty}$.

Proof: Suppose that $\frac{1}{M} \le \|y_n\| \le M$, that K is the basis constant of $\{x_n\}_{n=1}^{\infty}$ and that $y_n = \sum_{i=1}^{\infty} b_i^n x_i$. Let $0 < \varepsilon < \frac{1}{2KM}$ and let $0 < \varepsilon_i$ such that $\sum_{i=1}^{\infty} \varepsilon_i < \varepsilon$. We will proceed by induction:
Let $p_1 = 0$ and $n_1 = 1$. There exists p_2 such that

$$\|y_1 - \sum_{i=1}^{p_2} b_i^1 x_i\| < \varepsilon_1.$$

Let $y'_1 = \sum_{i=1}^{p_2} b_i^1 x_i$. Suppose we have constructed $y'_1,\dots,y'_{k-1}$, $p_1<\dots<p_k$ and $n_1<\dots<n_{k-1}$ such that for $j = 1,\dots,k-1$

$$y'_j = \sum_{i=p_j+1}^{p_{j+1}} b_i^{n_j} x_i \text{ and } \|y_{n_j} - y'_j\| < \varepsilon_j.$$

Let $n_k > n_{k-1}$ be such that $\sum_{i=1}^{p_k} |b_i^{n_k}| < \varepsilon_k/2$. This can be done because $\{x_m^*(y_n)\}_{n=1}^{\infty}$ converges to zero for $m = 1,\dots,p_k$. Now let $p_{k+1} > p_k$ be such that

$$\|\sum_{i=p_{k+1}+1}^{\infty} b_i^{n_k} x_i\| < \varepsilon_k/2.$$

Define $y'_k = \sum_{i=p_k+1}^{p_{k+1}} b_i^{n_k} x_i$. Then $\|y_{n_k} - y'_k\| < \varepsilon_k$ and by Proposition 1.8 we are done.

Another result often used in this work is the following proposition about the associativity and other properties of the ℓ_2 direct sum; for the proof, the reader is referred to Singer [1] and Beauzamy [1].
Recall that if $\{X_n\}_{n=1}^{\infty}$ is a sequence of Banach spaces then

$$(\sum_n X_n)_{\ell_2} = \{x = \{x_n\} : x_n \in X_n \text{ and } \sum_{n=1}^{\infty} \|x_n\|_{X_n}^2 < \infty\},$$

with the norm $\|x\| = (\sum_{n=1}^{\infty} \|x_n\|_{X_n}^2)^{1/2}$.

Proposition 1.10. (a) Let X and Y be Banach spaces; then

$$(\sum_n X \oplus Y)_{\ell_2} \approx (\sum_n X)_{\ell_2} \oplus (\sum_n Y)_{\ell_2},$$

where $X \approx Y$ means that X is isomorphic to Y.

(b) If $\{X_n\}$ and $\{Y_n\}$ are sequences of Banach spaces such that $X_n \approx Y_n$ and there is a constant $C \ge 1$ such that for $n = 1, 2,\dots,$ there are onto isomorphisms $T_n : X_n \longrightarrow Y_n$ with $\|T_n\|\|T_n^{-1}\| \le C$, then there exists an onto isomorphism

$$T : (\Sigma_n X_n)_{\ell_2} \longrightarrow (\Sigma_n Y_n)_{\ell_2}$$

with $\|T\|\,\|T^{-1}\| \leq C$.

(c) If $\{X_n\}$ and $\{Y_n\}$ are sequences of Banach spaces such that Y_n is a closed subspace of X_n and there is a constant $C \geq 1$ such that for $n = 1, 2, \ldots$ there are onto projections $P_n : X_n \longrightarrow Y_n$ with $\|P_n\| \leq C$, then there exists an onto projection $P : (\Sigma_n X_n)_{\ell_2} \longrightarrow (\Sigma_n Y_n)_{\ell_2}$ with $\|P\| \leq C$.

(d) If $\{X_n\}$ is a sequence of Banach spaces then $(\Sigma_n X_n)^*_{\ell_2} \approx (\Sigma_n X^*_n)_{\ell_2}$.

In general a space X is not complemented in X^{**}; however, the dual of every Banach space X is complemented in X^{***}; to see this, we first need the following definition:

Definition 1.11. Let X be a Banach space, let Y be a subspace of X and let Z be a subspace of X^*. We define

$$Y^{\perp} = \{f \in X^* : fy = 0 \quad \text{for every} \quad y \in Y\}$$

and

$$Z_{\perp} = \{x \in X : gx = 0 \quad \text{for every} \quad g \in Z\}.$$

Lemma 1.12. Let X be a Banach space and let j_X denote the canonical embedding of X in X^{**}, then

$$X^{***} = j_{X^*}(X^*) \oplus (j_X(X))^{\perp}.$$

Proof: Let $P : X^{***} \longrightarrow j_{X^*}(X^*)$ be defined as $P = j_{X^*} \circ (j_X)^*$, where $(j_X)^*$ denotes the transpose of j_X. Then it is easy to see that $(j_X)^* \circ j_{X^*} = Id_{X^*}$ and hence P is an onto projection satisfying $(I - P)X^{***} = (j_X(X))^{\perp}$.

The fact that the James space J is of codimension one in its double dual, as will be shown in Chapter two, is one of its most outstanding features. A generalization of this property, called quasi-reflexivity, is presented in Definition 1.13, and Theorem 1.14, due to Civin and Yood

[1], states that, similarly to the property of reflexivity, a space is quasi-reflexive if and only if its dual shares this characteristic.

Definition 1.13. A Banach space X is said to be quasi-reflexive (of order k) if the quotient of X^{**} by the natural image of X in X^{**} has finite dimension (dimension k).

The existence of such spaces for every k was proved by Singer [2] and will be shown in Theorem 2.a.4.

Theorem 1.14. A Banach space X is quasi-reflexive of order n if and only if X^* is quasi-reflexive of order n.

Proof: By Lemma 1.12 $X^{***}/j_{X^*}(X^*)$ is isomorphic to $(j_X(X))^\perp$ and it is a well known fact (see e.g. Beauzamy [1]) that $(j_X(X))^\perp$ is isomorphic to $(X^{**}/j_X(X))^*$.

CHAPTER 2. THE JAMES SPACE J

Nous approchons
Dans les forêts
Prenez la rue du matin
Montez les marches de la brume

Paul Eluard

In the appendix of his pioneering treatise [1], S. Banach stated a number of questions about the properties of Banach spaces, some of which remained unanswered for decades, providing a strong stimulus for research. Among them, those regarding the reflexivity of spaces with "enough good properties" were of special importance. In his basic paper [1] in 1950, James introduced the famous space which now bears his name and is called the J space for short; this space was constructed as a counterexample to a question posed by James himself, regarding the reflexivity of spaces not containing subspaces isomorphic to c_0 or ℓ_1 but proved to be also a counterexample to several of the questions posed by Banach. It soon became a rich source of conjectures, examples and counterexamples which brought about an enormous amount of research around its strange "pathological" behavior.

In this chapter we study the basic properties of J in considerable detail, starting with its fundamental property of quasi-reflexivity and going as far as a description of its isometries. We also treat the dual space J^* which in many ways resembles J and yet is completely different; in fact, as we shall see in Section 2.i, J and J^* are incomparable, that is, they have no isomorphic infinite dimensional non-reflexive subspaces.

The results and their proofs presented here are extracted from numerous sources, but whenever possible and convenient, it is the original proof that is discussed.

2.a. Definition and fundamental properties of J

We start with the definition of J, with the norm originally introduced in James [1] as well as with some equivalent norms of J which are more convenient in order to prove some of its properties.

Definition 2.a.1. J is the space of all real sequences $x = (a_1, a_2, \ldots)$ such that $\lim_{n\to\infty} a_n = 0$ for which

$$\| |x| \| = \sup \left(\sum_{i=1}^{n} (a_{p_{2i-1}} - a_{p_{2i}})^2 \right)^{1/2} < \infty,$$

where the supremum is taken over all choices of n and positive integers $p_1 < p_2 < \ldots < p_{2n}$. It is easy to verify that $(J, \| | \ | \|)$ is a Banach space.

We note that if

$$\|\|x\|\| = \sup \left(\frac{1}{2} \left(\sum_{i=1}^{n} (a_{p_{i+1}} - a_{p_i})^2 + (a_{p_{n+1}} - a_{p_1})^2 \right) \right)^{1/2}$$

$$\text{and } \|x\| = \sup \left(\frac{1}{2} \sum_{i=0}^{n} (a_{p_{i+1}} - a_{p_i})^2 \right)^{1/2},$$

where $a_0 = 0$ and the sup is taken over all choices of n and all positive integers $0 = p_0 < p_1 < \ldots < p_{n+1}$, then $\|\| \ \|\|$ and $\| \ \|$ are equivalent norms to $\| | \ | \|$ on J.

The unit vectors $\{e_n\}_{n=1}^{\infty}$ in J form a monotone basis with respect to all three norms and hence J is separable.

Unless we specify one of the above norms the results we are going to give are true for all of them.

The next theorem by James [1] shows that the space J is a counter-example to two of Banach's questions, namely whether every Banach space with a separable dual must be reflexive and whether every Banach space isomorphic to its double dual is reflexive.

Theorem 2.a.2. J is a separable Banach space such that:

(i) The unit vector basis $\{e_n\}_{n=1}^{\infty}$ is shrinking.

(ii) J is quasi-reflexive of order one, in fact $J^{**} = j(J) \oplus [1]$ where 1 is the functional given by $\mathbb{1}(e_n^*) = 1$ for all n and j is the canonical injection of J into J^{**}.

(iii) $(J, \| \ \|)$ is isomorphic to J^{**}, furthermore $(J, \||\ \||)$ is isometrically isomorphic to J^{**}.

(iv) The successive duals J^*, J^{**},... are separable, and this implies that J cannot have subspaces isomorphic to either c_0 or ℓ_1.

(v) J is not isomorphic to a subspace of a space with an unconditional basis.

Proof: (i) Let $\{e_n\}_{n=1}^{\infty}$ be the unit vector basis. Then $\{e_n\}_{n=1}^{\infty}$ is shrinking, that is $\lim_n \|x^*\|_n = 0$ where $\|x^*\|_n$ is the norm of $x^*|_{[e_i]_{i=n}^{\infty}}$.

Suppose the contrary; then there exist $x^* \in J^*$, $\varepsilon > 0$ and a block basis $\{y_k\}_{k=1}^{\infty}$ of $\{e_n\}_{n=1}^{\infty}$ such that $x^*(y_k) > \varepsilon$ and $\|y_k\| = 1$ for each k, and this gives us that $\sum_{k=1}^{\infty}(y_k/k)$ converges in J, which is a contradiction.

The proof of the last statement is straightforward; in fact each term in the calculation of the norm of $\sum_{k=1}^{\infty}(y_k/k)$ is either of the type $\frac{1}{2}\left(e_j^*(y_k)/k - e_r^*(y_k)/k\right)^2$ or of the type $\frac{1}{2}\left(e_j^*(y_k)/k - e_r^*(y_{k+m})/(k+m)\right)^2$ for some k, m, j and r. Therefore, since

$$\frac{1}{2}\left(e_j^*(y_k)/k - e_r^*(y_{k+m})/(k+m)\right)^2 \le \left((e_j^*(y_k))^2/k^2\right) + (e_r^*(y_{k+m}))^2/(k+m)^2$$

it follows that

$$\|\textstyle\sum_{k=1}^{\infty}(y_k/k)\|^2 \le 2\sum_{k=1}^{\infty}(\|y_k\|^2/k^2) = 2\sum_{k=1}^{\infty}(1/k^2) < \infty.$$

Thus $\{e_n\}_{n=1}^{\infty}$ is shrinking, and hence $\{e_n^*\}_{n=1}^{\infty}$, the set of biorthogonal functionals associated to $\{e_n\}_{n=1}^{\infty}$, is a basis of J^*.

(ii) Since all the vectors $e_1 + e_2 + \ldots + e_n$, $n = 1,2,\ldots$, have norm one, but this sequence does not have a weak limit point in J, it follows that the unit vector basis $\{e_n\}_{n=1}^{\infty}$ is not boundedly complete and that the unit ball of J is not weakly compact; hence J is not reflexive.

By (i) and Theorem 1.5 J^{**} can now be characterized as the set of all sequences $x^{**} = (a_1, a_2,\ldots)$ such that $\sup_n \|\sum_{i=1}^{n} a_i e_i\| < \infty$ and

$$\|x^{**}\| = \sup_n \|\textstyle\sum_{i=1}^{n} a_i e_i\|.$$

Thus let $x^{**} = (a_1, a_2,\ldots) \in J^{**}$ and suppose

$$M = \sup_n \|\sum_{i=1}^{n} a_i e_i\| = \|x^{**}\|;$$

we will prove that $\lim_n a_n = \lambda$ exists. If this is not the case, there exist $\varepsilon > 0$ and a subsequence $\{a_{m_n}\}_{n=1}^{\infty}$ of $\{a_i\}_{i=1}^{\infty}$ such that $|a_{m_{N+1}} - a_{m_N}| > \varepsilon$ for $N = 1, 2,\ldots$. Let $L \in \mathbb{N}$ be such that $L\varepsilon^2 > 2M^2$ and $n > m_{L+1}$, then

$$2M^2 \geq 2\,\|\sum_{i=1}^{n} a_i e_i\|^2 \geq \sum_{N=1}^{L}(a_{m_{N+1}} - a_{m_N})^2 > L\varepsilon^2 > 2M^2,$$

which is a contradiction.

Furthermore, by the definition of the norm in J, if $\lambda = 0$ then the sequence $(a_1, a_2,\ldots) \in J$ and if $\lambda \neq 0$, $x^{**} = (x^{**} - \lambda\mathbb{1}) + \lambda\mathbb{1}$. Therefore J^{**} is the direct sum of J and the space generated by $\mathbb{1}$.

(iii) Let $T : J^{**} \longrightarrow J$ be defined by

$$Tx^{**} = (-\lambda,\ x^{**}(e_1^*) - \lambda,\ x^{**}(e_2^*) - \lambda,\ldots),$$

where $\lambda = \lim_n x^{**}(e_n^*)$.

Due to the definition of $|||\ |||$ and to the form of the norm in J^{**} mentioned above, it is easy to see that T is an isometry.

(iv) This follows immediately from (iii).

(v) Since J cannot have subspaces isomorphic to c_0 or to ℓ_1, by (ii) and Theorem 1.7, (v) follows.

J was the first quasi-reflexive space ever constructed and Singer in [2] made use of it to construct examples of quasi-reflexive spaces of order k for every k (see Definition 1.13). Singer also defined the concept of a k-shrinking basis and in order to show the existence of such a basis for every k, he again resorted to the space J, proving that the canonical basis of $J^* \oplus J^* \oplus \cdots \oplus J^*$ is k-shrinking.

Definition 2.a.3. Let k be a non-negative integer. A basis $\{x_n\}_{n=1}^{\infty}$ of a Banach space X is called k-shrinking if $\text{codim}_{X^*} [x_n^*] = k$.

Observe that the notion of 0-shrinking basis coincides with that of a shrinking basis.

Theorem 2.a.4. Let $\{e_n^*\}_{n=1}^{\infty}$ in J^* be the biorthogonal basis to the basis $\{e_i\}_{i=1}^{\infty}$. This basis is 1-shrinking. Let $\{f_n\}_{n=1}^{\infty} \subset J^* \oplus J^*$ be defined by

$$f_{2n-1} = (e_n^*, 0),\ f_{2n} = (0, e_n^*).$$

Then $\{f_n\}_{n=1}^{\infty}$ is a 2-shrinking basis in $J^* \oplus J^*$. Continuing in this manner we obtain a k-shrinking basis in $J^* \oplus J^* \oplus \cdots \oplus J^*$.

Proof: By Theorem 2.a.2 $\{e_n^*\}_{n=1}^{\infty}$ is a basis of J^* and since for every m, $\langle j(e_n), e_m^*\rangle = e_m^*(e_n) = \delta_m^n$,

$$j(e_n) = (e_n^*)^*.$$

Thus, also by Theorem 2.a.2, $\text{codim}_{J^{**}} [(e_n^*)^*] = 1$ and $\{e_n^*\}_{n=1}^{\infty}$ is 1-shrinking.

The basis $\{f_n\}_{n=1}^{\infty}$ of $J^* \oplus J^*$ is 2-shrinking and so on: indeed by Theorem 2.a.2 $\mathbb{1} = (1, 1, \ldots) \in J^{**}$ and each $x^{**} \in J^{**}$ can be uniquely expressed as $x^{**} = j(x) + \lambda\mathbb{1}$ where $x \in J$, j is the canonical injection of J into J^{**} and $\lambda \in \mathbb{R}$. Hence every $(x^{**}, y^{**}) \in J^{**} \oplus J^{**}$ can be uniquely written as $(j(x), j(y)) + (\lambda\mathbb{1}, \mu\mathbb{1})$ with $x, y \in J$ and $\lambda, \mu \in \mathbb{R}$. Thus

$$\text{codim}_{J^{**} \oplus J^{**}}\ j(J) \oplus j(J) = 2,$$

that is $J \oplus J$ is quasi-reflexive of order 2. Hence the basis $\{f_n\}_{n=1}^{\infty}$ of $J^* \oplus J^*$ is 2-shrinking. For $k > 2$ the procedure is analogous.

Among several questions asked by Banach in his book [1] was the following: is it true that every infinite dimensional Banach space X is isomorphic to $X \oplus X$?

The first ones to give a counterexample to this, were Bessaga and Pelczynski, showing in [1] that if X and Y are quasi-reflexive Banach spaces of order n and m respectively, then $X \oplus Y$ is quasi-reflexive of order $n + m$. The above theorem is a simple proof of this when X and Y are equal to J.

Corollary 2.a.5. $J \oplus J$ is not isomorphic to J.

We now turn our attention to another of Banach's questions, namely the conjecture that every infinite dimensional real Banach space is the real underlying space of some complex Banach space. This was answered, this time by Dieudonné [1], also negatively and also using the James space, proving that J is not the real underlying space of any complex Banach space, as will be seen in Theorem 2.a.9, after we give the necessary concepts for properly understanding this problem.

Definition 2.a.6. Let F be a topological vector space over $\mathbb{C}$, the restriction of scalar multiplication to $\mathbb{R}$ turns F into a topological vector space over $\mathbb{R}$, called the real underlying space of F.

Definition 2.a.7. Let E be a topological vector space over $\mathbb{R}$. A topological vector space E_0 over $\mathbb{C}$ is called a complexification of E if the real underlying space of E_0 coincides with E.

It is known (see e.g. Schaefer [1]) that a topological vector space E over $\mathbb{R}$ has a complexification E_0 if and only if there exists an automorphism u of E such that $u^2(x) = -x$ for all $x \in E$. If u exists then multiplication by i in E_0 is defined as $ix = u(x)$ for every $x \in E$.

Let E be a Banach space over $\mathbb{R}$ with complexification E_0. We denote by E_0^* the dual of E_0 over $\mathbb{C}$, by E_0^{**} the dual of the real underlying space of E_0^* and by E_{00}^{**} the dual of E_0^* over $\mathbb{C}$.

Proposition 2.a.8. Let E be a real Banach space with complexification E_0. Then E^{**}/E is isomorphic to the real underlying space of E_{00}^{**}/E_0.

Proof: Since the element $u(x) \in E$ can be identified with $ix \in E_0$ and since, considered as sets, E is equal to E_0, we see that the function $\phi : E^* \longrightarrow E_0^*$ defined by

$$\phi(f)(x) = f(x) - if(u(x)), \text{ for } f \in E^* \text{ and } x \in E,$$

is an isomorphism from E^* onto the real underlying space of E_0^*. For $g \in E_0^*$ let Reg denote its real part. Then the inverse function ψ of ϕ, $\psi : E_0^* \longrightarrow E^*$, is

$$\psi(g)(x) = \text{Re}g(x) \text{ for } g \in E_0 \text{ and } x \in E.$$

As above, the function $\Phi : E_0^{**} \longrightarrow E_{00}^{**}$ defined by

$$\Phi(G)(h) = G(h) - iG(ih) \text{ for } G \in E_0^{**} \text{ and } h \in E_0$$

is an isomorphism between E_0^{**} and the real underlying space of E_{00}^{**}. Let $\Lambda = \Phi \circ \psi^*$, where $\psi^* : E^{**} \to E_0^{**}$ denotes the transpose of ψ. Then Λ is an isomorphism from E^{**} onto the real underlying space of E_{00}^{**}. Let j and j^0 be the canonical injections from E into E^{**} and E_0 into E_{00}^{**}, respectively. We will see that for $x \in E$, $\Lambda(j(x)) = j^0(x)$, which means that under Λ, E considered as a subspace of E^{**} is transformed into E_0 considered as a subspace of E_{00}^{**}. For that, let $f \in E_0^*$, $x \in E$; then

$$\langle \Lambda(j(x)), f\rangle = \langle \Phi\circ\psi^*(j(x)), f\rangle = \langle \psi^*(j(x)), f\rangle - i\langle \psi^*(j(x)), if\rangle =$$

$$= \langle j(x), \psi(f)\rangle - i\langle j(x), \psi(if)\rangle = \mathrm{Re}f(x) - i\mathrm{Re}\, if(x) = f(x) = \langle j^0(x), f\rangle.$$

This gives us the desired result.

Theorem 2.a.9 J is not the real underlying space of any complex Banach space.

Proof: Suppose there exists a complex Banach space J_0 such that J is the real underlying space of J_0. Then by Proposition 2.a.8 we have that J^{**}/J is isomorphic to the real underlying space of J_{00}^{**}/J_0. By Theorem 2.a.2, J^{**}/J has dimension one over $\mathbb{R}$, but the real underlying space of J_{00}^{**}/J_0 is a vector space whose dimension is either even or infinite over $\mathbb{R}$. This contradiction proves the assertion.

2.b. Finite representability

When trying to prove that J might be a counterexample to the conjecture that a Banach space is reflexive if and only if it is not uniformly ℓ_1^n, Giesy and James established in fact that ℓ_1 is finitely representable in J and that a Banach space X is not uniformly ℓ_1^n if and only if ℓ_1 is not finitely representable in X, that is if and only if X is B-convex. Indeed they showed even more, namely that every Banach space is finitely representable in J, so J is not the desired counterexample; however, James himself found a counterexample to this conjecture (see Chapter 5 **(ii)**).

The important concept of finite representability was introduced, also by James, in [10], and roughly speaking means the following: a space F is finitely representable in a space E, if the finite dimensional subspaces of F can be found in E with a degree of approximation as good as one wishes. Thus, although the spaces E and F may be very different, the finite dimensional subspaces of F can be found in E.

The proof of the finite representability of an arbitrary Banach space in J is done in two steps, the first and more difficult is to show that c_0 is finitely representable in J. It is done here along the lines of Giesy and James [1] whose proof has the interest of being constructive, in the sense that it exhibits an explicit isomorphism between $c_0^{(n)}$ and an n-dimensional subspace of J. This however requires several complicated lemmata which follow after the definition of finite representability and the introduction of some terminology. The second step is then to show that every Banach space is finitely representable in c_0, a result of Bessaga and Pelczynski [1] which is proved here following Bombal [1].

Definition 2.b.1. Let E and F be two Banach spaces. We will say that F is finitely representable in E (F f.r. in E) if for every $\varepsilon > 0$ and every finite dimensional subspace M of F, there exist a finite dimensional subspace N of E, with dimM = dimN, and an isomorphism $T : M \longrightarrow N$, with $\|T\|\|T^{-1}\| \le 1 + \varepsilon$.

Clearly if F is finitely representable in E and E is isomorphic to G, then F is finitely representable in G.

The following technical lemmata will enable us to show that there exist in J subspaces as close as one wishes to the span of the first n elements of the unit vector basis of c_0.

For notational convenience we need the following definition.

Definition 2.b.2. (a) For $x \in$ J and n $\in \mathbb{N}$, let $x(n) = e_n^*(x)$.
(b) $\mathfrak{z}_{2k}$ denotes the sequence given by

$$\mathfrak{z}_{2k}(2n) = 0 \text{ for } n = 0, 1, 2, \ldots,$$

$$\mathfrak{z}_{2k}(2n + 1) = (2k)^{-1/2} \quad \text{for } n = 0, 1, \ldots, k - 1,$$

$$\mathfrak{z}_{2k}(2n + 1) = 0 \text{ for } n \geq k.$$

(c) For any real sequence x and every positive integer n, denote by $T_n x$ the sequence y such that $y(kn + j) = \frac{n - j}{n} x(k) + \frac{j}{n} x(k + 1)$ for $k \geq 0$ and $j = 0, 1, \ldots, n$.

Lemma 2.b.3. Let m be a positive integer and γ, δ, ε_1, ε_2 and N be positive numbers for which $\gamma < \delta$ and

(1) $$\gamma^{1/2}N^{-1/2} + \gamma^{1/2} < \delta^{1/2}, \qquad \varepsilon_1 + 4\gamma N^{-1/2} < \varepsilon_2.$$

If $n > N$, if $x \in J$ has the properties

(i) $x(i) = 0$ if $i \geq 2m$,

(ii) $2\|x\|^2 < \sum_{i=0}^{2m-1} |x(i) - x(i + 1)|^2 + \varepsilon_1$

(iii) $|x(i) - x(i + 1)| < \left(\frac{\gamma}{2m}\right)^{1/2}$ if $i \geq 0$,

and if $y = T_n x$, $z = \gamma^{1/2}\mathfrak{z}_{2mn}$, and $w = y + z$ or $w = y - z$, then

(i') $w(i) = 0$ if $i \geq 2mn$,

(ii') $2\|w\|^2 < \sum_{i=0}^{2mn-1} |w(i) - w(i + 1)|^2 + \varepsilon_2$,

(iii') $|w(i) - w(i + 1)| < \left(\frac{\delta}{2mn}\right)^{1/2}$ if $i \geq 0$.

Proof: Suppose $n > N$. Clearly (i') is satisfied. Also, it follows from (iii), the definition of z, and (1) that

$$|w(i) - w(i + 1)| \leq \frac{1}{n}\left(\frac{\gamma}{2m}\right)^{1/2} + \left(\frac{\gamma}{2mn}\right)^{1/2} = \frac{\gamma^{1/2}n^{-1/2} + \gamma^{1/2}}{(2mn)^{1/2}} < \left(\frac{\delta}{2mn}\right)^{1/2}$$

for every i, so (iii') is satisfied.

If $\{p_i : 0 \leq i \leq k\}$ is an arbitrary increasing sequence of positive integers for which $p_0 = 0$ and $p_k = 2mn$, then (iii) implies

$$|y(p_i) - y(p_{i+1})| < \frac{p_{i+1} - p_i}{n}\left(\frac{\gamma}{2m}\right)^{1/2},$$

and it follows from $z = \gamma^{1/2}\mathfrak{z}_{2mn}$ that

(2) $$\sum_{i=0}^{k-1} |y(p_i)-y(p_{i+1})|\,|z(p_i)-z(p_{i+1})| \leq \sum_{i=0}^{k-1} |y(p_i)-y(p_{i+1})| \left(\frac{\gamma}{2mn}\right)^{1/2} <$$

$$< \frac{2mn}{n}\left(\frac{\gamma}{2m}\right)^{1/2}\left(\frac{\gamma}{2mn}\right)^{1/2} < \gamma N^{-1/2}.$$

Now suppose $\{p_i\}$ also has the property that

$$(3) \qquad 2\|w\|^2 = \sum_{i=0}^{k-1} |w(p_i) - w(p_{i+1})|^2.$$

Such a sequence exists because the support of w is finite. It follows from (2) that

$$(4) \quad 2\|w\|^2 < \sum_{i=0}^{k-1} |y(p_i) - y(p_{i+1})|^2 + \sum_{i=0}^{k-1} |z(p_i) - z(p_{i+1})|^2 + 2\gamma N^{-1/2}.$$

Observe that if $p_i < u < p_{i+1}$ then

$$|z(p_i) - z(p_{i+1})|^2 \le |z(p_i) - z(u)|^2 + |z(u) - z(p_{i+1})|^2.$$

Let $A = \{0, n, 2n, 3n, \ldots\}$.

For i fixed suppose that there does not exist $r \in A$ with $p_i < r < p_{i+1}$ and that $p_{i-1} < kn \le p_i < p_{i+1} < \ldots < p_{i+s} \le (k+1)n < p_{i+s+1}$. From the linearity of y on $[kn,(k+1)n]$ it follows that

$$\sum_{j=i}^{i+s-1} |y(p_j) - y(p_{j+1})|^2 \le |y(p_i) - y(p_{i+s})|^2,$$

hence s is at most 1. Thus assume that $p_{i-1} < kn \le p_i < p_{i+1} \le (k+1)n$. If $y(p_{i-1}) < y(p_i) < y(p_{i+1})$, then

$$|y(p_{i-1}) - y(p_i)|^2 + |y(p_i) - y(p_{i+1})|^2 < |y(p_{i-1}) - y(p_{i+1})|^2,$$

which is impossible by (3), and thus in this case $s = 0$.

If $y(p_{i-1}) > y(p_i) < y(p_{i+1}) < y(p_{i+2})$, then

$$|y(p_{i-1}) - y(p_i)|^2 \le |y(p_{i-1}) - y(kn)|^2$$

$$\text{and } |y(p_i) - y(p_{i+1})|^2 + |y(p_{i+1}) - y(p_{i+2})|^2 < |y(kn) - y(p_{i+2})|^2,$$

which is again a contradiction.

Now, if $y(p_{i-1}) > y(p_i) < y(p_{i+1}) > y(p_{i+2})$, then

$$|y(p_{i-1}) - y(p_i)|^2 \le |y(p_{i-1}) - y(kn)|^2,$$

$$|y(p_i) - y(p_{i+1})|^2 \le |y(kn) - y((k+1)n)|^2$$

$$\text{and } |y(p_{i+1}) - y(p_{i+2})|^2 \le |y((k+1)n) - y(p_{i+2})|^2;$$

thus $p_i = kn$ and $p_{i+1} = (k+1)n$. If $y(p_i) > y(p_{i+1})$ similarly we obtain that either $s = 0$ or if $s = 1$, $p_i = kn$ and $p_{i+1} = (k+1)n$.

Now suppose that there exists $r \in A$ such that $p_i < r < p_{i+1}$ and such that $y(r)$ is not between $y(p_i)$ and $y(p_{i+1})$. Let $r_1 \in A$ be the first one with this property. Then the insertion of r_1 in the sequence $\{p_i\}$ does not decrease the right hand side of (4). In fact, if $y(r_1) < y(p_i) < y(p_{i+1})$ then $(y(p_i) - y(p_{i+1}))^2 \le (y(r_1) - y(p_{i+1}))^2$,

if $y(r_1) < y(p_{i+1}) < y(p_i)$ then $(y(p_i) - y(p_{i+1}))^2 \le (y(p_i) - y(r_1))^2$

and the other two cases are similar.

We repeat this argument with the sequence $\{p_i\} \cup \{r_1\}$ and so forth until we arrive at a sequence $\{s_j\}$ such that

(a) $s_1 < s_2 <\ldots< s_t$,

(b) $\{p_i\} \subset \{s_j\}$,

(c) if $s_j \ne p_i$ for every i then $s_j \in A$,

(d) if there is no $r \in A$ with $s_j < r < s_{j+1}$, then $s_j = kn$ and $s_{j+1} = (k+1)n$ for some k,

(e) if $r \in A$, $s_j < r < s_{j+1}$, then $y(r)$ is between $y(s_j)$ and $y(s_{j+1})$.

By our previous remarks it follows that

$$(5) \qquad 2\|w\|^2 < \textstyle\sum_{i=1}^{t-1}|y(s_i) - y(s_{i+1})|^2 + \sum_{i=1}^{t-1}|z(s_i) - z(s_{i+1})|^2 + 2\gamma N^{-1/2}.$$

Let i be fixed and suppose that

$$m_i n \le s_i < (m_i + 1)n <\ldots< (m_i + \lambda_i)n < s_{i+1} \le (m_i + \lambda_i + 1)n.$$

Recalling that y is linear on $[kn, (k+1)n]$, if $y(s_i) < y(s_{i+1})$ then

$$(6) \qquad y(m_i n) \le y(s_i) \le y((m_i + j)n) \le y(s_{i+1}) \le y((m_i + \lambda_i + 1)n)$$

for $1 \le j \le \lambda_i$.

In fact, if $r \in A$ and $s_i < r < s_{i+1}$ then by d)

$$y(s_i) \le y(r) \le y(s_{i+1}).$$

In particular $y(s_i) \le y((m_i+1)n)$ and since y is linear on $[m_i n,(m_i+1)n]$, $y(m_i n) \le y(s_i)$. Also since $(m_i+\lambda_i)n < s_{i+1}$ we have $y((m_i+\lambda_i)n) \le y(s_{i+1})$ and by the linearity of y on $[(m_i + \lambda_i)n, (m_i + \lambda_i + 1)n]$ we get $y(s_{i+1}) \le y((m_i + \lambda_i + 1)n)$.

If $y(s_i) \ge y(s_{i+1})$ we get the reverse of the inequalities in (6) and the proof is similar. In both cases

$$(7) \qquad \Big(|y(s_i) - y((m_i+1)n)|^2 + \textstyle\sum_{j=1}^{\lambda_i-1} |y((m_i+j)n) - y((m_i+j+1)n)|^2 +$$
$$+ |y((m_i+\lambda_i)n) - y(s_{i+1})|^2 - |y(s_i) - y(s_{i+1})|^2\Big) -$$
$$-\Big(\textstyle\sum_{j=0}^{\lambda_i} |y((m_i+j)n) - y((m_i+j+1)n)|^2 - |y(m_i n) - y((m_i+\lambda_i+1)n)|^2\Big) \ge 0$$

since it is equal to twice

$$(y((m_i + \lambda_i)n) - y(m_i n))(y((m_i + \lambda_i + 1)n) - y((m_i + 1)n)) -$$

$$- (y((m_i + \lambda_i)n) - y(s_i))(y(s_{i+1}) - y((m_i + 1)n)) \geq 0.$$

Let $B=\{i\colon$ there is $r\in A$ with $s_i<r<s_{i+1}\}$ and $C=\{i\colon s_i=m_i n,\ s_{i+1}=(m_{i+1}+1)n\}$. Then by d)

$$\sum_{i\in B}\sum_{j=0}^{\lambda_i}|y((m_i+j)n) - y((m_i+j+1)n)|^2 - \sum_{i\in B}|y(m_i n) - y((m_i+\lambda_i+1)n)|^2 =$$

$$= \sum_{i\in B}\sum_{j=0}^{\lambda_i}|x(m_i+j) - x(m_i+j+1)|^2 - \sum_{i\in B}|x(m_i) - x(m_i+\lambda_i+1)|^2 =$$

$$= \sum_{i=0}^{2m-1}|x(i) - x(i+1)|^2 - \sum_{i\in C}|x(m_i) - x(m_i+1)|^2 -$$

$$- \sum_{i\in B}|x(m_i) - x(m_i+\lambda_i+1)|^2 \geq \sum_{i=0}^{2m-1}|x(i) - x(i+1)|^2 - 2\|x\|^2,$$

and by (ii) this expression is greater than $2\|x\|^2 - \varepsilon_1 - 2\|x\|^2 = -\varepsilon_1$. Hence, using (7),

$$\sum_{i\in B}|y(s_i) - y(s_{i+1})|^2 \leq \sum_{i\in B}|y(s_i) - y((m_i+1)n)|^2 +$$

$$+ \sum_{i\in B}\sum_{j=1}^{\lambda_i-1}|y((m_i+j)n) - y((m_i+j+1)n)|^2 +$$

$$+ \sum_{i\in B}|y((m_i+\lambda_i)n) - y(s_{i+1})|^2 + \varepsilon_1.$$

Therefore, if $\{q_i : 1 \leq i \leq \kappa\}$ is $\{s\} \cup \{kn : kn \leq 2mn\}$, then

$$(8)\quad 2\|w\|^2 < \sum_{i=1}^{\kappa-1}|y(q_i) - y(q_{i+1})|^2 + \sum_{i=1}^{\kappa-1}|z(q_i) - z(q_{i+1})|^2 + 2\gamma N^{-1/2} + \varepsilon_1.$$

For each i, there exists ℓ with $\ell n \leq q_i < q_{i+1} \leq (\ell+1)n$ and y is linear between q_i and q_{i+1}. Therefore

$$|y(q_i) - y(q_{i+1})|^2 + |z(q_i) - z(q_{i+1})|^2 -$$

$$- \sum_{j=q_i}^{q_{i+1}-1}\left(|y(j) - y(j+1)|^2 + |z(j) - z(j+1)|^2\right) =$$

$$= \left(\frac{q_{i+1} - q_i}{n}\right)^2 |x(\ell) - x(\ell+1)|^2 + |z(q_i) - z(q_{i+1})|^2 -$$

$$- \frac{q_{i+1} - q_i}{n^2}|x(\ell) - x(\ell+1)|^2 - (q_{i+1} - q_i)\frac{\gamma}{2mn} \leq$$

$$\leq ((q_{i+1} - q_i)^2 - (n+1)(q_{i+1} - q_i) + n)\frac{\gamma}{2mn^2} =$$

$$= (q_{i+1} - q_i - 1)(q_{i+1} - q_i - n)\frac{\gamma}{2mn^2} \leq 0.$$

From this, (8) and (2) we get

$$2\|w\|^2 < \sum_{i=1}^{\kappa}\sum_{j=q_i}^{q_{i+1}-1}(|y(j) - y(j+1)|^2 + |z(j) - z(j+1)|^2) + \varepsilon_1 + 2\gamma N^{-1/2} =$$

$$= \sum_{j=1}^{2mn-1}(|y(j) - y(j+1)|^2 + |z(j) - z(j+1)|^2) + \varepsilon_1 + 2\gamma N^{-1/2} \le$$

$$\le \sum_{j=1}^{2mn-1}|w(j) - w(j+1)|^2 + \varepsilon_1 + 4\gamma N^{-1/2}.$$

This and (1) imply (ii').

Lemma 2.b.4. Let $x_1,\dots,x_k$ be elements of J and

$$\Theta = \{\theta(k) = \{\theta_1,\dots,\theta_k\} \text{ with } \theta_i = \pm 1 \text{ for } i = 1,\dots,k\}.$$

Then there is some $\vartheta(k) \in \Theta$ such that for every $\theta(k) \in \Theta$,

$$\|\textstyle\sum_{i=1}^{k} a_i\theta_i x_i\| \le \max_i |a_i| \,\|\vartheta_1 x_1 + \vartheta_2 x_2 + \dots + \vartheta_k x_k\|.$$

Proof: Let $\theta(k) \in \Theta$. Since $x_i = \sum_{j=1}^{\infty} e_j^*(x_i)e_j$, we have

$$\textstyle\sum_{i=1}^{k} a_i\theta_i x_i = \sum_{j=1}^{\infty}(\sum_{i=1}^{k} a_i\theta_i e_j^*(x_i))e_j.$$

Let $0 = p_0 < p_1 <\dots< p_{n+1}$ and $\theta_i = \pm 1$. If $\theta'_i = \mathrm{sgn}(e_{p_r}^*(x_i) - e_{p_{r+1}}^*(x_i))$,

$$\frac{1}{2}\sum_{r=0}^{n}(\sum_{i=1}^{k} a_i\theta_i e_{p_r}^*(x_i) - \sum_{i=1}^{k} a_i\theta_i e_{p_{r+1}}^*(x_i))^2 \le$$

$$\le \frac{1}{2}\sum_{r=0}^{n}(\max_i a_i^2)(\sum_{i=1}^{k}|e_{p_r}^*(x_i)-e_{p_{r+1}}^*(x_i)|)^2 =$$

$$= \frac{1}{2}\sum_{r=0}^{n}(\max_i a_i^2)\Big(\sum_{i=1}^{k}\theta'_i(e_{p_r}^*(x_i)-e_{p_{r+1}}^*(x_i))\Big)^2 \le (\max_i a_i^2)\|\textstyle\sum_{i=1}^{k}\theta'_i x_i\|^2 \le$$

$$\le (\max_i a_i^2)\max_{\theta(k)\in\Theta}\|\textstyle\sum_{i=1}^{k}\theta_i x_i\|^2 = (\max_i a_i^2)\,\|\sum_{i=1}^{k}\vartheta_i x_i\|^2.$$

The key for the proof of the finite representability of c_0 in J lies in the following result.

Lemma 2.b.5. For each $k \ge 1$ and $\varepsilon > 0$ there exist elements $x_1,\dots,x_k$ in J with $\|x_i\| = 1$ for every i and such that

$$\max_{\theta(k)}\|\theta_1 x_1 + \theta_2 x_2 + \dots + \theta_k x_k\| < 1 + \varepsilon$$

where the max is taken over all sequences $\theta(k) \in \Theta$.

Proof: Let $\varepsilon > 0$ and $k \ge 1$. Let $\gamma_1,\dots,\gamma_{k+1}$ and $\varepsilon_1,\dots,\varepsilon_{k+1}$ satisfy

$$2 = \gamma_1 < \gamma_2 <\ldots< \gamma_{k+1} = 2(1 + 2\varepsilon), \quad 0 < \varepsilon_1 < \varepsilon_2 <\ldots< \varepsilon_{k+1} = 2\varepsilon^2.$$

Choose N so that for $1 \le i < k + 1$

$$\gamma_i^{1/2}N^{-1/2} + \gamma_i^{1/2} < \gamma_{i+1}^{1/2}, \quad \varepsilon_i + 4\gamma_i N^{-1/2} < \varepsilon_{i+1}.$$

We will proceed by induction using Lemma 2.b.3 repeatedly.

Let $\mathfrak{z}_{2k}$ and T_n be as in Definition 2.b.2 and observe that $T_nT_m = T_{nm}$.

Let $x^{(1)} = \gamma_1^{1/2}\mathfrak{z}_2$, $y^{(1)} = T_nx^{(1)}$, $z^{(1)} = \gamma_2^{1/2}\mathfrak{z}_{2n}$ and $w^{(1)} = y^{(1)} + z^{(1)}$.

Then by Lemma 2.b.3

(i) $w^{(1)}(i) = 0$ for $i \ge 2n$,

(ii) $2\|w^{(1)}\|^2 < \sum_{i=0}^{2n-1}|w^{(1)}(i) - w^{(1)}(i + 1)|^2 + \varepsilon_3$,

(iii) $|w^{(1)}(i) - w^{(1)}(i + 1)| < \left(\frac{\gamma_3}{2n}\right)^{1/2}$ if $i \ge 0$.

Remark that if we take $-x^{(1)}$ instead of $x^{(1)}$, $(y^{(1)})' = T_n(-x^{(1)})$ and $(w^{(1)})' = (y^{(1)})' + z^{(1)}$ we get that (i), (ii) and (iii) also hold for $(w^{(1)})'$.

Next we apply Lemma 2.b.3 to $x^{(2)} = w^{(1)}$, $y^{(2)} = T_nx^{(2)}$, $z^{(2)} = \gamma_3^{1/2}\mathfrak{z}_{2n^2}$ and $w^{(2)} = y^{(2)} + z^{(2)}$. Then

$$y^{(2)} = T_ny^{(1)} + T_nz^{(1)} = T_{n^2}\gamma_1^{1/2}\mathfrak{z}_2 + T_n\gamma_2^{1/2}\mathfrak{z}_{2n}$$

and

(i) $w^{(2)}(i) = 0$ for $i \ge 2n^2$,

(ii) $2\|w^{(2)}\|^2 < \sum_{i=0}^{2n^2-1}|w^{(2)}(i) - w^{(2)}(i + 1)|^2 + \varepsilon_4$,

(iii) $|w^{(2)}(i) - w^{(2)}(i + 1)| < \left(\frac{\gamma_4}{2n^2}\right)^{1/2}$ if $i \ge 0$.

Let $x^{(3)} = w^{(2)}$, $y^{(3)} = T_nx^{(3)}$, $z^{(3)} = \gamma_4^{1/2}\mathfrak{z}_{2n^3}$ and $w^{(3)} = y^{(3)} + z^{(3)}$.

Then

$$y^{(3)} = T_{n^3}\gamma_1^{1/2}\mathfrak{z}_2 + T_{n^2}\gamma_2^{1/2}\mathfrak{z}_{2n} + T_n\gamma_3^{1/2}\mathfrak{z}_{2n^2}.$$

....

Continuing in this fashion, if we write $u_i = T_{n^{k-i}}\gamma_i^{1/2}\mathfrak{z}_{2n^{i-1}}$, we finally arrive at

$$y^{(k-1)} = T_{n^{k-1}}\gamma_1^{1/2}\mathfrak{z}_2 + T_{n^{k-2}}\gamma_2^{1/2}\mathfrak{z}_{2n} + \ldots + T_n\gamma_{k-1}^{1/2}\mathfrak{z}_{2n^{k-2}} = u_1 + \ldots + u_{k-1},$$

$$z^{(k-1)} = \gamma_k^{1/2}\mathfrak{z}_{2n^{k-1}} = u_k, \quad w^{(k-1)} = y^{(k-1)} + z^{(k-1)} = u_1 + \ldots + u_{k-1} + u_k$$

and

(ii) $2\|w^{(k-1)}\|^2 < \sum_{i=0}^{2n^{k-1}-1} |w^{(k-1)}(i) - w^{(k-1)}(i+1)|^2 + \varepsilon_{k+1}$,

(iii) $|w^{(k-1)}(i) - w^{(k-1)}(i+1)| < \left(\dfrac{\gamma_{k+1}}{2n^{k-1}}\right)^{1/2}$ if $i \geq 0$.

Using (ii) and (iii) we get

$$\|u_1 + \ldots + u_k\|^2 = \|w^{(k-1)}\|^2 < (\gamma_{k+1} + \varepsilon_{k+1})/2 = (1+\varepsilon)^2.$$

Since Lemma 2.b.3 is also true for $w = y - z$ and by the above remark it holds that

$$\|\theta_1 u_1 + \ldots + \theta_k u_k\| < 1 + \varepsilon \quad \text{for every } \theta(k) \in \Theta.$$

Since $\|\mathfrak{z}_{2m}\| = 2^{-1/2}$, it is easy to see that $\|T_n\mathfrak{z}_{2m}\| = 2^{-1/2}$ for all m. Thus $\|u_i\| = (\gamma_i/2)^{1/2} \geq 1$ for $1 \leq i \leq k$ and if we let $x_i = u_i/\|u_i\|$, we get using Lemma 2.b.4, that there exists $\vartheta(k) \in \Theta$ such that for every $\theta(k) \in \Theta$

$$\|\theta_1 x_1 + \ldots + \theta_k x_k\| \leq \max_i(1/\|u_i\|)\|\vartheta_1 u_1 + \ldots + \vartheta_k u_k\| \leq$$
$$\leq \|\vartheta_1 u_1 + \ldots + \vartheta_k u_k\| < 1 + \varepsilon.$$

Notation: Let X be a Banach space with a basis $\{x_n\}$. We will denote by $X^{(m)}$ the subspace $[x_1,\ldots,x_m]$ of X.

For spaces X with a basis $\{x_n\}$, in order to prove that X is finitely representable in Y, it is enough to check for the n-dimensional subspaces of X of the form $[x_i]_{i=1}^n$ that they are $(1+\varepsilon)$-embedded in the space Y, instead of analyzing all the finite dimensional subspaces.

Lemma 2.b.6. Let F be an n-dimensional subspace of a Banach space $(X, \|\ \|)$ with a basis $\{x_n\}$ and let $\varepsilon > 0$. Then there exist an integer m and an isomorphism T from F into $X^{(m)}$ such that

$$(1-\varepsilon)\|x\| \leq \|Tx\| \leq (1+\varepsilon)\|x\|$$

for every $x \in F$ and

$$\|T - I_F\| \le \varepsilon n^{-1/2}$$

where I_F is the identity map.

Proof: Let $\varepsilon > 0$ and F be a subspace of dimension n of X and f_2, $f_4,\dots,f_{2n}$ be an algebraic basis of F. Complete this basis so that $f_1, f_2,\dots,f_{2n}$ span a subspace of dimension 2n of X. Let $K > 1$ and $\delta > 0$ be such that

$$\text{(1)} \qquad \begin{cases} (\sum_{i=1}^{2n} a_i^2)^{1/2} \le K\|\sum_{i=1}^{2n} a_i x_i\|, \\ \delta\|\sum_{i=1}^{2n} a_i x_i\| \le \|\sum_{i=1}^{2n} a_i f_i\|, \\ \delta\sum_{i=1}^{n} |\alpha_i| \le \|\sum_{i=1}^{n} \alpha_i f_{2i}\|, \end{cases}$$

for all sequences of scalars $a_1,\dots,a_{2n}$ and $\alpha_1,\dots,\alpha_n$.

For $1 \le i \le n$, let $f'_i \in X$ with only finitely many non-zero coefficients with respect to the basis $\{x_n\}$ be such that

$$\|f'_i - f_{2i}\| \le (\varepsilon\delta/K)n^{-1/2}.$$

Then for any sequence of scalars $b_1,\dots,b_n$, using Hölder's inequality and (1),

$$\|\sum_{i=1}^{n} b_i(f'_i - f_{2i})\| \le \frac{\varepsilon\delta}{K}n^{-1/2}\sum_{i=1}^{n} |b_i| \le \frac{\varepsilon\delta}{K}(\sum_{i=1}^{n} b_i^2)^{1/2} \le$$

$$\le \varepsilon\delta\|\sum_{j=1}^{n} b_j x_{2j}\| \le \varepsilon\|\sum_{i=1}^{n} b_i f_{2i}\|.$$

Hence

$$\|\sum_{i=1}^{n} b_i f'_i\| \le \|\sum_{i=1}^{n} b_i f_{2i}\| + \varepsilon\|\sum_{i=1}^{n} b_i f_{2i}\|$$

and

$$\|\sum_{i=1}^{n} b_i f_{2i}\| \le \|\sum_{i=1}^{n} b_i f'_i\| + \varepsilon\|\sum_{i=1}^{n} b_i f_{2i}\|.$$

Therefore

$$\text{(2)} \qquad (1 - \varepsilon)\|\sum_{i=1}^{n} b_i f_{2i}\| \le \|\sum_{i=1}^{n} b_i f'_i\| \le (1 + \varepsilon)\|\sum_{i=1}^{n} b_i f_{2i}\|.$$

If $f'_i = \sum_{j=1}^{n_i}\langle x_j^*, f'_i\rangle x_j$ for $i = 1,\dots,n$, let $m = \max n_i$.

Then $[f'_1,\dots,f'_n]$ is a subspace of $X^{(m)}$. Let T from F into $X^{(m)}$ be the isomorphism given by $Tf_{2i} = f'_i$.

Using (1) we obtain

$$\|T(\sum_{i=1}^{n} \alpha_i f_{2i}) - \sum_{i=1}^{n} \alpha_i f_{2i}\| \le \sum_{i=1}^{n} |\alpha_i|\,\|f_{2i} - f'_i\| \le$$

$$\le \frac{\varepsilon\delta}{K} n^{-1/2}(\sum_{i=1}^{n}|\alpha_i|) \le \frac{\varepsilon}{K} n^{-1/2}\|\sum_{i=1}^{n}\alpha_i f_{2i}\|.$$

Hence $\|T - I_F\| \le \varepsilon n^{-1/2}$ and by (2) $(1 - \varepsilon)\|x\| \le \|Tx\| \le (1 + \varepsilon)\|x\|$ for every $x \in F$.

Corollary 2.b.7. Let $(E, \| \|_E)$ be a Banach space. In order to prove that X is finitely representable in E where X is a Banach space with a basis $\{x_n\}$, it suffices to prove that for every m and every $\varepsilon > 0$ there exists an isomorphism T from $X^{(m)}$ into E such that for every $x \in X^{(m)}$

$$(1 - \varepsilon)\|x\|_X \le \|Tx\|_E \le (1 + \varepsilon)\|x\|_X.$$

After all the tiresome but necessary technical lemmata, finally wc are able to prove the main result of this section.

Theorem 2.b.8. The space c_0 is finitely representable in $(J, \| \|)$.

Proof: Let $\varepsilon > 0$, $k \in \mathbb{N}$ and $x_1,\dots,x_k$ be the elements of J given by Lemma 2.b.5 with $\|x_i\| = 1$ for all i and such that

$$\max_{\theta(k)} \|\theta_1 x_1 + \theta_2 x_2 +\dots+ \theta_k x_k\| < 1 + \varepsilon$$

where the sup is taken over all sequences $\theta(k) \in \Theta$. Then using Lemma 2.b.4 there is $\vartheta(k) = \{\vartheta_1,\dots,\vartheta_k\}$ such that for every $\theta(k) \in \Theta$

(1) $$\|\sum_{i=1}^{k} a_i\theta_i x_i\| \le (\sup_i|a_i|)\|\sum_{i=1}^{k}\vartheta_i x_i\| < (1 + \varepsilon)\sup_i|a_i|.$$

If $\sup_i|a_i| = |a_j|$, since $\|x_i\| = 1$, then

$$2|a_j| \le \|\sum_{i=1}^{k} a_i x_i\| + \|2a_j x_j - \sum_{i=1}^{k} a_i x_i\| < \|\sum_{i=1}^{k} a_i x_i\| + (1 + \varepsilon)\sup_i|a_i|,$$

so that

(2) $$\|\sum_{i=1}^{k} a_i x_i\| > (1 - \varepsilon)\sup_i|a_i|.$$

By (1), (2) and Corollary 2.b.7 c_0 is finitely representable in $(J, \| \|)$.

The by now well known theorem of Dvoretzky and Rogers (see for example Day [1]) implies that ℓ_2 is finitely representable in every Banach space; on the other side Bessaga and Pelczynski [2] proved that every Banach space is f.r. in c_0. Here we give an easy proof of the second fact, which can be found in Bombal [1]. This result in turn implies that every Banach space is f.r. in J; thus J and c_0, while being very different spaces, have the "same" finite dimensional subspaces.

Proposition 2.b.9. Every Banach space X is finitely representable in c_0.

Proof: Let E be a finite dimensional subspace of X and $\varepsilon > 0$. Since the unit sphere in E^* is compact, there exist $f_1, f_2,\dots,f_n$ in E^* with $\|f_i\| = 1$ for every $i = 1,\dots,n$, such that for every $f \in E^*$ with $\|f\| = 1$ there is $i \le n$ with $\|f - f_i\| \le \varepsilon$.
Define $T : E \longrightarrow c_0$ by $Tx = (f_1x,\dots,f_nx, 0, 0,\dots)$. Then for every $x \in E$

$$\|Tx\|_\infty = \max_{1\le i\le n}|f_ix| \le \|x\|.$$

Let $x \in E$ with $\|x\| = 1$. Then there is $f \in E^*$ with $\|f\| = 1$ and $fx = 1$. Let i satisfy $\|f - f_i\| \le \varepsilon$; then

$$f_ix = fx - (f - f_i)x \ge 1 - \varepsilon.$$

Therefore $\|Tx\|_\infty \ge 1 - \varepsilon = (1 - \varepsilon)\|x\|$. Hence for every $x \in E$

$$(1 - \varepsilon)\|x\| \le \|Tx\|_\infty \le \|x\|$$

and this proves the proposition.

Corollary 2.b.10. Every Banach space is finitely representable in J.

Proof: By Proposition 2.b.9 every Banach space is f.r. in c_0, by Theorem 2.b.8 c_0 is f.r. in J, and finite representability is transitive (see e.g. Beauzamy [1]).

This corollary shows that J contains "all" finite dimensional Banach spaces almost isometrically and also that J is not B-convex, according to the following definition.

Definition 2.b.11. A Banach space X is B-convex if and only if ℓ_1 is not finitely representable in X.

The above results are related to the concepts of type and cotype, which we will introduce in Section 2.i where we will show that the above corollary implies that J has no type other than 1 and no cotype other than ∞.

2.c. Complemented subspaces of J and the space $\mathfrak{J}$

One of the central topics in Banach space theory is the study of complemented subspaces, since this gives the closest analogy with the important orthogonal decomposition for Hilbert spaces.

In the present section we study the complemented subspaces of J, and to this end we introduce the important space $\mathfrak{J} = (\sum_n J_n)_{\ell_2}$, defined in Casazza, Lin and Lohman [1], which plays a central role in the proof of the primarity of J that will be given in Section 2.d. The key result of this section, namely the fact that every reflexive complemented subspace of J is isomorphic to a complemented subspace of $\mathfrak{J}$ (Proposition 2.c.18), requires considerable elaboration, and along the way we need to discuss several interesting notions such as superreflexivity and Schauder decompositions which generalize the concepts of reflexivity and Schauder bases respectively.

Definition 2.c.1. Let $\{e_n\}_{n=1}^{\infty}$ be the unit vector basis of J. We define $J_0 = \{0\}$, where 0 is the 0-vector in J, and $J_n = [e_1, e_2, \ldots, e_n]$. Let $\{t_n\}_{n=1}^{\infty}$ in $\mathbb{N}$. $(\sum_n J_{t_n})_{\ell_2}$ is the space of sequences $x = \{x_n\}_{n=1}^{\infty}$ with $x_n \in J_{t_n}$ and $\|x\|^2 = \sum_{n=1}^{\infty} \|x_n\|^2 < \infty$ and we denote the space $(\sum_n J_n)_{\ell_2}$ by $\mathfrak{J}$.

The following inequalities, giving an estimate of the norm of the sum of elements in J having strictly disjoint supports, are essential in the rest of the chapter.

Lemma 2.c.2. Let $\{e_n\}_{n=1}^{\infty}$ be the unit vector basis of J. Let $p_1 < q_1 < p_2 < q_2 < \ldots$ be an increasing sequence of positive integers such that $p_{k+1} - q_k \geq 2$ for all k. Then for any real sequence $\{a_n\}_{n=1}^{\infty}$,

$$\sum_{j=1}^{k} \|\sum_{n=p_j}^{q_j} a_n e_n\|^2 \leq \|\sum_{j=1}^{k} \sum_{n=p_j}^{q_j} a_n e_n\|^2 \leq 2\sum_{j=1}^{k} \|\sum_{n=p_j}^{q_j} a_n e_n\|^2.$$

Proof: The first inequality follows easily from the fact that

$\|\sum_{n=p_j}^{q_j} a_n e_n\|$ can be calculated for every j by using only indices in the interval $[p_j - 1, q_j + 1]$. The second inequality follows from the fact that $(a - b)^2 \le 2(a^2 + b^2)$, applied to the appropriate summands of the middle member of the inequality.

The spaces generated by the subsequences of the unit vector basis in J are very few, in fact we will see that these spaces are isomorphic to either J, $\mathfrak{J}$ or ℓ_2 and are complemented in J. These results are taken from Casazza, Lin and Lohman [1].

Definition 2.c.3. Let $\{n_i\}$ be an increasing sequence in $\mathbb{N}$. The sequence $\mathbb{N} \setminus \{n_i\}$ is called the complementary sequence to $\{n_i\}$. $\{n_i\}$ is called proper if the complement $\mathbb{N} \setminus \{n_i\}$ of $\{n_i\}$ in $\mathbb{N}$ is infinite.

As the following theorem shows, the spaces $(\sum_n J_{t_n})_{\ell_2}$ are isomorphic to some particular subspaces of J which will later be proved to be complemented in J.

Theorem 2.c.4. For any sequence $\{t_n\}_{n=1}^{\infty}$ in $\mathbb{N}$, the natural basis of $(\sum_n J_{t_n})_{\ell_2}$ is equivalent to $\{e_{n_i}\}_{i=1}^{\infty}$ where $\{n_i\}$ is the complementary sequence to $\{\sum_{i=1}^{n} t_i + n\}_{n=1}^{\infty}$.

Proof: Define $m_0 = 0$ and for $n = 1, 2, \ldots,$ $m_n = \sum_{i=1}^{n} t_i + n$. Let $\{n_i\}$ be the complementary sequence to $\{m_n\}$. Let $T : (\sum_n J_{t_n})_{\ell_2} \to [e_{n_i}]$ be as follows:

For $x = (\sum_{i=1}^{t_1} a_i^{(1)} e_i, \ldots, \sum_{i=1}^{t_n} a_i^{(n)} e_i, 0, \ldots)$ in $(\sum_n J_{t_n})_{\ell_2}$, let

$$Tx = \sum_{k=1}^{n} \sum_{i=1}^{t_k} a_i^{(k)} e_{m_{k-1}+i}.$$

Here $\sum_{i=1}^{t_k} a_i^{(k)} e_i$ means the zero vector in J if $t_k = 0$.

Then by Lemma 2.c.2,

$$\|x\| \le \|Tx\| \le \sqrt{2}\ \|x\|.$$

From this we get that $\{e_{n_i}\}_{i=1}^{\infty}$ is equivalent to the natural basis of $(\sum_n J_{t_n})_{\ell_2}$.

Corollary 2.c.5. If $\{n_i\}_{i=1}^{\infty}$ is a proper sequence, then $[e_{n_i}]_{i=1}^{\infty}$ is a reflexive subspace of J isomorphic to $(\sum_n J_{t_n})_{\ell_2}$ for some sequence $\{t_n\}$. In particular $[e_{n_i}]_{i=1}^{\infty}$, where $\{n_i\} = \{m : m \ne n(n+3)/2 \text{ for every } n\}$, is isomorphic to $\mathfrak{J}$ and thus this space is isomorphic to a reflexive subspace of J.

Proof: Let $m_0 = 0$, and let $\{m_k\}$ be the complementary sequence to $\{n_i\}$ in $\mathbb{N}$. Define $t_n = m_n - m_{n-1} - 1$; then by the proof of Theorem 2.c.4, $\{e_{n_i}\}_{i=1}^{\infty}$ is equivalent to the natural basis of $(\sum_n J_{t_n})_{\ell_2}$. As $(\sum_n J_{t_n})_{\ell_2}$ is reflexive (see Proposition 1.10) we have the first assertion, and if in Theorem 2.c.4 we take $t_n = n$ we get that $\mathfrak{J}$ is isomorphic to $[e_{n_i}]_{i=1}^{\infty}$ where $\{n_i\} = \{m : m \ne n(n+3)/2 \text{ for every } n\}$, and this proves the second assertion.

Theorem 2.c.6. For every subsequence $\{e_{n_i}\}_{i=1}^{\infty}$ of $\{e_n\}_{n=1}^{\infty}$, $[e_{n_i}]_{i=1}^{\infty}$ is complemented in J by a projection P with $\|P\| \le 2$.

Proof: If $\{n_i\}_{i=1}^{\infty}$ is not a proper sequence the result is obvious. So let $\{n_i\}_{i=1}^{\infty}$ be a proper sequence. Let $m_0 = 0$, let $\{m_k\}_{k=1}^{\infty}$ be the complementary sequence to $\{n_i\}_{i=1}^{\infty}$ in $\mathbb{N}$ and let $X = [e_{n_i}]_{i=1}^{\infty}$. Define $P : J \longrightarrow X$ as follows: for $y = \sum_{n=1}^{\infty} a_n e_n \in J$ let

$$Py = \sum_{i=1}^{\infty} (a_{n_i} - a_{r_i}) e_{n_i}$$

where $r_i = m_s$ if $m_{s-1} < n_i < m_s$; observe that $r_i \le r_{i+1}$ for every i. Then P is a projection onto X, and if $0 = p_0 < p_1 < \ldots < p_k$ we have

$$\frac{1}{2}\sum_{j=0}^{k-1}\left(a_{n_{p_j}} - a_{r_{p_j}} - a_{n_{p_{j+1}}} + a_{r_{p_{j+1}}}\right)^2 \le$$

$$\le \sum_{j=0}^{k-1}\left((a_{n_{p_j}} - a_{n_{p_{j+1}}})^2 + (a_{r_{p_j}} - a_{r_{p_{j+1}}})^2\right) \le 4\|\sum_{n=1}^{\infty} a_n e_n\|^2.$$

Hence $\|P\| \le 2$.

In fact, if $\{n_i\}$ is a proper sequence, not only is $[e_{n_i}]_{i=1}^{\infty}$ complemented in J, but its complementary space under P is isometric to J.

Corollary 2.c.7. Let $\{n_i\}$ be a proper sequence and $X = [e_{n_i}]_{i=1}^{\infty}$. Let P be as above and W = (I - P)J. Then J is isometric to W.

Proof: Let $m_0 = 0$, $\{m_k\}$ be the complementary sequence to $\{n_i\}$ and

$$f_s = \sum_{j=m_{s-1}+1}^{m_s} e_j.$$

We will see that W is the closed linear span of $\{f_s\}_{s=1}^{\infty}$ in J. For $y \in J$, $y = \sum_{n=1}^{\infty} a_n e_n$, let $Qy = (I - P)y$. Then $Qy = \sum_{n=1}^{\infty} b_n e_n$ where

$$b_n = \begin{cases} a_n & \text{if } n \in \{m_k\}, \\ a_{m_s} & \text{if } n_j = n \text{ and } m_{s-1} < n_j < m_s. \end{cases}$$

Hence $Qy \in [f_s]_{s=1}^{\infty}$, $Qw = w$ for $w \in [f_s]_{s=1}^{\infty}$ and $\|Q\| = 1$. Therefore $W = [f_s]_{s=1}^{\infty}$ and it is easily seen that $T : J \longrightarrow W$ given by $Te_i = f_i$ is an isometry.

In order to be able to classify the spaces $(\sum_i J_{m_i})_{\ell_2}$, we need to prove a combinatorial lemma which appears in Casazza, Kottman and Lin [1].

Lemma 2.c.8. Let $M = \{m_i\}_{i=1}^{\infty}$ be a sequence of positive integers such that $\limsup_i m_i = \infty$. Then there exist sequences $N' = \{n'_i\}_{i=1}^{\infty}$, $N'' = \{n''_i\}_{i=1}^{\infty}$, $M' = \{m'_i\}_{i=1}^{\infty}$ and $M'' = \{m''_i\}_{i=1}^{\infty}$ such that

(a) $N' \cap N'' = \emptyset$ and if $n \in \mathbb{N}$ then either $n = n'_i$ or $n = n''_i$ for exactly one $i \in \mathbb{N}$,

(b) $M' \cap M'' = \emptyset$ and if $m_k \in M$ then either $m_k = m'_i$ or $m_k = m''_i$ for exactly one $i \in \mathbb{N}$,

(c) $n'_{2i-1} + n'_{2i} = m'_i$ and $m''_{2i-1} + m''_{2i} = n''_i$ for $i = 1, 2, \ldots$.

Proof: We proceed by induction: Let $n'_1 = 1$ and

$$n'_2 = \min\{n \in \mathbb{N} : n \neq n'_1 \text{ and } n'_1 + n \in M\}.$$

Let

$$\gamma_1 = \min\{i \in \mathbb{N} : n'_1 + n'_2 = m_i \in M\} \text{ and let } m'_1 = m_{\gamma_1}.$$

Now let

$$\alpha_1 = \min\{i \in \mathbb{N} : m_i \in M \setminus \{m'_1\}\}$$

and

$$\beta_1 = \min\{i \in \mathbb{N} : i \neq \alpha_1,\ m_i \in M \setminus \{m'_1\} \text{ and } m_i + m_{\alpha_1} \in \mathbb{N} \setminus \{n'_1, n'_2\}\}.$$

Define $m''_1 = m_{\alpha_1}$, $m''_2 = m_{\beta_1}$ and $n''_1 = m''_1 + m''_2$.

Suppose that $n'_1, n'_2, \ldots, n'_{2k}$; $n''_1, n''_2, \ldots, n''_k$; $m'_1, m'_2, \ldots, m'_k$ and m''_1, m''_2, $m''_3, \ldots, m''_{2k}$ have been chosen such that for $i = 1, 2, \ldots, k$

$$n'_{2i-1} + n'_{2i} = m'_i \quad \text{and} \quad m''_{2i-1} + m''_{2i} = n''_i.$$

Let

$$n'_{2k+1} = \min\{n \in \mathbb{N} : n \neq n'_i \text{ for } i = 1,2,\ldots,2k \text{ and } n \neq n''_i \text{ for } i = 1,2,\ldots,k\}$$

and

$$n'_{2k+2} = \min\{n \in \mathbb{N} : n \neq n'_i \text{ for } i = 1,2,\ldots,2k+1,\ n \neq n''_i \text{ for } i = 1,\ldots,k \text{ and } n'_{2k+1} + n \in M \setminus \{m'_1, \ldots, m'_k,\ m''_1, \ldots, m''_{2k}\}\}.$$

This is well defined because $\limsup_i m_i = \infty$. Now let

$$\gamma_{k+1} = \min\{j \in \mathbb{N} : m_j = n'_{2k+1} + n'_{2k+2}\}.$$

Define $m'_{k+1} = m_{\gamma_{k+1}}$. Finally let

$$\alpha_{k+1} = \min\{i \in \mathbb{N} : m_i \in M \setminus \{m'_1, \ldots, m'_{k+1},\ m''_1, \ldots, m''_{2k}\}\}$$

and

$$\beta_{k+1} = \min\{i \in \mathbb{N} : i \neq \alpha_{k+1},\ m_i \in M \setminus \{m'_1, \ldots, m'_{k+1},\ m''_1, \ldots, m''_{2k}\} \text{ and } m_i + m_{\alpha_{k+1}} \in \mathbb{N} \setminus \{n'_1, \ldots, n'_{2k+2},\ n''_1, \ldots, n''_k\}\}.$$

Let $m''_{2k+1} = m_{\alpha_{k+1}}$, $m''_{2k+2} = m_{\beta_{k+1}}$ and $n''_{k+1} = m''_{2k+1} + m''_{2k+2}$.

This step completes the proof of the lemma.

By virtue of Corollary 2.c.5 the following theorem leads to the characterization of the subspaces of J spanned by subsequences of the canonical basis.

Theorem 2.c.9. Let $\{m_i\}$ be a sequence of positive integers. If
(a) $\sup_i m_i = t < \infty$, then $(\sum_i J_{m_i})_{\ell_2}$ is isomorphic to ℓ_2.
(b) $\sup_i m_i = \infty$, then $(\sum_i J_{m_i})_{\ell_2}$ is isomorphic to $\mathfrak{J}$.

Proof: (a) Let $t = \sup_i m_i$ and δ, $\mu > 0$ be such that

$$(\sum_{j=1}^t a_j^2)^{1/2} \le \delta\|\sum_{j=1}^t a_j e_j\| \le \delta\mu(\sum_{j=1}^t a_j^2)^{1/2}.$$

Let $S_i : J_{m_i} \longrightarrow \ell_2^{(m_i)}$ be defined as

$$S_i\left(\sum_{j=1}^{m_i} a_j e_j\right) = (a_1,\dots,a_{m_i}).$$

Then clearly $\|S_i\|\|S_i^{-1}\| \le \delta\mu$ and by Proposition 1.10 the result follows because $(\sum_i \ell_2^{(m_i)})_{\ell_2}$ is isomorphic to ℓ_2.
(b) Let $\sup_i m_i = \infty$ and $M = \{m_i\}$. Then by Lemma 2.c.8 there exist sequences $N' = \{n'_i\}$, $N'' = \{n''_i\}$, $M' = \{m'_i\}$ and $M'' = \{m''_i\}$ such that

$$N' \cap N'' = \emptyset$$

and if $n \in \mathbb{N}$ then either $n = n'_i$ or $n = n''_i$ for exactly one $i \in \mathbb{N}$; also

$$M' \cap M'' = \emptyset$$

and if $m_k \in M$ then either $m_k = m'_i$ or $m_k = m''_i$ for exactly one $i \in \mathbb{N}$. Furthermore $n'_{2i-1} + n'_{2i} = m'_i$ and $m''_{2i-1} + m''_{2i} = n''_i$ for $i = 1, 2,\dots$.
On the other hand it is easy to see, using the definition of the norm in J, that for every n and m

$$\|\sum_{i=1}^n a_i e_i\|^2 + \|\sum_{j=1}^m b_j e_j\|^2 \le 2\|\sum_{i=1}^n a_i e_i + \sum_{j=1}^m b_j e_{j+n}\|^2 \le$$

$$\le 4(\|\sum_{i=1}^n a_i e_i\|^2 + \|\sum_{j=1}^m b_j e_j\|^2).$$

Hence there is an isomorphism T between $J_n \oplus J_m$ and J_{n+m} with $\|T\|\|T^{-1}\| \le 2$. Applying Proposition 1.10 we get

$$\mathfrak{J} = (\Sigma_n J_n)_{\ell_2} \approx (\Sigma_i J_{n'_i})_{\ell_2} \oplus (\Sigma_i J_{n''_i})_{\ell_2} \approx$$

$$\approx (\Sigma_i (J_{n'_{2i-1}} \oplus J_{n'_{2i}}))_{\ell_2} \oplus (\Sigma_i J_{m''_{2i-1}+m''_{2i}})_{\ell_2} \approx$$

$$(\Sigma_i J_{m'_i})_{\ell_2} \oplus (\Sigma_i (J_{m''_{2i-1}} \oplus J_{m''_{2i}}))_{\ell_2} \approx (\Sigma_i J_{m'_i})_{\ell_2} \oplus (\Sigma_i J_{m''_i})_{\ell_2} \approx (\Sigma_i J_{m_i})_{\ell_2}.$$

Corollary 2.c.10. Let $\{e_{n_i}\}$ be a subsequence of the unit vector basis $\{e_n\}$ in J. Then the closed linear subspace spanned by $\{e_{n_i}\}$ is isomorphic to either J, $\mathfrak{J}$ or ℓ_2.

Proof: If $\{n_i\}$ is not a proper sequence, then it is easy to see that $[e_{n_i}]$ is isomorphic to J. If it is proper, by Corollary 2.c.5 and Theorem 2.c.9, $[e_{n_i}]$ is isomorphic either to $\mathfrak{J}$ or to ℓ_2.

To show that the above result is meaningful we will prove that $\mathfrak{J}$ is not superreflexive, which implies that $\mathfrak{J}$ is not isomorphic to ℓ_2. The notion of superreflexivity was introduced by James in [10] when studying uniformly convexifiable spaces.

Definition 2.c.11. A Banach space X is superreflexive if, whenever F is finitely representable in X, F is reflexive.

Definition 2.c.12. A Banach space X is uniformly convex if for every $\varepsilon > 0$ there exists $\delta > 0$ such that if $x, y \in X$, $\|x\| = \|y\| = 1$, $\|x - y\| \geq \varepsilon$, then $\|\frac{x+y}{2}\| \leq 1 - \delta$.

X is uniformly convexifiable if there exists an equivalent norm on X making X into a uniformly convex space.

The notion of uniform convexifiability is closely related to that of superreflexivity, in fact they are equivalent as the next theorem shows. It is stated as in Diestel [1], and a detailed proof can be found spread throughout Beauzamy [1]. The proofs of (a) implies (c) and of (b) equivalent to (c) are due to James, and the proof of (c) implies (a) is due to Enflo.

Theorem 2.c.13. Let X be a Banach space. The following are equivalent:

(a) X is uniformly convexifiable.

(b) There exist p, q > 1 such that for every normalized basic sequence $\{x_n\}$ in X there is K such that for every $a_1,\ldots,a_n \in \mathbb{R}$

$$K^{-1}(\sum_{i=1}^{n}|a_i|^p)^{1/p} \le \|\sum_{i=1}^{n}a_ix_i\| \le K(\sum_{i=1}^{n}|a_i|^q)^{1/q}.$$

(c) X is superreflexive.

Proposition 2.c.14. The space $\mathfrak{J}$ is not superreflexive.

Proof: Let K > 0, p > 1 and n be large enough such that $n^{1/p} > K^{-1}$. If for $i = 1,\ldots,n$, $n = 1, 2,\ldots$, $x_n^i = (0,\ldots,0,e_i,0,\ldots)$, where $e_i \in J_n$, then

$$1 = \|(0,\ldots,0,\sum_{i=1}^{n}e_i,0,\ldots)\| = \|\sum_{i=1}^{n}x_n^i\| < Kn^{1/p},$$

and applying Theorem 2.c.13 we get that $\mathfrak{J}$ is not superreflexive.

Corollary 2.c.15. $\mathfrak{J}$ is not isomorphic to ℓ_2.

Proof: ℓ_2 is obviously uniformly convex.

In this part of the section we will use the concepts of finite dimensional decompositions and π_λ-spaces and several results related to those topics, as a tool for proving that every complemented reflexive subspace of J is isomorphic to a complemented subspace of $\mathfrak{J}$. Since they are general results not directly related to the James spaces, to avoid an unnecessary detour, the definitions of these notions and the theorems connected to them that we will use in the rest of this section can be found in Appendix 2.j.

Concretely, the techniques explained in Appendix 2.j are used in the next proposition in order to see that certain complemented subspaces of J always have a shrinking finite dimensional decomposition, and in Theorem 2.c.17 to prove that reflexive subspaces of J with a shrinking finite dimensional decomposition can be decomposed in a very special form.

All of the results in the rest of this section are by Casazza [1].

Proposition 2.c.16. Let $\mathfrak{J} = (\sum_n J_n)_{\ell_2}$. If X is a complemented subspace of J, then $X \oplus \mathfrak{J}$ has a shrinking finite dimensional decomposition and is isomorphic to a complemented subspace of J.

Proof: Let $P : J \longrightarrow X$ be a projection. We will show first that $X \oplus \mathfrak{J}$ is a π_λ-space. Let F be a finite dimensional subspace of $X \oplus \mathfrak{J}_n$, where

$$\mathfrak{J}_n = \{(x_1,\dots,x_n,0,0,\dots) : x_i \in J_i \text{ for } i = 1,\dots,n\}.$$

Then $F \subset E \oplus \mathfrak{J}_n$ for some finite dimensional subspace E of X. Since X is complemented in J, we may apply the theorem saying that a separable Banach space has the bounded approximation property (B.A.P.) if and only if it is isomorphic to a complemented subspace of a space with a basis (see e.g. Lindenstrauss and Tzafriri [1]); thus X has the B.A.P. Hence if $\varepsilon > 0$ and $(1 - \varepsilon)^{-1}\varepsilon(\dim E) < 1$, there exists a finite rank operator $S_1 : X \longrightarrow X$ with $\|S_1\| \le \beta$, with β independent of E, such that $\|S_1x - x\| \le \varepsilon$ for every $x \in B_E$, the closed unit ball in E. Then by Lemma 2.j.2 we can choose an operator $S : X \longrightarrow X$ with finite dimensional range such that $S|_E = I_E$ and $\|S\| \le 2\beta$. Since J has a basis it is a π_λ-space and by Lemma 2.j.3 there exists a projection $Q' : J \longrightarrow J$ with finite dimensional range such that $\|Q'\| < \lambda$ and $S(X) \cup (I-S)S(X) \subset Q'J$. Let $H = Q'(J)$, $\varepsilon > 0$ and $\{f_1,\dots,f_k\}$ be a normalized basis of H with basis constant M. By Lemma 2.b.6 there exist $m > n$ and an isomorphism L from H into J_m with $\|Lh - h\| \le \varepsilon\|h\|/(2\sqrt{k}M\|Q'\|)$ and $\|L^{-1}\| < 2$. Hence by Proposition 1.8 there exists an onto projection $Q : J \longrightarrow LH$ with $\|Q\| \le 2\|Q'\|$. Then $Q|_{J_m} : J_m \longrightarrow LH$ is also a projection.

Let i_n be the inclusion from J_n into $\mathfrak{J}$, let P_n be the natural projection from $\mathfrak{J}$ onto i_nJ_n and let P_0 be the natural projection of $\mathfrak{J}$ onto $\mathfrak{J}_n$. Clearly $\|P_i\| = 1$ for $i = 0, 1, \dots$. Define $R : X \oplus \mathfrak{J} \longrightarrow X \oplus \mathfrak{J}$ by

$$R(x,y) = (Sx + PL^{-1}Qi_m^{-1}P_m y,\ i_mL(I - S)Sx + i_mL(I - SP)L^{-1}Qi_m^{-1}P_m y + P_0y).$$

Then R has finite dimensional range and there is a uniform bound μ for all such operators R. We will show that R is a projection on $X \oplus \mathfrak{J}$.

(1) If $x \in E$ then $Sx = x$ and $(I - S)Sx = 0$, so that

$$R(x, 0) = (Sx, i_m L(I - S)Sx) = (x, 0).$$

(2) If $y \in \mathfrak{J}_n$ then $P_m y = 0$ and $P_0 y = y$ so that

$$R(0, y) = (PL^{-1}Qi_m^{-1}P_m y, i_m L(I - SP)L^{-1}Qi_m^{-1}P_m y + P_0 y) = (0, y).$$

(3) Since $P_0 i_m L = 0$, $P_m i_m L = i_m L$, $QL = L$ and $PS = S$, we have

$$R^2(x, 0) = R(Sx, i_m L(I - S)Sx) =$$

$$= (S^2 x + (I - S)Sx, i_m L(I - S)S^2 x + i_m L(I - SP)(I - S)Sx) =$$

$$= (Sx, i_m L(I - S)Sx) = R(x, 0).$$

(4) Since $P_m i_m L = i_m L$, $QL = L$, $P_m P_0 = 0$, $P_0 i_m L = 0$ and $PS = S$, for every $y \in \mathfrak{J}$

$$R^2(0, y) = R(PL^{-1}Qi_m^{-1}P_m y, i_m L(I - SP)L^{-1}Qi_m^{-1}P_m y + P_0 y) =$$

$$= (SPL^{-1}Qi_m^{-1}P_m y + P(I - SP)L^{-1}Qi_m^{-1}P_m y, i_m L(I - S)SPL^{-1}Qi_m^{-1}P_m y +$$

$$+ i_m L(I - SP)(I - SP)L^{-1}Qi_m^{-1}P_m y + P_0 y) =$$

$$= (PL^{-1}Qi_m^{-1}P_m y, i_m LSPL^{-1}Qi_m^{-1}P_m y - i_m LS^2PL^{-1}Qi_m^{-1}P_m y +$$

$$+ i_m L(I - SP)L^{-1}Qi_m^{-1}P_m y - i_m LSPL^{-1}Qi_m^{-1}P_m y + i_m LSPSPL^{-1}Qi_m^{-1}P_m y + P_0 y) =$$

$$= (PL^{-1}Qi_m^{-1}P_m y, i_m L(I - SP)L^{-1}Qi_m^{-1}P_m y + P_0 y) = R(0, y).$$

Therefore R is a projection with $R(x, y) = (x, y)$ for $(x, y) \in E \oplus \mathfrak{J}_n$ and in particular $R(x, y) = (x, y)$ if $(x, y) \in F$.

Now, if F is any finite dimensional subspace of $X \oplus \mathfrak{J}$, then $F \subseteq E \oplus G$, where E and G are finite dimensional subspaces of X and $\mathfrak{J}$ respectively. Let $\varepsilon > 0$ with $\varepsilon\dim(E \oplus G) < 1 - \varepsilon$. By Lemma 2.b.6, there exists an integer n such that G is isomorphic to a subspace of $\mathfrak{J}_n$ via an isomorphism ϕ with

$$\|\phi - I_G\| \le \frac{\varepsilon}{\mu + 1},$$

where μ is the bound for the operators R found above.

Therefore, if $T : E \oplus G \longrightarrow E \oplus \mathfrak{J}_n$ is given by $T(e, g) = (e, \phi g)$, then

$$\|T - I_{E\oplus G}\| \le \frac{\varepsilon}{\mu + 1}.$$

Let R be the projection found above for the space $E \oplus \phi G$; then for $e \in E$ and $g \in G$, since $RT = T$, and $\|R\| \le \mu$,

$$\|R(e, g) - (e, g)\| \le \|R(e, g) - R(T(e, g))\| + \|R(T(e, g)) - T(e, g)\| +$$
$$+ \|T(e, g) - (e, g)\| \le \|T(e, g) - (e, g)\|(\|R\| + 1) < \varepsilon\|(e, g)\|.$$

Thus, we may apply Lemma 2.j.2 to obtain a projection P on $X \oplus \mathfrak{J}$, so that $E \oplus G$ and therefore F, are contained in its range and $\|P\| \le 2\|R\| \le 2\mu$. Hence $X \oplus \mathfrak{J}$ satisfies condition (ii) of Lemma 2.j.3, and therefore is a π_λ-space.

By Corollary 2.c.5 and Theorem 2.c.6 J is isomorphic to $\mathfrak{J} \oplus W$ for some space W, and by Corollary 2.c.7 W is isomorphic to J. Thus J is isomorphic to $J \oplus \mathfrak{J}$. Since by assumption X is complemented in J, it follows that $X \oplus \mathfrak{J}$ is isomorphic to a complemented subspace of J. Therefore, using the theorem mentioned at the beginning of this proof, $(X \oplus \mathfrak{J})^*$ has the bounded approximation property. It follows from Theorem 2.j.10 that $X \oplus \mathfrak{J}$ has a shrinking finite dimensional decomposition.

Theorem 2.c.17. Suppose X is a reflexive subspace of J with a shrinking finite dimensional decomposition. Then X is isomorphic to $(\sum_n X_n)_{\ell_2}$ for some sequence of finite dimensional spaces.

Proof: Let $\{H_n\}$ be a shrinking F.D.D. for X. Since $\{e_n\}$ is a shrinking basis for J, as was shown in Theorem 2.a.2, applying Lemma 2.j.12, we get that for every $\varepsilon > 0$ there exist blockings $\{E_n\}$ of $\{e_n\}$ and $\{F_n\}$ of $\{H_n\}$ such that for any $x_n \in F_n$ there is a $z_n \in E_n \oplus E_{n+1}$ with

(1) $$\|x_n - z_n\| \le (\varepsilon/2^{n+2})\|x_n\|.$$

By Lemma 2.j.11 there exist natural numbers $0 = p_0 < p_1 < p_2 < \ldots$ such that if $x = \sum_{i=1}^{\infty} x_i$ in X with $x_i \in F_i$ and $\|x\| \le 1$, then for any n there is some $p_n + 2 < i_n < p_{n+1}$ such that

(2) $$\max\{\|x_{i_n}\|, \|x_{i_n+1}\|\} \le \varepsilon/2^{n+1}.$$

Let $x = \sum_{i=1}^{\infty} x_i$ with $x_i \in F_i$, $\|x\| \le 1$, $\{p_n\}$, $\{i_n\}$ as above and define $i_{-1} = 0$. If $\{q_n\} \subset \mathbb{N}$ is such that $1 \le q_n < q_{n+1}$ and C is the decomposition constant of $\{F_n\}$ then $\|x_n\| \le 2C$ for n = 1, 2,... and

(3) $$\left(\sum_{n=0}^{\infty}\left\|\sum_{j=q_n}^{q_{n+1}}(z_j - x_j)\right\|^2\right)^{1/2} \le \left(\sum_{n=0}^{\infty}\left(\sum_{j=q_n}^{q_{n+1}}\|(z_j - x_j)\|\right)^2\right)^{1/2} \le$$

$$\le \left(\sum_{n=0}^{\infty}\left(\sum_{j=q_n}^{q_{n+1}}(2C\varepsilon/2^{j+2})\right)^2\right)^{1/2} \le \frac{\varepsilon C}{2}\left\{\sum_{n=0}^{\infty}\left((1/2)^{q_n-1} - (1/2)^{q_{n+1}}\right)^2\right\}^{1/2} \le$$

$$\le \frac{\varepsilon C}{2}\left(\sum_{n=0}^{\infty}(1/2)^{2n}\right)^{1/2} \le \varepsilon C.$$

Define

$$X_n = F_{p_n+1} \oplus F_{p_n+2} \oplus \ldots \oplus F_{p_{n+1}}.$$

We will show that $(\sum_n X_n)_{\ell_2}$ satisfies the conclusion of the theorem.

Using (2) and (3) we obtain

(4) $$(\sum_{n=0}^{\infty}\|\sum_{j=p_n+1}^{p_{n+1}}x_j\|^2)^{1/2} \le (\sum_{n=0}^{\infty}\|\sum_{j=p_n+1}^{i_n-1}x_j\|^2)^{1/2} + (\sum_{n=0}^{\infty}\|x_{i_n}\|^2)^{1/2} +$$

$$+ (\sum_{n=0}^{\infty}\|x_{i_n+1}\|^2)^{1/2} + (\sum_{n=0}^{\infty}\|\sum_{j=i_n+2}^{p_{n+1}}x_j\|^2)^{1/2} \le$$

$$\le (1 + C)(\sum_{n=0}^{\infty}\|\sum_{j=i_{n-1}+2}^{i_n-1}x_j\|^2)^{1/2} + 2\varepsilon + C(\sum_{n=0}^{\infty}\|\sum_{j=i_n+2}^{i_{n+1}-1}x_j\|^2)^{1/2} \le$$

$$\le 2\varepsilon + (1 + 2C)(\sum_{n=0}^{\infty}\|\sum_{j=i_{n-1}+2}^{i_n-1}x_j\|^2)^{1/2} \le$$

$$\le 2\varepsilon + (1+2C)\left((\sum_{n=0}^{\infty}\|\sum_{j=i_{n-1}+2}^{i_n-1}(x_j-z_j)\|^2)^{1/2} + (\sum_{n=0}^{\infty}\|\sum_{j=i_{n-1}+2}^{i_n-1}z_j\|^2)^{1/2}\right) \le$$

$$\le \varepsilon(2 + C + 2C^2) + (1 + 2C)(\sum_{n=0}^{\infty}\|\sum_{j=i_{n-1}+2}^{i_n-1}z_j\|^2)^{1/2}.$$

But $\left\{\sum_{j=i_{n-1}+2}^{i_n-1}z_j\right\}_{n=1}^{\infty}$ is a sequence in J of disjoint blocks of $\{e_n\}$ so that $(i_n + 2) - (i_n - 1) > 2$. Hence by Lemma 2.c.2

(5) $$(\sum_{n=0}^{\infty}\|\sum_{j=i_{n-1}+2}^{i_n-1}z_j\|^2)^{1/2} \le \|\sum_{n=0}^{\infty}\sum_{j=i_{n-1}+2}^{i_n-1}z_j\|.$$

Combining (4), (5) (1) and (2) we get

$$(\sum_{n=0}^{\infty}\|\sum_{j=p_n+1}^{p_{n+1}}x_j\|^2)^{1/2} \le \varepsilon(2 + C + 2C^2) + (1 + 2C)\|\sum_{n=0}^{\infty}\sum_{j=i_{n-1}+2}^{i_n-1}z_j\| \le$$

$$\le \varepsilon(2 + C + 2C^2) + (1 + 2C)\sum_{n=0}^{\infty}\sum_{j=i_{n-1}+2}^{i_n-1}\|z_j - x_j\| +$$

$$+ (1 + 2C)\|\sum_{n=0}^{\infty}\sum_{j=i_{n-1}+2}^{i_n-1}x_j\| \le$$

$$\leq \varepsilon(2 + C + 2C^2) + (1 + 2C)(2C\varepsilon) + (1 + 2C)\left\|\sum_{n=0}^{\infty}\sum_{j=i_{n-1}+2}^{i_n-1} x_j\right\| \leq$$

$$\leq \varepsilon(2+3C+6C^2) + (1+2C)\left(\left\|\sum_{n=0}^{\infty}\sum_{j=i_{n-1}+1}^{i_n} x_j\right\| + \sum_{n=0}^{\infty}\|x_{i_{n-1}+1}\| + \sum_{n=0}^{\infty}\|x_{i_n}\|\right) \leq$$

$$\leq \varepsilon(4 + 7C + 6C^2) + (1 + 2C)\|x\|.$$

Therefore, if ε is such that $\varepsilon < 1/(4 + 7C + 6C^2)$, for all $x \in X$ with $\|x\| = 1$ we have

$$(6) \qquad (\sum_{n=0}^{\infty}\|\sum_{j=p_n+1}^{p_{n+1}} x_j\|^2)^{1/2} \leq 1 + (1 + 2C)\|x\| = 2(1 + C)\|x\|.$$

On the other hand, for all $x \in X$ with $\|x\| = 1$, using the triangle inequality, (1), (3) and Lemma 2.c.2, we get

$$\|x\| \leq \sum_{i=0}^{2}\left\|\sum_{n=0}^{\infty}\sum_{j=p_{3n+i}+1}^{p_{3n+i+1}} x_j\right\| \leq C\varepsilon + \sum_{i=0}^{2}\left\|\sum_{n=0}^{\infty}\sum_{j=p_{3n+i}+1}^{p_{3n+i+1}} z_j\right\| \leq$$

$$\leq C\varepsilon + \sqrt{2}\sum_{i=0}^{2}\left(\sum_{n=0}^{\infty}\left\|\sum_{j=p_{3n+i}+1}^{p_{3n+i+1}} z_j\right\|^2\right)^{1/2} \leq C\varepsilon + 3\sqrt{2}\left(\sum_{n=0}^{\infty}\left\|\sum_{j=p_n+1}^{p_{n+1}} z_j\right\|^2\right)^{1/2} \leq$$

$$\leq C\varepsilon+3\sqrt{2}C\varepsilon+3\sqrt{2}\left(\sum_{n=0}^{\infty}\left\|\sum_{j=p_n+1}^{p_{n+1}} x_j\right\|^2\right)^{1/2} \leq 6C\varepsilon+3\sqrt{2}\left(\sum_{n=0}^{\infty}\left\|\sum_{j=p_n+1}^{p_{n+1}} x_j\right\|^2\right)^{1/2}.$$

Therefore, if $\varepsilon < 1/(12C)$ then

$$(7) \qquad 1/2 = (1/2)\|x\| \leq \|x\| - 6C\varepsilon \leq 3\sqrt{2}\left(\sum_{n=0}^{\infty}\left\|\sum_{j=p_n+1}^{p_{n+1}} x_j\right\|^2\right)^{1/2}.$$

Inequalities (6) and (7) show that X is isomorphic to $(\sum_n X_n)_{\ell_2}$.

Finally, combining the two previous results, we can give Casazza's proof which characterizes the complemented reflexive subspaces of J as complemented subspaces of $\mathfrak{J}$. In Section 2.g we will show that not only can the complemented reflexive subspaces of J be viewed as subspaces of $\mathfrak{J}$, but all reflexive subspaces of J are isomorphic to subspaces of $\mathfrak{J}$.

Proposition 2.c.18. Every complemented reflexive subspace X of J is isomorphic to a complemented subspace of $\mathfrak{J}$.

Proof: By Proposition 2.c.16, $X \oplus \mathfrak{J}$ has a shrinking F.D.D. and is isomorphic to a complemented subspace W of J. Since both X and $\mathfrak{J}$ are reflexive, it follows from Theorem 2.c.17 that $X \oplus \mathfrak{J}$ is isomorphic to $(\sum_n X_n)_{\ell_2}$ for some sequence of finite dimensional subspaces of J. Thus as

$(\sum_n X_n)_{\ell_2}$ is isomorphic to W, let S be an isomorphism between $(\sum_n X_n)_{\ell_2}$ and W, let Q be the projection from J onto W and P_n the natural projection from $(\sum_n X_n)_{\ell_2}$ onto $i_n X_n$ where $i_n : X_n \to (\sum_n X_n)_{\ell_2}$ is the inclusion map. Then $\|P_n\| \le 1$ and clearly $Q_n = SP_nS^{-1}Q$ is a projection from J onto $S(i_n X_n) = Y_n$ with $\|Q_n\| \le \|S^{-1}\|\|S\|\|Q\|$ for every $n \in \mathbb{N}$. Let k_n be the dimension of Y_n and $1 > \varepsilon_n > 0$ such that

(1) $$\delta_n = \left(\varepsilon_n k_n^{-1/2}\|Q_n\|(1 + \varepsilon_n)/(1 - \varepsilon_n)\right) < 1/(k_n + 1).$$

By Lemma 2.b.6 there exist J_{m_n} and an isomorphism T_n from Y_n into J_{m_n} with

$$\|T_n\| \le (1 + \varepsilon_n),\ \|T_n^{-1}\| \le 1/(1 - \varepsilon_n) \text{ and } \|T_n - I_{Y_n}\| \le \varepsilon_n k_n^{-1/2}.$$

Let $R_n : J \to T_n Y_n$ be defined by $R_n = T_n Q_n$. Then, if $z = T_n y$ with $y \in Y_n$, keeping in mind that $Q_n y = y$ we have

$$\|R_n z - z\| \le \|T_n\|\|Q_n T_n y - Q_n y\| \le \|T_n\|\|Q_n\|\varepsilon_n k_n^{-1/2}\|y\| \le$$

$$\le \|T_n\|\|T_n^{-1}\|\|Q_n\|\varepsilon_n k_n^{-1/2}\|z\| \le \left(\varepsilon_n k_n^{-1/2}\|Q_n\|(1 + \varepsilon_n)/(1 - \varepsilon_n)\right)\|z\| = \delta_n\|z\|.$$

Applying Lemma 2.j.2 there exists an operator S_n from J onto H where H is a subspace of dimension k_n of J such that $S_n|_{T_nY_n} = I_{T_nY_n}$ and

$$\|S_n - R_n\| \le (k_n\delta_n/(1 - \delta_n))\|R_n\| < \|R_n\|,$$

where this inequality follows by (1). Therefore, since $\dim H = \dim Y_n$, we obtain that $H = T_nY_n$, and hence S_n is a projection from J onto T_nY_n. Since $T_nY_n \subset J_{m_n}$, for every $n \in \mathbb{N}$, $S_n|_{J_{m_n}}$ is a projection from J_{m_n} onto T_nY_n with

$$\|S_n\| \le 2\|R_n\| \le 2\|T_n\|\|Q_n\| \le 2\|T_n\|\|S^{-1}\|\|S\|\|Q\| \le 4\|S^{-1}\|\|S\|\|Q\|.$$

Observing that we may take $m_1 < m_2 < \ldots$, we conclude by Proposition 1.10 that there exists a projection from $(\sum_n J_{m_n})_{\ell_2}$ onto $(\sum_n T_nY_n)_{\ell_2}$. Since

$$\|T_nSi_n\| \le (1 + \varepsilon_n)\|S\| \text{ and } \|i_n^{-1}S^{-1}T_n^{-1}\| \le 1/(1 - \varepsilon_n),$$

again by Proposition 1.10, $(\sum_n X_n)_{\ell_2}$ is isomorphic to a complemented subspace of $(\sum_n J_{m_n})_{\ell_2}$, and thus by Theorem 2.c.9, to a complemented subspace of $\mathfrak{J}$.

We finish with three results related to the topics of this section, which are needed later, in Section 2.d, for the proof of the primarity of J.

Proposition 2.c.19. The spaces $\mathfrak{J} \oplus \mathfrak{J}$ and $(\mathfrak{J} \oplus \mathfrak{J} \oplus \ldots)_{\ell_2}$ are isomorphic to $\mathfrak{J}$.

Proof: Using Theorem 2.c.9 and Proposition 1.10 we see that

$$\mathfrak{J} \oplus \mathfrak{J} = (\Sigma_n J_n)_{\ell_2} \oplus (\Sigma_n J_n)_{\ell_2} \approx (\Sigma_n J_{2n})_{\ell_2} \approx \mathfrak{J}.$$

On the other hand suppose that $\mathbb{N} = \bigcup_{i=1}^{\infty} N_i$ where N_i is infinite for every i and $N_i \cap N_j = \emptyset$ for $i \neq j$. For instance let $\{p_i\}_{i=1}^{\infty}$ be the set of primes, $N_i = \{p_i^j\}_{j=1}^{\infty}$, $i = 1, 2, \ldots$, and $N_0 = \mathbb{N} \setminus \bigcup_{i=0}^{\infty} N_i$. Then by Theorem 2.c.9, for every i, $(\Sigma_{n \in N_i} J_n)_{\ell_2} \approx \mathfrak{J}$. Let $x_i \in J_i$; since

$$\Sigma_{i=1}^{\infty} \|x_i\|^2 = \Sigma_{j=0}^{\infty} \Sigma_{i \in N_j} \|x_i\|^2,$$

we get

$$\mathfrak{J} \approx (\mathfrak{J} \oplus \mathfrak{J} \oplus \ldots)_{\ell_2}.$$

Corollary 2.c.20. If X is a complemented reflexive subspace of J, then $(\Sigma_n X)_{\ell_2}$ is isomorphic to a complemented subspace of J.

Proof: By Proposition 2.c.18, X is isomorphic to a complemented subspace of $\mathfrak{J}$, and by Corollary 2.c.5 and Theorem 2.c.6, $\mathfrak{J}$ is isomorphic to a complemented subspace of J. Using Proposition 1.10 and the fact that $(\Sigma_n \mathfrak{J})_{\ell_2}$ is isomorphic to $\mathfrak{J}$, we get the desired result.

Corollary 2.c.21. If X is a complemented reflexive subspace of J, then J is isomorphic to $J \oplus X$.

Proof: By Corollary 2.c.20, J is isomorphic to $(\Sigma_n X)_{\ell_2} \oplus W$ for some W. Hence

$$J \approx (\Sigma_n X)_{\ell_2} \oplus W \approx X \oplus (\Sigma_n X)_{\ell_2} \oplus W \approx X \oplus J.$$

Until now we have only dealt with the reflexive complemented subspaces of J; in Section 2.d we will show that the only non-reflexive complemented subspace of J is J itself.

2.d. Basic sequences and the primarity of J

A central topic in the geometry of Banach spaces is the study of complemented subspaces and from this point of view, the simplest case besides prime spaces is that of primary spaces. It is known that most of the classical spaces are of this type, and in this section we will study the behavior of block basic sequences, unconditional, spreading and symmetric basic sequences in J, ultimately showing that J is also primary.

Furthermore, along the way, several properties of these basic sequences which are interesting in their own right will be established, notably Corollary 2.d.3 and Theorem 2.d.18, characterizing the unconditional block basic sequences of $\{e_n\}$ and the spreading basic sequences of J.

The following results about block basic sequences of the unit vector basis of J are by Casazza, Lin and Lohman [1] and will be used to prove that J has uncountably many non-equivalent unconditional basic sequences and that J is somewhat reflexive.

Theorem 2.d.1. Suppose $p_1 \le q_1 < p_2 \le q_2 <\dots$. For $n = 1, 2,\dots$ let $y_n = \sum_{i=p_n}^{q_n} e_i$. Then

(i) If $p_{n+1} = q_n + 1$ for all except finitely many n, $[y_n]$ is isomorphic to J.

(ii) If $p_{n+1} > q_n + 1$ for infinitely many n, there exists a sequence of positive integers $\{t_k\}_{k=1}^{\infty}$ such that $[y_n]$ is isomorphic to $(\sum_k J_{t_k})_{\ell_2}$.

(iii) $[y_n]$ is complemented in J.

Proof: (i) Suppose $p_{n+1} = q_n + 1$ for $n \ge N$. Let δ be such that for every sequence of real numbers $\alpha_1,\dots,\alpha_N$,

$$|\sum_{i=1}^{N}\alpha_i^2| \leq \delta\|\sum_{i=1}^{N}\alpha_i e_i\|^2.$$

Then, by the definition of the norm in J, for every $m \in \mathbb{N}$ we have

$$\|\sum_{i=1}^{m}\alpha_i e_i\| \leq \|\sum_{i=1}^{m}\alpha_i y_i\| \leq \sqrt{1+\delta}\ \|\sum_{i=1}^{m}\alpha_i e_i\|.$$

(ii) Let $\{n_k\}$ be the sequence of all positive integers n such that $p_{n+1} > q_n + 1$. Let $n_0 = 0$ and $t_k = n_k - n_{k-1}$ for $k = 1, 2, \ldots$. Then for each fixed k and for every sequence of real numbers $\alpha_1, \ldots, \alpha_{t_k}$,

$$\|\sum_{i=1}^{t_k}\alpha_i e_i\| = \|\sum_{i=1}^{t_k}\alpha_i y_{n_{k-1}+i}\|.$$

For $x = (\sum_{i=1}^{t_1} a_i^{(1)} e_i, \ldots, \sum_{i=1}^{t_m} a_i^{(m)} e_i, 0, 0, \ldots)$ in $(\sum_k J_{t_k})_{\ell_2}$ let

$$y = \sum_{j=1}^{m}\sum_{i=1}^{t_j} a_i^{(j)} y_{n_{j-1}+i}.$$

Then

$$y = \sum_{j=1}^{m}\sum_{r=p_{n_{j-1}+1}}^{q_{n_j}} \beta_r^{(j)} e_r$$

where $\beta_r^{(j)} = a_i^{(j)}$ if $p_{n_{j-1}+i} \leq r \leq q_{n_{j-1}+i}$. Therefore, using the above and Lemma 2.c.2,

$$\|x\|^2 = \sum_{j=1}^{m}\|\sum_{i=1}^{t_j} a_i^{(j)} e_i\|^2 = \sum_{j=1}^{m}\|\sum_{i=1}^{t_j} a_i^{(j)} y_{n_{j-1}+i}\|^2 =$$

$$= \sum_{j=1}^{m}\|\sum_{r=p_{n_{j-1}+1}}^{q_{n_j}} \beta_r^{(j)} e_r\|^2 \leq \|\sum_{j=1}^{m}\sum_{r=p_{n_{j-1}+1}}^{q_{n_j}} \beta_r^{(j)} e_r\|^2 =$$

$$= \|y\|^2 \leq 2\sum_{j=1}^{m}\|\sum_{r=p_{n_{j-1}+1}}^{q_{n_j}} \beta_r^{(j)} e_r\|^2 = 2\|x\|^2.$$

(iii) Let W be the closed linear span of

$$[y_n] \cup \{e_i : q_k < i < p_{k+1} \text{ for some } k \geq 0\} = [f_s],$$

where $q_0 = 0$ and

$$\{f_s\} = \{e_1, \ldots, e_{p_1-1}, y_1, e_{q_1+1}, \ldots, e_{p_2-1}, y_2, e_{q_2+1}, \ldots\}.$$

Then, by the proof of Corollary 2.c.7, there is a projection Q from J onto W with $\|Q\| = 1$, and J is isometric to W via an isometry T with $Te_i = f_i$. If $f_{m_i} = y_i$, by Theorem 2.c.6 there is a projection P of J onto

$[T^{-1}y_i] = [e_{m_i}]$ with $\|P\| \le 2$. Therefore $R = TPT^{-1}$ is a projection of W onto $[y_n]$ with $\|R\| \le 2$, and since $Qy_n = y_n$, RQ is a projection of J onto $[y_n]$ with $\|RQ\| \le 2$.

Since c_0 and ℓ_1 do not contain infinite dimensional reflexive subspaces, it was believed that every non-reflexive Banach space contained a closed infinite dimensional subspace with this property. In 1967 Herman and Whitley [1] proved this conjecture to be false, by showing that every closed infinite dimensional subspace of J contains a subspace isomorphic to ℓ_2. The following theorem and its two corollaries lead to this result; the proof given here comes from Casazza, Lin and Lohman [1].

Theorem 2.d.2. For $j = 1, 2, \ldots$ let $y_j = \sum_{n=p_j}^{q_j} \alpha_n e_n$ be a block basic sequence of $\{e_n\}_{n=1}^{\infty}$. If $p_{j+1} - q_j > 1$ then $[y_j]_{j=1}^{\infty}$ is complemented in J. If in addition $\{y_j\}_{j=1}^{\infty}$ is seminormalized, then it is equivalent to the unit vector basis of ℓ_2.

Proof: Let $\{n_i\}_{i=1}^{\infty}$ denote the sequence

$$p_1, p_1 + 1, \ldots, q_1, p_2, p_2 + 1, \ldots, q_2, \ldots.$$

By the nature of the norm in J, $[e_{n_i}]$ is isometric to $[e_{r_i}]$ where

$$r_{s_k + j} = s_k + j + 2k$$

for $k = 0, 1, \ldots,$ $j = 1, \ldots, q_{k+1} - p_{k+1} + 1$, $s_0 = 0$ and $s_k = \sum_{i=1}^{k}(p_i - q_i)$.

By Theorem 2.c.4, $\{e_{r_i}\}_{i=1}^{\infty}$ is equivalent to the natural basis of $(\sum_n J_{t_n})_{\ell_2}$, where $t_n = q_n - p_n + 1$, via an isomorphism T_1 with $\|T_1\|\|T_1^{-1}\| \le \sqrt{2}$, such that

$$T_1 \sum_{n=p_j}^{q_j} \beta_n e_n = (0, \ldots, 0, \sum_{n=1}^{t_j} \beta_{p_j+n-1} e_n, 0, 0, \ldots).$$

Thus $\{e_{n_i}\}$ is also equivalent to $(\sum_n J_{t_n})_{\ell_2}$, via an isomorphism T with

$$T \sum_{n=p_j}^{q_j} \beta_n e_n = T_1 \sum_{n=p_j}^{q_j} \beta_n e_n.$$

For each $n \in \mathbb{N}$, choose $f_n \in (i_n J_{t_n})^*$, with $\|f_n\| = 1$, such that

$$f_n(Ty_n) = \|Ty_n\|, \tag{1}$$

where i_n is the inclusion of J_{t_n} into $(\sum_n J_{t_n})_{\ell_2}$. For each $x \in [e_{n_i}]_{i=1}^{\infty}$, $x = \sum_{i=1}^{\infty} b_{n_i} e_{n_i}$, we can write $x = \sum_{n=1}^{\infty}\sum_{m=p_n}^{q_n} b_m e_m$ and define $P : J \longrightarrow [y_n]$ by

$$P(x) = T^{-1}\left(\sum_{n=1}^{\infty}\|Ty_n\|^{-1} f_n(T(\sum_{m=p_n}^{q_n} b_m e_m))Ty_n\right).$$

Since $Ty_n \in i_n J_{t_n}$, we get from (1) and the definition of the norm in $(\sum_n J_{t_n})_{\ell_2}$ that

$$\|\sum_{n=1}^{\infty}\|Ty_n\|^{-1} f_n(T(\sum_{m=p_n}^{q_n} b_m e_m))Ty_n\| = \left(\sum_{n=1}^{\infty}\left(f_n(T(\sum_{m=p_n}^{q_n} b_m e_m))\right)^2\right)^{1/2}.$$

Using Lemma 2.c.2 we obtain

$$\|P(x)\| \le \|T\|\|T^{-1}\|\left(\sum_{n=1}^{\infty}\|\sum_{m=p_n}^{q_n} b_m e_m\|^2\right)^{1/2} \le \sqrt{2}\|x\|.$$

Since $Py_n = y_n$, it follows that P is a bounded projection of $[e_{n_i}]_{i=1}^{\infty}$ onto $[y_n]_{n=1}^{\infty}$. But by Theorem 2.c.6, $[e_{n_i}]_{i=1}^{\infty}$ is complemented in J by a projection of norm less than or equal to 2. Hence $[y_j]_{j=1}^{\infty}$ is complemented in J by a projection of norm at most $2\sqrt{2}$. Now suppose $L \le \|y_n\| \le M$ for $n = 1, 2,\ldots$. Let $y = \sum_{m=1}^{\infty}\beta_m y_m \in [y_m]_{m=1}^{\infty}$. Then

$$y = \sum_{m=1}^{\infty}\sum_{i=p_n}^{q_n} \beta_m \alpha_i e_i,$$

and by Lemma 2.c.2,

$$\sum_{m=1}^{\infty}\|\sum_{i=p_n}^{q_n} \beta_m \alpha_i e_i\|^2 \le \|y\|^2 \le 2\sum_{m=1}^{\infty}\|\sum_{i=p_n}^{q_n} \beta_m \alpha_i e_i\|^2.$$

But $\|\sum_{i=p_n}^{q_n} \beta_m \alpha_i e_i\|^2 = \beta_m^2\|y_m\|^2$. Therefore

$$L^2\sum_{m=1}^{\infty}\beta_m^2 \le \|y\|^2 \le 2M^2\sum_{m=1}^{\infty}\beta_m^2$$

and thus $\{y_n\}$ is equivalent to the unit vector basis of ℓ_2.

Corollary 2.d.3. Every seminormalized unconditional block basic sequence $\{y_n\}_{n=1}^{\infty}$ of $\{e_n\}_{n=1}^{\infty}$ is equivalent to the unit vector basis of ℓ_2.

Proof: This follows from Theorem 2.d.2 and the fact that since $\{y_n\}_{n=1}^{\infty}$ is an unconditional basic sequence, $\sum_{n=1}^{\infty} a_n y_n$ converges if and only if $\sum_{n=1}^{\infty} a_{2n} y_{2n}$ and $\sum_{n=1}^{\infty} a_{2n-1} y_{2n-1}$ converge (see e.g. Beauzamy [1]).

The following result analyzes the subspaces of J with regard to their relationship to ℓ_2. Herman and Whitley [1] proved that every quasi-reflexive Banach space is somewhat reflexive, that is every closed infinite dimensional subspace contains a reflexive space. The next proof of this fact for the space J is direct and says more.

Corollary 2.d.4. Every closed infinite dimensional subspace E of J has a subspace F such that F is complemented in J and F is isomorphic to ℓ_2. In particular, J is a somewhat reflexive space.

Proof: By the well known theorem stating that if Y is a closed infinite dimensional subspace of a Banach space X with a Schauder basis $\{u_n\}_{n=1}^{\infty}$, then there is a subspace Z of Y having a basis which is equivalent to a block basis of $\{u_n\}_{n=1}^{\infty}$ (see e.g. Lindenstrauss and Tzafriri [1]), there exists in E a normalized basic sequence $\{x_n\}_{n=1}^{\infty}$ equivalent to a block basis of $\{e_n\}_{n=1}^{\infty}$. Then $\{x_{2n}\}_{n=1}^{\infty}$ satisfies the conditions of Theorem 2.d.2. Therefore $\{x_{2n}\}_{n=1}^{\infty}$ is equivalent to the unit vector basis of ℓ_2, and $[x_{2n}]_{n=1}^{\infty}$ is complemented in J.

Although, as we saw in Theorem 2.a.2, J has no unconditional bases, it has uncountably many mutually non-equivalent unconditional basic sequences. This is a consequence of the next results.

Proposition 2.d.5. Given $\lambda > 1$, $1 \leq p \leq \infty$ and a positive integer n there exist an integer m and an isomorphism T from $\ell_p^{(n)}$ into J_m such that for all $x \in \ell_p^{(n)}$

$$\lambda^{-1}\|x\| \leq \|Tx\| \leq \lambda\|x\|.$$

Proof: Since by Corollary 2.b.10 ℓ_p is finitely representable in J, for every $\varepsilon > 0$ there exists an isomorphism S from $\ell_p^{(n)}$ into a subspace of J with $(1 - \varepsilon)^{-1}\|x\| \leq \|Sx\| \leq (1 + \varepsilon)\|x\|$.

On the other hand, by Lemma 2.b.6 there are an integer m and an isomorphism U from $S(\ell_p^{(n)})$ into J_m such that $(1-\varepsilon)^{-1}\|x\| \le \|Ux\| \le (1+\varepsilon)\|x\|$.

Theorem 2.d.6. Let $\{n_i\}_{i=1}^{\infty}$ be a sequence of positive integers. Then

(i) the Banach space $\left(\sum_{i=1}^{\infty} \ell_{\infty}^{(n_i)}\right)_{\ell_2}$ is isomorphic to a complemented subspace of J,

(ii) the Banach space $\left(\sum_{i=1}^{\infty} \ell_p^{(n_i)}\right)_{\ell_2}$, $1 \le p < \infty$, is isomorphic to a subspace of J.

Proof: (i) By Proposition 2.d.5 there exist a sequence $\{m_i\}_{i=1}^{\infty}$ and isomorphisms T_i from $\ell_{\infty}^{(n_i)}$ into J_{m_i} with $\sup_i\|T_i\| \le \sqrt{2}$ and $\sup_i\|T_i^{-1}\| \le \sqrt{2}$.

Then by Proposition 1.10, $\left(\sum_{i=1}^{\infty} \ell_{\infty}^{(n_i)}\right)_{\ell_2}$ is isomorphic to the subspace $\left(\sum_{i=1}^{\infty} T_i\ell_{\infty}^{(n_i)}\right)_{\ell_2}$ of $\left(\sum_{i=1}^{\infty} J_{m_i}\right)_{\ell_2}$. It is easy to see that there exists an extension $R_i : J_{(m_i)} \to \ell_{\infty}^{(n_i)}$ of T_i^{-1} with $\|R_i\| = \|T_i^{-1}\|$ for $i = 1,2,\ldots$.

Hence $P_i = T_iR_i$ is a projection from J_{m_i} onto $T_i\left(\ell_{\infty}^{(n_i)}\right)$ with $\|P_i\| \le 2$.

By Theorems 2.c.4 and 2.c.6, $\left(\sum_{i=1}^{\infty} J_{m_i}\right)_{\ell_2}$ is isomorphic to a complemented subspace of J. Using Proposition 1.10 again, $\left(\sum_{i=1}^{\infty} T_i\ell_{\infty}^{(n_i)}\right)_{\ell_2}$, and therefore $\left(\sum_{i=1}^{\infty} \ell_{\infty}^{(n_i)}\right)_{\ell_2}$, are isomorphic to a complemented subspace of J.

(ii) By Proposition 2.d.5 there exist a sequence $\{m_i\}_{i=1}^{\infty}$ and isomorphisms T_i from $\ell_p^{(n_i)}$ into J_{m_i} with $\sup_i\|T_i\| \le \sqrt{2}$ and $\sup_i\|T_i^{-1}\| \le \sqrt{2}$.

Applying Proposition 1.10, we get an isomorphism from $\left(\sum_{i=1}^{\infty} \ell_p^{(n_i)}\right)_{\ell_2}$ into $\left(\sum_{i=1}^{\infty} J_{m_i}\right)_{\ell_2}$, and this concludes the proof.

Corollary 2.d.7. There are uncountably many mutually non-equivalent unconditional basic sequences in J.

Proof: The canonical basis of $(\sum_{n=1}^{\infty} \ell_p^{(n)})_{\ell_2}$ is unconditional for $p \ge 1$.

Another natural basis of J is the so-called summing basis $\{\xi_n\}_{n=1}^{\infty}$, which will be shown to be the only spreading basic sequence of J besides those sequences equivalent to the unit vector basis in ℓ_2.

Definition 2.d.8. The summing basis $\{\xi_n\}_{n=1}^{\infty}$ in J is given by $\xi_n = \sum_{i=1}^{n} e_i$ where $\{e_n\}_{n=1}^{\infty}$ is the unit vector basis in J.

Lemma 2.d.9. $\{\xi_n\}_{n=1}^{\infty}$ is a monotone, boundedly complete basis of J.

Proof: $e_1 = \xi_1$, $e_n = \xi_n - \xi_{n-1}$ for $n = 2, 3, \ldots$, and if $x = \sum_{i=1}^{\infty} a_i \xi_i \in J$, then

$$\|x\| = \|(\textstyle\sum_{i=1}^{\infty} a_i)e_1 + (\sum_{i=2}^{\infty} a_i)e_2 + (\sum_{i=3}^{\infty} a_i)e_3 + \ldots\| =$$

$$= \sup_n \left(\frac{1}{2} \sum_{k=0}^{n} (\sum_{i=p_k}^{p_{k+1}-1} a_i)^2 \right)^{1/2}$$

where the sup is taken over all positive integers n and all sequences of integers such that $0 = p_0 < p_1 < \ldots < p_{n+1}$. From this the result follows.

Definition 2.d.10. A basis $\{x_n\}_{n=1}^{\infty}$ of a Banach space X is called spreading if for every increasing sequence of integers $\{n_i\}_{i=1}^{\infty}$, $\{x_{n_i}\}_{i=1}^{\infty}$ is equivalent to $\{x_n\}_{n=1}^{\infty}$. If in addition $\{x_n\}_{n=1}^{\infty}$ is unconditional, then the basis is said to be subsymmetric.

A sequence $\{x_n\}_{n=1}^{\infty}$ is invariant under spreading (IS) if

$$\|\textstyle\sum_{i=1}^{k} a_i x_i\| = \|\sum_{i=1}^{k} a_i x_{n_i}\|$$

for every finite real sequence $\{a_i\}_{i=1}^{k}$ and for every $n_1 < n_2 < \ldots < n_k$.

Clearly if $\{x_n\}_{n=1}^{\infty}$ is an IS basis, it is spreading.

We will list several inequalities involving the canonical and summing bases in J, which come in handy when estimating norms and all of which are direct consequences of the definition of the norm in J.

Proposition 2.d.11. Let $\{e_n\}_{n=1}^{\infty}$ be the canonical basis and $\{\xi_n\}_{n=1}^{\infty}$ the summing basis of J; let $\{a_i\}_{i=1}^{\infty}$ be a sequence of real numbers. Then:

(1) $\|\sum_{i=1}^{n} a_i e_i\| \le \|\sum_{i=1}^{m} a_i e_i\|$ if $n \le m$.

(2) $\sup_i |a_i| \le \|\sum_{i=1}^{\infty} a_i e_i\|$.

(3) $\|\sum_{i=1}^{\infty} a_{n_i} e_i\| \le \|\sum_{i=1}^{\infty} a_i e_i\|$ for any $n_1 < n_2 <\dots$.

(4) $\|\sum_{i=1}^{\infty} a_i e_i\| = \|\sum_{i=n+1}^{\infty} a_{i-n} e_i\|$ for every n.

(5) $\|\sum_{i=1}^{\infty} a_i e_i\| \le \sqrt{2}(\sum_{i=1}^{\infty} a_i^2)^{1/2}$.

(6) $\|\sum_{i=1}^{\infty} a_i \xi_i\| = \|\sum_{i=1}^{\infty} a_i \xi_{n_i}\|$ for any $n_1 < n_2 <\dots$.

(7) $|\sum_{i=1}^{\infty} a_i| \le \|\sum_{i=1}^{\infty} a_i \xi_i\|$.

(8) $(\sum_{i=1}^{\infty} a_i^2)^{1/2} \le \sqrt{2}\|\sum_{i=1}^{\infty} a_i \xi_i\|$.

An immediate consequence from Proposition 2.d.11(6) is that the summing basis $\{\xi_n\}_{n=1}^{\infty}$ of J is invariant under spreading.

The next lemma is similar to Theorem 2.c.6, with the role of the unit vector basis played by the summing basis.

Lemma 2.d.12. For any $p_1 < p_2 <\dots$, the space $[\xi_{p_i}]$ is complemented in J.

Proof: The operator $P : J \longrightarrow [\xi_{p_i}]$ defined by

$$P(\sum_{i=1}^{\infty} a_i \xi_i) = \sum_{i=1}^{\infty}(\sum_{j=p_{i-1}+1}^{p_i} a_j)\xi_{p_i},$$

where $p_0 = 0$, is a projection. In fact it is clear that $P^2 = P$, and by (6) and (3) of Proposition 2.d.11,

$$\|P(\sum_{i=1}^{\infty} a_i \xi_i)\| = \|\sum_{i=1}^{\infty}(\sum_{j=p_{i-1}+1}^{p_i} a_j)\xi_{p_i}\| = \|\sum_{i=1}^{\infty}(\sum_{j=p_{i-1}+1}^{p_i} a_j)\xi_i\| =$$

$$= \|\sum_{i=1}^{\infty}(\sum_{j=p_{i-1}+1}^{\infty} a_j)e_i\| \le \|\sum_{i=1}^{\infty}(\sum_{j=i}^{\infty} a_j)e_i\| = \|\sum_{i=1}^{\infty} a_i \xi_i\|.$$

Hence $\|P\| = 1$.

We will see that the summing basis plays an important role in the study of seminormalized sequences which don't have weak cluster points. Indeed, the key technical tool for the proof of the primarity of J is the characterization of its spreading basic sequences and these fall naturally into two classes: those having no weak cluster point and those having zero as a weak cluster point. The former will be analyzed in Proposition 2.d.15 and the latter in Proposition 2.d.16.

The following results are by Andrew [2].

Proposition 2.d.13. Let $y \in J^{**} \setminus jJ$, where j is the canonical embedding of J into J^{**}, and suppose $n_1 < n_2 <\dots$ is a sequence of natural numbers. Let $y_i = \sum_{j=1}^{n_i} y(e_j^*)e_j \in J$ and suppose that $\{n_i\}$ is such that $y_i \neq y_{i+1}$ for every i. Then

(a) $\{y_i\}$ is a spreading basic sequence equivalent to $\{\xi_i\}$, and the constant of this equivalence depends only on y and not on $\{n_i\}$,

(b) $[y_i]$ is complemented in J by a projection P with $\|P\| < K$, where K depends on y but not on $\{n_i\}$.

Proof: Recall that if $y \in J^{**}$, using Theorems 1.5 and 2.a.2, we may identify y with the sequence $(y(e_1^*), y(e_2^*),\dots)$ and that $y \in J$ if and only if $\lim_n y(e_n^*) = 0$. Also from Theorem 1.5 it follows that

$$\|y\|_{J^{**}} = \sup_n \|\textstyle\sum_{i=1}^{n} y(e_i^*)e_i\|.$$

Thus let $y \in J^{**} \setminus jJ$ and $y_i = \sum_{j=1}^{n_i} y(e_j^*)e_j$.

First we prove the case where $y(e_j^*) \neq 0$ for all j. Define $T : J \longrightarrow J$ by

$$T\textstyle\sum_{i=1}^{\infty} a_i e_i = \sum_{i=1}^{\infty} y(e_i^*)a_i e_i.$$

Let $p_1 < p_2 < p_3 <\dots$. By the triangle inequality and (2) of Proposition 2.d.11,

$$\left(\frac{1}{2}\sum_{i=1}^{n}(y(e_{p_i}^*)a_{p_i} - y(e_{p_{i-1}}^*)a_{p_{i-1}})^2\right)^{1/2} \le$$

$$\le \left(\frac{1}{2}\sum_{i=1}^{n}(y(e_{p_i}^*)(a_{p_i} - a_{p_{i-1}}))^2\right)^{1/2} + \left(\frac{1}{2}\sum_{i=1}^{n}\left((y(e_{p_i}^*) - y(e_{p_{i-1}}^*))a_{p_{i-1}}\right)^2\right)^{1/2} \le$$

$$\le (\sup_i|y(e_i^*)|)\|\textstyle\sum_{i=1}^{\infty}a_ie_i\| + \sup_i|a_i|\,\|y\|_{J^{**}} \le$$

$$\le (\sup_i|y(e_i^*)| + \|y\|_{J^{**}})\|\textstyle\sum_{i=1}^{\infty}a_ie_i\| \le 2\|y\|_{J^{**}}\|\sum_{i=1}^{\infty}a_ie_i\|.$$

Therefore

$$\|T\textstyle\sum_{i=1}^{\infty}a_ie_i\| \le 2\|y\|_{J^{**}}\|\sum_{i=1}^{\infty}a_ie_i\|,$$

$$\|T\| \le 2\|y\|_{J^{**}},$$

and T is continuous. Moreover $T\xi_{n_i} = y_i$.

Hence if $\sum_{i=1}^{\infty} a_i\xi_i \in J$,

(1) $$\|\sum_{i=1}^{\infty} a_i y_i\| \le \|T\|\|\sum_{i=1}^{\infty} a_i\xi_{n_i}\| = \|T\|\|\sum_{i=1}^{\infty} a_i\xi_i\| \le 2\|y\|_{J^{**}}\|\sum_{i=1}^{\infty} a_i\xi_i\|.$$

Since $y \in J^{**} \setminus \tilde{j}J$, $\lim_j y(e_j^*)$ exists and is different from zero, and since also $y(e_j^*) \ne 0$ for every j, $\sup_j(|y(e_j^*)|^{-1}) < \infty$. It follows that the sequence $y' = \{y(e_j^*)^{-1}\}_{j=1}^{\infty} \in J^{**}$, and using the definition of norm in J^{**} we get

$$\|y'\|_{J^{**}} = \sup_n\|\sum_{i=1}^{n} y(e_i^*)^{-1}e_i\| \le \sup_j(|y(e_j^*)|^{-2})\sup_n\|\sum_{i=1}^{n} y(e_i^*)e_i\| =$$

$$= \sup_j(|y(e_j^*)|^{-2})\|y\|_{J^{**}}.$$

Therefore, as was done before for T, we get that T^{-1} is a bounded operator and $\|T^{-1}\| \le 2\|y'\|_{J^{**}} \le 2\sup_j(|y(e_j^*)|^{-2})\|y\|_{J^{**}}$.
Hence

$$\|\sum_{i=1}^{\infty} a_i\xi_i\| = \|\sum_{i=1}^{\infty} a_i\xi_{n_i}\| = \|\sum_{i=1}^{\infty} a_i T^{-1}y_i\| \le \|T^{-1}\|\|\sum_{i=1}^{\infty} a_i y_i\|.$$

Thus $\{\xi_i\}$ is equivalent to $\{y_i\}$ and the constant of the equivalence is

$$\|T\|\|T^{-1}\| \le 4\|y\|_{J^{**}}^2 \sup_j(|y(e_j^*)|^{-2}).$$

By Lemma 2.d.12 $[\xi_{n_i}]$ is complemented by a projection P' with $\|P'\| \le 1$.
But then $TP'T^{-1}$ is a projection P from J onto $[y_i]$ with norm

$$\|P\| \le 4\|y\|_{J^{**}}^2 \sup_j(|y(e_j^*)|^{-2}).$$

Since $y \in J^{**} \setminus \tilde{j}J$, it follows that $\lim_j y(e_j^*) \ne 0$, and therefore the set $\{j : y(e_j^*) = 0\}$ is finite. Suppose it is not empty and name it $\{j_i\}_{i=1}^{r}$, where $j_i < j_{i+1}$, $i = 1,\dots,r-1$.
Let

$$\{m_i : i = 1,\dots,j_r - r\} = \{j \le j_r : j \ne j_i \text{ for } i = 1,\dots,r\}$$

and suppose $m_i < m_{i+1}$ for $i = 1,\dots,j_r - r$. Consider the permutation v of $\mathbb{N}$ defined by $v(i) = j_i$, $i = 1,\dots,r$, $v(r+i) = m_i$, $i = 1,\dots,j_r - r$ and $v(j) = j$ for $j > j_r$.

Let $\sum_{i=1}^{\infty} a_i e_i \in J$ and define $U(\sum_{i=1}^{\infty} a_i e_i) = \sum_{i=1}^{\infty} a_{v(i)} e_i$.

Then U is an automorphism of J with $\|U\| \le \sqrt{3}$. In fact, let $p_1<p_2<\dots<p_n$ and let $p_k = \max\{p_i : p_i \le r\}$. Since $v(i) < v(i+1)$ for $i \ne r$,

$$\frac{1}{2}\sum_{j=1}^{n-1}(a_{v(p_j)} - a_{v(p_{j+1})})^2 = \frac{1}{2}\Big(\sum_{j=1}^{k-1}(a_{v(p_j)} - a_{v(p_{j+1})})^2 +$$

$$+ (a_{v(p_k)} - a_{v(p_{k+1})})^2 + \sum_{j=k+1}^{n-1}(a_{v(p_j)} - a_{v(p_{j+1})})^2\Big) \le 3\|\sum_{i=1}^{\infty}a_ie_i\|^2.$$

Similarly $v^{-1}(i) < v^{-1}(i+1)$ for $i \ne 1,\dots,j_r$; hence

$$\|U^{-1}\| \le (j_r + 1)^{1/2}.$$

Let $S : J \longrightarrow J$ be defined by

$$S(\sum_{i=1}^{\infty}a_ie_i) = \sum_{i=1}^{\infty}a_ie_{i+r}.$$

Then $\|S\| = \|S^{-1}\| = 1$, and the sequence $\{Uy_i\} \subset SJ$. The sequence $\{y_i\}$ is equivalent to the sequence $\{S^{-1}Uy_i\}$, and $S^{-1}Uy_i$ has no non-zero coordinates in J. Hence we are in the first case and $\{S^{-1}Uy_i\}$ is equivalent to $\{\xi_i\}$. Also $[S^{-1}Uy_i]$ is complemented in J via a projection P. It follows that $\{y_i\}$ is equivalent to $\{\xi_i\}$ and $[y_i]$ is complemented in J via the projection $Q = U^{-1}SPS^{-1}U$ and $\|Q\|$ depends on y but is independent of $\{n_i\}$.

In the following lemma we construct a projection from J^{**} into J^{**} which leaves jJ invariant and will be used in the proof of Proposition 2.d.15.

Lemma 2.d.14. Let $\{q_i\}$ and $\{r_i\}$ be sequences in $\mathbb{N}$ such that for every i, $q_i < r_i < r_i + 1 < q_{i+1}$. Let $y \in J^{**}$ and define $Q : J^{**} \longrightarrow J^{**}$ by

$$Qy(e_j^*) = \begin{cases} y(e_{r_1+1}^*) & \text{if } 1 \le j \le r_1, \\ y(e_{r_i+1}^*) & \text{if } q_i \le j \le r_i, \\ y(e_j^*) & \text{otherwise.} \end{cases}$$

Then Q is a projection with $\|Q\| = 1$, and $Qy \in jJ$ if and only if $y \in jJ$.

Proof: Q is obviously a projection. Recall that for $y \in J^{**}$,

$$\|y\|_{J^{**}} = \sup_n\|\sum_{i=1}^{n}y(e_i^*)e_i\|,$$

and hence it follows using Proposition 2.d.11(3) that

$$\|Qy\|_{J^{**}} \le \|y\|_{J^{**}} \text{ and } \|Q\| = 1.$$

Using Theorem 2.a.2, we know that $y \in J^{**}$ is also in jJ if and only if $\lim_j y(e_j^*) = 0$. Since $\lim_j Qy(e_j^*) = \lim_j y(e_j^*)$, it follows that $Qy \in jJ$ if and only if $y \in jJ$.

The next proposition analyzes the seminormalized sequences of J which don't have weak cluster points.

From now on we will frequently use the notation w^* for the term weak star.

Proposition 2.d.15. Let $\{z_n\} \subset J$ be a sequence with $\|z_n\| \le M$ for all n, such that $\{jz_n\}$ has a w^* limit $y \in J^{**} \setminus jJ$. Let $a = \frac{1}{2} \lim_{j\to\infty} |y(e_j^*)|$. Then $\{z_n\}$ has a subsequence $\{z_{n_k}\}$ equivalent to $\{\xi_k\}$ such that $[z_{n_k}]$ is complemented in J and the norm of the equivalence and the projection depend only on a and M.

Proof: Let $j_0 \in \mathbb{N}$ be such that for $j \ge j_0$, $|y(e_j^*)| > a$. We know by Theorem 2.a.2 that $a > 0$. Since for $x^* \in J^*$ with $\|x^*\| = 1$ we have $|\langle x^*,z_n\rangle| \le M$ and $\lim_{n\to\infty} |\langle x^*,z_n\rangle| = |\langle y,x^*\rangle|$, we obtain that $a < \|y\| \le M$ and thus we may assume, possibly by omitting a finite number of terms, that $\|z_k\| > a$. Let $P_n : J^{**} \longrightarrow J$ be defined by

$$P_n x = \textstyle\sum_{j=1}^n x(e_j^*)e_j \text{ for } x \in J^{**}$$

and let $\{\varepsilon_k\}$ be a sequence of reals decreasing to zero with $\varepsilon_1 < a/2$. Choose an increasing sequence of integers $\{n_k\}$ such that

(1) $$\|P_{n_k} jz_k - z_k\| < \varepsilon_k.$$

Then

$$(1/2)a \le \|P_{n_k} jz_k\| \le M.$$

Let $z'_k = P_{n_k} jz_k$; then the sequence $\{jz'_k\} \subset J^{**}$ converges in the weak star topology to $y \in J^{**}$ and we may construct an increasing sequence of natural numbers $\{m_k\}$ such that

(2) $$\|P_{n_{m_{k-1}}+2} (jz'_{m_k} - y)\| < \varepsilon_k.$$

and $n_{m_k} > n_{m_{k-1}} + 3$. This is done as follows: Let $m_0 = 1$ and m_1 be such that $n_{m_1} > \max(j_0, 4)$ and for $i = 1,\dots,n_{m_0} + 2$

$$|\langle e_i^*, z'_{m_1}\rangle - y(e_i^*)| < \varepsilon_1/(n_{m_0} + 2).$$

Then

$$\|P_{n_{m_0}+2}(jz'_{m_1} - y)\| < \varepsilon_1.$$

Proceeding inductively we get the desired sequence.
Define

$$z''_k = P_{n_{m_{k-1}}+2}y + (I - P_{n_{m_{k-1}}+2}j)z'_{m_k} = y_k + w_k.$$

Then $w_k = \sum_{j=q_k}^{r_k} b_j e_j$ where $q_k = n_{m_{k-1}} + 3$ and $r_k = n_{m_k}$. Also

$$\|w_k\| \leq 2\|z'_{m_k}\| \leq 2M.$$

The sequences $\{q_i\}$ and $\{r_i\}$ satisfy the hypotheses of Lemmas 2.c.2 and 2.d.14. It follows from Lemma 2.c.2 and (8) of Proposition 2.d.11 that

$$\|\textstyle\sum_{k=1}^{\infty} a_k w_k\| \leq 2\sqrt{2}M(\sum_{k=1}^{\infty}|a_k|^2)^{1/2} \leq 4M\|\sum_{k=1}^{\infty} a_k \xi_k\|$$

for all sequences $\{a_k\}$ with $\|\sum_{k=1}^{\infty} a_k \xi_k\| < \infty$. Hence by (1) in the proof of Proposition 2.d.13,

(3) $$\|\textstyle\sum_{k=1}^{\infty} a_k z''_k\| \leq \|\sum_{k=1}^{\infty} a_k y_k\| + \|\sum_{k=1}^{\infty} a_k w_k\| \leq$$

$$\leq 2\|y\|_{J^{**}} \|\textstyle\sum_{k=1}^{\infty} a_k \xi_k\| + 4M\|\sum_{k=1}^{\infty} a_k \xi_k\| \leq 6M\|\sum_{k=1}^{\infty} a_k \xi_k\|.$$

Let $Q : J^{**} \longrightarrow J^{**}$ be the norm one projection defined in Lemma 2.d.14 for the given sequences $\{q_i\}$ and $\{r_i\}$. Let $y' = Qy$ and $y'_k = P_{n_{m_{k-1}}+2}y'$. Then y' also belongs to $J^{**}\setminus jJ$ and since $r_1 > j_0$, $|y'(e_m)| > a$ for every m; thus

$$a < \|y'\| \leq M.$$

But $jw_k(e^*_{r_i+1}) = 0$ for every i and k and $QjP_n = jP_nQ$ for $r_i < n < q_{i+1}$; hence it follows that

$$Qjw_k = 0 \text{ and } Qjz''_k = QjP_{n_{m_{k-1}}+2}y = jP_{n_{m_{k-1}}+2}Qy = jP_{n_{m_{k-1}}+2}y' = jy'_k.$$

Since $\|Q\| = 1$ and $y'_k \neq y'_{k+1}$, we have by Proposition 2.d.13 and (3) that

$$6M\|\textstyle\sum_{k=1}^{\infty} a_k \xi_k\| \geq \|\sum_{k=1}^{\infty} a_k z''_k\| \geq \|Qj\sum_{k=1}^{\infty} a_k z''_k\| = \|\sum_{k=1}^{\infty} a_k y'_k\| \geq$$

$$\geq (2\sup_j(|y(e_j^*)|^{-2})\|y'\|)^{-1} \|\sum_{k=1}^{\infty} a_k \xi_k\| \geq (a^2/2M)\|\sum_{k=1}^{\infty} a_k \xi_k\|.$$

Hence $\{z''_k\}$ and $\{y'_k\}$ are equivalent to $\{\xi_k\}$.

Let $S = j^{-1}Qj$. Then $\|S\| \leq 1$ and $S|_{[z''_k]}$ is an isomorphism from $[z''_k]$ onto $[y'_k]$ with $Sz''_k = y'_k$. Again by Proposition 2.d.13, the subspace spanned by any subsequence of $\{y'_k\}$ is complemented in J by a projection that depends on $\{y'_k\}$, but that we will call P nevertheless with

$$\|P\| \leq 4\|y'\|^2_{J^{**}} \sup_j(|y'(e_j^*)|^{-2}) \leq 4M^2/a^2.$$

Thus any subsequence of $\{z''_k\}$ is complemented by a projection $S^{-1}PS$ with

$$\|S^{-1}PS\| \leq \|S^{-1}\|\|S\|\|P\| \leq (2M/a^2)(6M)(4M^2/a^2) = 48M^4/a^4.$$

Now by (1) and (2)

$$\|z''_k - z_{m_k}\| \leq \|z''_k - z'_{m_k}\| + \|z'_{m_k} - z_{m_k}\| =$$

$$= \|P_{n_{m_{k-1}}+2}(jz'_{m_k} - y)\| + \|z'_{m_k} - z_{m_k}\| < \varepsilon_k + \varepsilon_{m_k} < 2\varepsilon_k.$$

Hence, if $\sum_{k=1}^{\infty}\varepsilon_k$ is small enough it follows from Proposition 1.8 that $\{z_{m_k}\}$ is equivalent to $\{\xi_k\}$,

$$(a^2/4M)\|\sum_{k=1}^{\infty} a_k \xi_k\| \leq \|\sum_{k=1}^{\infty} a_k z_{m_k}\| \leq 12M\|\sum_{k=1}^{\infty} a_k \xi_k\|,$$

and $\left[z_{m_{k_s}}\right]$ is complemented in J for any subsequence of $\{z_{m_k}\}$ by a projection P' with $\|P'\| \leq 96M^2/a^2$.

We now analyze the case in which zero is a weak cluster point of a sequence in J.

Proposition 2.d.16. Let $\{x_n\} \subset J$ with $L^{-1} \leq \|x_n\| \leq L$ for $n = 1, 2,\ldots$ such that $\{x_n\}$ has a subsequence weakly convergent to zero. Then there is a subsequence $\{x_{n_i}\}$ equivalent to the unit vector basis of ℓ_2 such that $[x_{n_i}]$ is complemented in J.

Proof: We may assume that $\|x_n\| = 1$ for $n = 1, 2,\ldots$ and $x_n \longrightarrow 0$ weakly. Let $\{\varepsilon_k\}$ be a sequence of positive real numbers with $\sum_{k=1}^{\infty}\varepsilon_k < 2^{-7}$. As in Corollary 1.9, choose two increasing sequences of integers $\{n_k\}$, $\{m_k\}$ so that $m_0 = 0$, $n_1 < m_1 < n_2 < m_2 < n_3 < m_3 <\dots$ and if

$$x'_{n_k} = \sum_{i=m_{k-1}+2}^{m_k} e_i^*(x_{n_k})e_i,$$

then

$$\|x'_{n_k} - x_{n_k}\| < \varepsilon_k.$$

Thus $1/2 \le \|x'_{n_k}\| \le 2$, and since the supports of the x'_{n_k} are separated by one integer, by Theorem 2.d.2 $[x'_{n_k}]$ is complemented in J by a projection P of norm at most $2\sqrt{2}$ and $\{x'_{n_k}\}$ is equivalent to the unit vector basis of ℓ_2. In fact for every real sequence $\{b_n\}$,

$$(1/4)\sum_{j=1}^k b_j^2 \le \|\sum_{j=1}^k b_j x'_{n_j}\|^2 \le 8\sum_{j=1}^k b_j^2.$$

Hence $\{x'_{n_k}\}$ has basis constant K at most $4\sqrt{2}$. We conclude that

$$\sum_{k=1}^{\infty}\|x'_{n_k} - x_{n_k}\| < \sum_{k=1}^{\infty}\varepsilon_k < 2^{-7} \le \frac{1}{8K\|P\|},$$

and from Proposition 1.8 it follows that $[x_{n_k}]$ is complemented in J and $\{x_{n_k}\}$ is equivalent to the unit vector basis of ℓ_2.

To characterize the spreading basic sequences in J, first we will see that every spreading basic sequence is seminormalized.

Lemma 2.d.17. Let $\{z_k\}_{k=1}^{\infty}$ be a spreading basic sequence in a Banach space X. Then there exists $M > 0$ such that $\frac{1}{M} \le \|z_k\| \le M$ for every $k = 1, 2,\dots$.

Proof: Suppose that $\{z_k\}$ is not seminormalized and that there is a subsequence $\{z_{k_m}\}_{m=1}^{\infty}$ such that $\|z_{k_m}\| \ge m$. Let $x = \sum_{n=1}^{\infty} z_n^*(x)z_n \in X$ with $z_n^*(x) \ne 0$, and let m_n be such that $|z_n^*(x)|m_n \ge 1$. Then $\sum_{n=1}^{\infty} z_n^*(x)z_{k_{m_n}}$ diverges because $\|z_n^*(x)z_{k_{m_n}}\| \ge |z_n^*(x)|m_n \ge 1$ and thus, since the sequence $\{z_k\}$ is spreading, $\sum_{n=1}^{\infty} z_n^*(x)z_n$ also diverges, which is a contradiction.
On the other hand, if there is a subsequence $\{z_{k_m}\}_{m=1}^{\infty}$ with $\|z_{k_m}\| \le \frac{1}{m}$,

taking $m_n \geq \frac{2^n}{\|z_n\|}$, we have that $\|z_{k_{m_n}}\| \leq \|z_n\|/2^n$. Then $\sum_{n=1}^{\infty} \frac{1}{\|z_n\|} z_{k_{m_n}} \in X$, but $\sum_{n=1}^{\infty} \frac{1}{\|z_n\|} z_n$ diverges, which is again a contradiction.

Theorem 2.d.18. Let $\{z_k\}$ be a spreading basic sequence in J.
(a) If $\{z_k\}$ converges weakly to zero, then $\{z_k\}$ is equivalent to the unit vector basis of ℓ_2.
(b) If $\{z_k\}$ does not converge weakly to zero, then $\{z_k\}$ is equivalent to $\{\xi_k\}$.

Proof: Since $\{z_k\}$ is a spreading basic sequence, by Lemma 2.d.17 there exists M such that $M^{-1} \leq \|z_k\| \leq M$. Also $\{z_k\}$ cannot have a non-zero weak cluster point, because z_n^* extended to J is such that $\lim_k z_n^*(z_k) = 0$. Thus by Proposition 2.d.16, if $\{z_k\}$ converges weakly to zero, $\{z_k\}$ has a subsequence equivalent to the unit vector basis of ℓ_2. Since $\{z_k\}$ is spreading, $\{z_k\}$ is itself equivalent to the unit vector basis of ℓ_2. In the case that $\{z_k\}$ does not converge weakly to zero, Proposition 2.d.15 implies that $\{z_k\}$ has a subsequence equivalent to $\{\xi_k\}$, from which it follows that $\{z_k\}$ is equivalent to $\{\xi_k\}$.

The following corollary showing the uniqueness of subsymmetric basic sequences in J is due to Casazza, Lin and Lohman [1], but this proof is by Andrew [2].

Corollary 2.d.19. J has a unique subsymmetric basic sequence up to equivalences.

Proof: Let $\{y_n\}$ be a subsymmetric basic sequence in J. Then $\{y_n\}$ is spreading, and therefore, by Lemma 2.d.17, there exists M such that $M^{-1} \leq \|y_n\| \leq M$. Let $Y = [y_n]$; since J^{**} is separable, Y does not have a subspace isomorphic to ℓ_1. Hence, using a theorem of James (see e.g. Lindenstrauss and Tzafriri [1]), which says that if $\{x_n\}$ is an unconditional basis in a Banach space X, then $\{x_n\}$ is shrinking if and only if X does not contain ℓ_1, we get that $\{y_n\}$ is a shrinking basis of Y. Therefore every $f \in Y^*$ may be written as $f = \sum_{i=1}^{\infty} a_i y_i^*$ with $f(y_n) = a_n$, and so $f(y_n) \to 0$. Hence $\{y_n\}$ is a spreading basic sequence which converges weakly to zero and by Theorem 2.d.18 it is equivalent to the unit vector basis of ℓ_2.

We now come to one of the main results of the section. First recall that a Banach space X is primary if $X = Y \oplus W$ implies that either Y or W is isomorphic to X. As mentioned, many of the classical Banach spaces such as c_0, ℓ_p (which are moreover prime), L_p, for $1 \le p \le \infty$, and C(K) are known to be primary, and in 1977 Casazza [1] proved that this is also the case for J. The proof we are going to give here is due in part to Andrew [2] and is based on the following characterization of the subspaces of J:

Theorem 2.d.20. If X is a subspace of J, then X is isomorphic to $Y \oplus W$, where W is reflexive, Y is complemented in J, and Y is either trivial or isomorphic to J.

Proof: If X is reflexive, we choose $Y = \{0\}$ and $W = X$.
If X is not reflexive there exists a sequence $\{z_n\} \subset X$ with $\|z_n\| = 1$ for all n, such that $\{z_n\}$ has no weak cluster point in J (see e.g. Beauzamy [1]). By Proposition 2.d.15, there is a subsequence $\{z_{n_k}\}$ equivalent to $\{\xi_n\}$ such that $[z_{n_k}]$ is complemented in J by a projection P. Let $Y = [z_{n_k}]$, then Y is isomorphic to J, and hence $(I - P)J$ is reflexive. This is easily obtained from the facts that Y is quasi-reflexive of order one and that the direct sum of a quasi-reflexive space of order one and a non-reflexive space is a non-reflexive space, and if it is quasi-reflexive, its order of quasi-reflexivity is greater than one. Since $Y \subset X$, $P|_X$ is a projection from X onto Y, and $W = (I - P)X$ is a reflexive space because it is contained in $(I - P)J$.

Corollary 2.d.21. If $X \subset J$ is non-reflexive then there exists a subspace $Y \subset X$ such that Y is isomorphic to J and Y is complemented in J.

Summarizing we get:

Theorem 2.d.22. J is primary.

Proof: Suppose that J is isomorphic to X ⊕ Y. Then either X or Y is reflexive and the other one is quasi-reflexive of order one. Assume X is reflexive. Then by Theorem 2.d.20, Y is isomorphic to J ⊕ W for some reflexive subspace W of Y and W is complemented in J, since Y is complemented in J. By Corollary 2.c.21

$$Y \approx J \oplus W \approx J,$$

and this finishes the proof.

Corollary 2.d.23. If J is isomorphic to Y ⊕ W with Y non-reflexive, then Y is isomorphic to J, and W is reflexive.

This last corollary finishes the classification of the complemented subspaces of J, since in Section 2.c we saw that every reflexive complemented subspace of J is isomorphic to a complemented subspace of $\mathfrak{J}$.

2.e. Isometries of the space J

Until now, except for the isometry between J and J^{**}, which is obviously norm dependent, we have described only properties of J which are independent of the specific equivalent norm one chooses to work with. In this section we focus our attention on the isometries of J which certainly depend on the norm of the space, since taking equivalent norms does not in general preserve the isometries.

The existence of Banach spaces with only the trivial isometries ± the identity was already known from 1971 (see Davis [1]), and in fact in 1986 Bellenot [5] showed that each separable Banach space has an equivalent norm with this property; but J is the first example where this is true for "natural" norms.

In this section we present Bellenot's proof of the fact that $(J, \|\ \|)$ only admits the trivial isometries, which although somewhat technical, requires only very simple tools.

In fact, it turns out that for all the norms we have been using, the groups of isometries are very small: Sersouri [1] proved a more general theorem yielding in particular that $(J, \||\ |\|)$ has also only the trivial isometries; as for the norm $\|\ |\ \ |\|$, Semenov and Skorik [1] showed that the group of isometries of J endowed with this norm consists of ± the identity and $\pm\sigma$, where σ is the permutation of the first two coordinates: $\sigma(e_1) = e_2$, $\sigma(e_2) = e_1$ and $\sigma(e_i) = e_i$ for $i > 2$. A final example of the crucial role played by the specific norm in the classification of the isometries in J is given by Corollary 2.c.21, showing that $J \oplus \ell_2$ is isomorphic to J and hence J can be renormed to have many isometries.

Since we will be working with the norm $\|\ \|$ in J, J^{**} will be considered in this section as

$$J^{**} = \{x = \{a_n\} \subset \mathbb{R} : \sup\textstyle\sum_{i=0}^{k}(a_{p_i} - a_{p_{i+1}})^2 < \infty\}$$

where the sup is taken over all sequences of integers $0 = p_0 <\ldots< p_{k+1}$. If $x \in J^{**}$, we will write $x = \sum_{n=1}^{\infty} a_n e_n$; observe that by Theorem 2.a.2 and its proof, both $\|x\| = \sup_m \|\sum_{n=1}^{m} a_n e_n\|$ and $\lim_n a_n$ exist.

We now recall the important notion of extreme points in a Banach space which will play an important role in the proof of the uniqueness (up to the sign) of the isometries in $(J, \|\ \|)$.

Definition 2.e.1. If X is a Banach space, we will say that $x \in X$ is extreme if x cannot be written as $\frac{1}{2}(y + z)$ with $y \neq x$ and $y,z \in \{u \in X : \|u\| \leq \|x\|\}$, or equivalently, if x cannot be written as $ty + (1-t)z$ with $t \in (0,1)$ and $y,z \in \{u \in X : \|u\| \leq \|x\|\}$.

The next two lemmata give sufficient conditions in order for a point $x \in J^{**}$ to be extreme.

Lemma 2.e.2. If $x = \sum_{n=1}^{\infty} a_n e_n \in J^{**}$ and for some integers n and m with $0 \leq n < m - 1$ we have

$$a_n > a_{n+1} =\ldots= a_{m-1} > a_m \quad \text{or} \quad a_n < a_{n+1} =\ldots= a_{m-1} < a_m,$$

where $a_0 = 0$, then x is not extreme.

Proof: Put

$$y = \sum_{i=1}^{n} a_i e_i + a_n \sum_{i=n+1}^{m-1} e_i + \sum_{i=m}^{\infty} a_i e_i,$$

$$z = \sum_{i=1}^{n} a_i e_i + a_m \sum_{i=n+1}^{m-1} e_i + \sum_{i=m}^{\infty} a_i e_i,$$

where $\sum_{i=1}^{n} a_i e_i$ means the zero vector if $n = 0$. Then by Proposition 2.d.11(3), $\|y\|$ and $\|z\|$ are both not bigger than $\|x\|$, and $x = ty+(1-t)z$, where t is such that $a_{n+1} = ta_n + (1-t)a_m$.

Lemma 2.e.3. If $x = \sum_{n=1}^{\infty} a_n e_n \in J^{**}$ and

$$2\|x\|^2 = \lim_{n\to\infty} a_n^2 + \sum_{n=0}^{\infty}(a_n - a_{n+1})^2,$$

then x is extreme.

Proof: Suppose this is not true. Then there exist $y = \sum_{n=1}^{\infty} b_n e_n$ and $z = \sum_{n=1}^{\infty} c_n e_n$ in J^{**} with $x \neq y$, $\|y\| \le \|x\|$ and $\|z\| \le \|x\|$ such that $x = \frac{1}{2}(y + z)$. Therefore

$$\|x\| = \|y\| = \|z\|.$$

Since for every n it is true that

$$2\|y\|^2 \ge b_n^2 + \sum_{i=0}^{n-1}(b_i - b_{i+1})^2$$

we get

(1) $$2\|y\|^2 \ge \lim_{n\to\infty} b_n^2 + \sum_{n=0}^{\infty}(b_n - b_{n+1})^2$$

and similarly

(2) $$2\|z\|^2 \ge \lim_{n\to\infty} c_n^2 + \sum_{n=0}^{\infty}(c_n - c_{n+1})^2.$$

Let $u_0 = \lim a_n$, $u_n = a_n - a_{n+1}$, $v_0 = \lim b_n$, $v_n = b_n - b_{n+1}$, $w_0 = \lim c_n$ and $w_n = c_n - c_{n+1}$ for $n = 1, 2, \ldots$. Then $u = \{u_n\}_{n=0}^{\infty}$, $v = \{v_n\}_{n=0}^{\infty}$ and $w = \{w_n\}_{n=0}^{\infty}$ are three sequences in ℓ_2 with $u = \frac{1}{2}(v + w)$ and by hypothesis, (1) and (2), $\|u\|_2 \ge \|v\|_2$, $\|u\|_2 \ge \|w\|_2$. But this would imply that u is not an extreme point in ℓ_2, which is a contradiction to the strict convexity of ℓ_2 (see e.g. Beauzamy [1]).

The fact that the elements $s_n = \sum_{i=n}^{\infty} e_i \in J^{**}$ are extreme, which is essential for the proof of Theorem 2.e.11, now follows easily.

Lemma 2.e.4. Let $s_n = \sum_{i=n}^{\infty} e_i \in J^{**}$. Then for each n, s_n is extreme and so are $s_n - ts_m$ for $n < m$ and $t \geq 0$ and $s_n - s_m + ts_q$ for $n < m < q$ and $t \geq 0$.

Proof: It is easy to see that all the points mentioned satisfy the conditions of Lemma 2.e.3.

The following technical lemmata are essential for the proof of the main theorem in this section.

Lemma 2.e.5.

(I) Suppose $0 \leq a_2 \leq a_1$ and $0 \leq b_2 \leq b_1$.

If $a_1 = \lambda b_1$ with $\lambda > 0$ but $a_2 \neq \lambda b_2$, then for small enough $\varepsilon > 0$ either

$a_2 > \lambda b_2$ and $0 < a_1 - (\lambda - \varepsilon)b_1 < a_2 - (\lambda - \varepsilon)b_2$ or

$a_2 < \lambda b_2$ and $0 > a_1 - (\lambda + \varepsilon)b_1 > a_2 - (\lambda + \varepsilon)b_2$.

(II) Suppose $0 \leq a_1, a_2, a_3, b_1, b_2, b_3$; $a_1 > a_2 \leq a_3$ and $b_1 > b_2 \leq b_3$.

If $a_1 = \lambda b_1$, $a_2 = \lambda b_2$ with $\lambda > 0$ but $a_3 \neq \lambda b_3$, then for small enough $\varepsilon > 0$ either

$a_3 > \lambda b_3$ and $a_1 - (\lambda + \varepsilon)b_1 < a_2 - (\lambda + \varepsilon)b_2 < a_3 - (\lambda + \varepsilon)b_3$ or

$a_3 < \lambda b_3$ and $a_1 - (\lambda - \varepsilon)b_1 > a_2 - (\lambda - \varepsilon)b_2 > a_3 - (\lambda - \varepsilon)b_3$.

(III) Suppose $0 \leq a_1, a_2, a_3, b_1, b_2, b_3$; $a_1 < a_2 \geq a_3$ and $b_1 < b_2 \geq b_3$.

If $a_1 = \lambda b_1$, $a_2 = \lambda b_2$ with $\lambda > 0$ but $a_3 \neq \lambda b_3$, then for small enough $\varepsilon > 0$ either

$a_3 > \lambda b_3$ and $a_1 - (\lambda - \varepsilon)b_1 < a_2 - (\lambda - \varepsilon)b_2 < a_3 - (\lambda - \varepsilon)b_3$ or

$a_3 < \lambda b_3$ and $a_1 - (\lambda + \varepsilon)b_1 > a_2 - (\lambda + \varepsilon)b_2 > a_3 - (\lambda + \varepsilon)b_3$.

The proof of this is direct.

Lemma 2.e.6. Let $T : J \longrightarrow J$ be an onto isometry, s_n as in Lemma 2.e.4 and $T^{**}s_n = \sum_{j=1}^{\infty} a_j^n e_j$. Then either $0 \leq a_j^n \leq 1$ for every j and every n and $\lim_{i\to\infty} a_i^n = 1$ for every n, or $0 \leq -a_j^n \leq 1$ for every j and every n and $\lim_{i\to\infty} a_i^n = -1$ for every n.

Proof: $T^{**} : J^{**} \longrightarrow J^{**}$ and the induced map $\tilde{T} : (J^{**}/J) \longrightarrow (J^{**}/J)$ are both onto isometries. But since J^{**}/J is isomorphic to $\mathbb{R}$, either $\tilde{T}$ = identity or $-\tilde{T}$ = identity. Therefore replacing T by -T if necessary, we may assume $\tilde{T}$ = identity. Hence, if $T^{**}(\sum_{i=1}^{\infty} a_i e_i) = \sum_{i=1}^{\infty} b_i e_i$ we get that $\lim_{i\to\infty} a_i = \lim_{i\to\infty} b_i$, and in particular applying this to $T^{**}s_n$,

$$\lim_{i\to\infty} a_i^n = 1.$$

Since $1 = \|s_n\|^2 = \|T^{**}s_n\|^2 \geq (a_j^n)^2$, and

$$1 = \|T^{**}s_n\|^2 \geq \frac{1}{2}((a_j^n)^2 + (a_j^n - a_i^n)^2 + (a_i^n)^2)$$

for every i and j, by passing to the limit as i tends to infinity we get $1 \geq (a_j^n)^2 - a_j^n + 1$ and hence

$$0 \leq a_j^n \leq 1.$$

Taking $\tilde{T}$ = -identity we get the other case.

Next we introduce some notation which will be used throughout this section.

Let T denote an isometry from J onto J, $s_n = \sum_{i=n}^{\infty} e_i$ and $T^{**}s_n = \sum_{j=1}^{\infty} a_j^n e_j$. Let

$$B_n = \{m : a_j^m = 0 \text{ for } j < n \text{ but } a_n^m \neq 0\},$$

and define $\Pi : \mathbb{N} \longrightarrow \mathbb{N}$ by $\Pi(m) = n$ if and only if $m \in B_n$, that is

$$\Pi(m) = n \text{ if and only if } T^{**}s_m = \sum_{j=n}^{\infty} a_j^m e_j \text{ and } a_n^m \neq 0.$$

Remark: Observe that since T^{**} is an isometry, $x \in J^{**}$ is an extreme point if and only if $T^{**}x$ is an extreme point.

From here on, we will suppose by virtue of Lemma 2.e.6 that T is such that $0 \leq a_j^n \leq 1$ for every j and every n and $\lim_{j\to\infty} a_j^n = 1$.

Lemma 2.e.7. The sets B_n are finite and Π is unbounded.

Proof: Suppose j, k $\in B_n$ and j < k; then $a_n^j > 0$ and $a_n^k > 0$. We will show that $a_n^j \leq a_n^k$.

Suppose the contrary, $a_n^j > a_n^k$, and let $\lambda > 1$ be such that $a_n^j = \lambda a_n^k$. We will show by an inductive procedure that $a^j_m = \lambda a^k_m$ for all m. If m < n, then $0 = a^j_m = a^k_m$ and thus $a^j_m = \lambda a^k_m$ for $m \le n$.

Step 1: Since $0 = a^j_0 =\ldots= a^j_{n-1} < a^j_n$ and $0 = a^k_0 =\ldots= a^k_{n-1} < a^k_n$ and s_j and s_k are extreme, by Lemma 2.e.2 $a^j_n \ge a^j_{n+1}$ and $a^k_n \ge a^k_{n+1}$. If $a^j_{n+1} \ne \lambda a^k_{n+1}$ by Lemma 2.e.5(I) there exists $\varepsilon > 0$ such that either

$$0 < a^j_n - (\lambda - \varepsilon)a^k_n < a^j_{n+1} - (\lambda - \varepsilon)a^k_{n+1} \quad \text{or}$$

$$0 > a^j_n - (\lambda + \varepsilon)a^k_n > a^j_{n+1} - (\lambda + \varepsilon)a^k_{n+1}.$$

But this, again by Lemma 2.e.2, contradicts the fact that both $s_j - (\lambda + \varepsilon)s_k$ and $s_j - (\lambda - \varepsilon)s_k$ are extreme points. Hence $a^j_{n+1} = \lambda a^k_{n+1}$. If $a^j_n > a^j_{n+1}$ go to step 2. Otherwise $a^j_n = a^j_{n+1}$, $a^k_n = a^k_{n+1}$, and since s_j, s_k are extreme, $a^j_{n+1} \ge a^j_{n+2}$ and $a^k_{n+1} \ge a^k_{n+2}$. Thus we reapply Lemma 2.e.5(I) getting $a^j_{n+2} = \lambda a^k_{n+2}$. Again, if $a^j_{n+1} > a^j_{n+2}$ go to step 2, otherwise repeat step 1 until $a^j_m > a^j_{m+1}$ for some m. If such an m does not exist $a^j_m = \lambda a^k_m$ for all m.

Step 2: When we enter this step we have $a^j_p > a^j_{p+1} \ge 0$ and $a^j_i = \lambda a^k_i$ for $i \le p + 1$. Again, since s_j and s_k are extreme, $a^j_{p+1} \le a^j_{p+2}$ and $a^k_{p+1} \le a^k_{p+2}$. If $a^j_{p+2} \ne \lambda a^k_{p+2}$ we can apply Lemma 2.e.5(II) and this would contradict the fact that $s_j - (\lambda + \varepsilon)s_k$ and $s_j - (\lambda - \varepsilon)s_k$ are both extreme points. Hence $a^j_{p+2} = \lambda a^k_{p+2}$.

If $a^j_{p+1} < a^j_{p+2}$ go to step 3. If $a^j_{p+1} = a^j_{p+2}$ reapply Lemma 2.e.5(II) until

$$a^j_{p+1} = a^j_{p+2} =\ldots= a^j_q < a^j_{q+1}.$$

Such a q exists since $a^j_{p+1} < a^j_p \le 1$ and since $\lim_{p\to\infty} a^j_p = 1$. Then go to step 3.

Step 3: When we enter this step we have $0 \le a^j_q < a^j_{q+1}$ and $a^j_i = \lambda a^k_i$ for $i \le q + 1$. Since s_j and s_k are extreme points, $a^j_{q+1} \ge a^j_{q+2}$ and $a^k_{q+1} \ge a^k_{q+2}$. If $a^j_{q+2} \ne \lambda a^k_{q+2}$ Lemma 2.e.5(III) applies, and this would contradict the fact that both $s_j - (\lambda + \varepsilon)s_k$ and $s_j - (\lambda - \varepsilon)s_k$ are extreme points. Hence $a^j_{q+2} = \lambda a^k_{q+2}$. We reapply Lemma 2.e.5(III) until

$$a^j_{q+1} = a^j_{q+2} = \ldots = a^j_p > a^j_{p+1}$$

and return to step 2. If such a p does not exist, then $a^j_m = \lambda a^k_m$ for all m.

Continuing in this fashion we obtain $a^j_n = \lambda a^k_n$ for n = 1, 2,....
However, this is a contradiction, since then

$$T^{**}s_j = \sum_{i=1}^{\infty} a^j_i e_i = \lambda \sum_{i=1}^{\infty} a^k_i e_i = \lambda T^{**}s_k;$$

but $\|T^{**}s_j\| = \|T^{**}s_k\| = 1$. Hence $a^j_n \le a^k_n$.
Next we will show that for each n,

$$\lim_i a^i_n = 0.$$

Let $\{e^*_n\}$ be the coefficient functionals of $\{e_n\} \subset J$. Since $\{e_n\}$ is shrinking, there is a sequence $\{b^n_j\}_{j=1}^{\infty}$ such that $T^*e^*_n = \sum_{j=1}^{\infty} b^n_j e^*_j$. If m > n,

$$\left|\sum_{j=i}^{m} b^n_j\right| = \left|\left\langle \sum_{j=i}^{\infty} b^n_j e^*_j, \sum_{k=1}^{m} e_k \right\rangle\right| \le \left\|\sum_{j=i}^{\infty} b^n_j e^*_j\right\|.$$

Hence

$$|a^i_n| = |\langle T^{**}s_i, e^*_n\rangle| = |\langle s_i, T^*e^*_n\rangle| = \left|\sum_{j=i}^{\infty} b^n_j\right| \le \left\|\sum_{j=i}^{\infty} b^n_j e^*_j\right\|,$$

which tends to zero as $i \longrightarrow \infty$. This, together with the fact that $j, k \in B_n$, $j < k$ imply $a^j_n \le a^k_n$, gives us that B_n is finite for each n and since $\bigcup_{n=1}^{\infty} B_n = \mathbb{N}$, Π is unbounded.

Lemma 2.e.8. Π is eventually non-decreasing, that is there exists $M \in \mathbb{N}$ such that $\Pi|_{(M,\infty)}$ is non-decreasing.

Proof: Let $i < j < k$, $p < q$, $p < r$, $p \in B_i$, $q \in B_j$, $r \in B_k$. We will show that $q \le r$.
Suppose the contrary: $r < q$. Then

$$T^{**}s_p = \sum_{u=i}^{\infty} a^p_u e_u,$$

$$T^{**}(s_p - ts_q) = \sum_{u=i}^{j-1} a^p_u e_u + \sum_{u=j}^{\infty}(a^p_u - ta^q_u)e_u$$

and

$$T^{**}(s_p - s_r + ts_q) = \sum_{u=i}^{j-1} a^p_u e_u + \sum_{u=j}^{k-1}(a^p_u + ta^q_u)e_u + \sum_{u=k}^{\infty}(a^p_u - a^r_u + ta^q_u)e_u$$

are extreme for $t \geq 0$.

Therefore, since $a_i^p > 0$ there exists $m < j - 1$ either such that

$$a_m^p < a_{m+1}^p = \ldots = a_{j-1}^p \geq a_j^p$$

or such that

$$a_m^p > a_{m+1}^p = \ldots = a_{j-1}^p \leq a_j^p,$$

where m is the largest index less than $j - 1$ such that $a_u^p \neq a_{j-1}^p$. (Remember $a_0^p = 0$). For the same reason there exists $\tilde{m}$ such that $m < j - 1$ and either

$$a_m^p < a_{m+1}^p = \ldots = a_{j-1}^p \geq a_j^p - ta_j^q \quad \text{or} \quad a_m^p > a_{m+1}^p = \ldots = a_{j-1}^p \leq a_j^p - ta_j^q.$$

But for t sufficiently large, $a_j^p - ta_j^q < a_{j-1}^p$. Hence there exists m with

$$a_m^p < a_{m+1}^p = \ldots = a_{j-1}^p \geq a_j^p.$$

Since $T^{**}(s_p - s_r + ts_q)$ is also extreme, this would imply

$$a_m^p < a_{m+1}^p = \ldots = a_{j-1}^p \geq a_j^p + ta_j^q,$$

which for large t gives us a contradiction. Hence $q \leq r$.

Let $M = \max\{j : \Pi(j) \leq \Pi(1)\}$. This M exists since B_n is finite for every n. If $i > j > M$, then $\Pi(i) > \Pi(1)$ and $\Pi(j) > \Pi(1)$.

If $\Pi(i) < \Pi(j)$, we have $\Pi(1) < \Pi(i) < \Pi(j)$, $1 < j$, $1 < i$, $1 \in B_{\pi(1)}$, $i \in B_{\pi(i)}$ and $j \in B_{\pi(j)}$. This, applying the first part of this lemma, gives us $j \geq i$ which is a contradiction. Hence $\Pi(j) \leq \Pi(i)$, that is, Π is eventually non-decreasing.

Lemma 2.e.9. Π is one to one restricted to (M,∞), where M is as in Lemma 2.e.8.

Proof: Let $i < j$, $p < q$, $p < r$, $p \in B_i$, $q, r \in B_j$. We will show that $q = r$.

Suppose $q < r$; since $T^{**}(s_p - ts_q)$ is extreme for $t > 0$, and for large t, $a_{j-1}^p \geq a_j^p - ta_j^q$, there exists $m < j - 1$ such that

$$a_m^p < a_{m+1}^p = \ldots = a_{j-1}^p \geq a_j^p - ta_j^q.$$

But

$$T^{**}(s_p - s_q + ts_r) = \sum_{u=i}^{j-1} a_u^p e_u + \sum_{u=j}^{\infty} (a_u^p - a_u^q + ta_u^r) e_u$$

is also extreme and hence we should have

$$a^p_m < a^p_{m+1} = \ldots = a^p_{j-1} \geq a^p_j - a^q_j + ta^r_j,$$

which for large t is false. Hence we get a contradiction and $r \leq q$. But by the same argument, interchanging the roles of q and r, we get $q \leq r$. Therefore $q = r$.

Let M be as in Lemma 2.e.8, then if $i > j > M$ we know by the said lemma that $\Pi(i) \geq \Pi(j)$. Suppose $\Pi(i) = \Pi(j)$. By the definition of M, $\Pi(1) < \Pi(i)$ and since $1 < i$, $1 < j$, $1 \in B_{\pi(1)}$, and $i, j \in B_{\pi(i)} = B_{\pi(j)}$, by the first part of this lemma, we get $i = j$, which is a contradiction. Therefore $\Pi(i) > \Pi(j)$. Hence Π is one to one on (M,∞).

Lemma 2.e.10. For $n \geq M$, $\Pi(n) \leq n$ and there exists K such that $\Pi(K + i) = \Pi(K) + i$ for $i \geq 0$.

Proof: First we will see that the cardinality of $\bigcup_{i<k} B_i$ is at least $k - 1$. The set $V_k = \{\sum_{j=k}^{\infty} b_j e_j \in J^{**}\}$ has codimension $k - 1$ in J^{**}. Since for $n = 1, 2, \ldots$, $e_n = s_n - s_{n+1}$ and $\{s_1\} \cup \{e_n\}_{n=1}^{\infty}$ is a basis for J^{**}, we get that $\{s_n\}_{n=1}^{\infty}$ is also a basis for J^{**}, and thus $\{T^{**}(s_n)\}_{n=1}^{\infty}$ is a basis for $T^{**}(J^{**}) = J^{**}$. If the number of elements in $\bigcup_{i<k} B_i$ were $r < k - 1$, then all but r of the $T^{**} s_n$ would belong to V_k, and $T^{**}(J^{**})/V_k = J^{**}/V_k$ would have dimension less than or equal to r, which is a contradiction.

In particular card $\bigcup_{j<\pi(i)} B_j \geq \Pi(i) - 1$.

Let $i > M$. By Lemma 2.e.9, if $i < u$, then $\Pi(i) < \Pi(u)$, thus

$$\bigcup_{j<\pi(i)} B_j = \{u : u \in B_j,\ j < \Pi(i)\} = \{u : \Pi(u) < \Pi(i)\} =$$
$$= \{u \leq M : \Pi(u) < \Pi(i)\} \cup \{M < u < i : \Pi(u) < \Pi(i)\}.$$

Hence card $\bigcup_{j<\pi(i)} B_j \leq M + (i - M - 1) = i - 1$. Therefore $\Pi(i) \leq i$. Let $f(i) = i - \Pi(i)$. If $i > M$ then by Lemmata 2.e.8 and 2.e.9 $\Pi(i+1) \geq \Pi(i)+1$. Therefore $0 \leq f(i + 1) \leq f(i)$. Since $f(i)$ is an integer, this implies that f is eventually constant, i.e. there exists $K \geq M$ such that $f(i) = f(j)$ if $i, j \geq K$. Thus $f(K + j) = f(K)$, that is $\Pi(K+j) = \Pi(K)+j$ for $j \geq 0$.

After the long preparation, we are finally ready to prove the main theorem.

Theorem 2.e.11. The only onto isometries of $(J, \|\ \|)$ are ± identity.

Proof: Let T be an onto isometry, T^{**}, $\tilde{T}$ as in Lemma 2.e.6 and its proof. Assume that $\tilde{T}$ = identity. From this we will show that T^{**} = identity and similarly if $\tilde{T}$ = -identity, T^{**} = -identity.

First we will prove that if n is large enough, then

$$T^{**} s_n = s_n.$$

Let K be as in Lemma 2.e.10 and let $n \geq K$, $m < n$. Suppose $a^m_{\pi(n)} < 1$. Let $\Pi(q)$ be the first index greater than $\Pi(n)$ such that $q > n$ and $a^m_{\pi(q)} > a^m_{\pi(n)}$. This q exists by Lemma 2.e.10 and because $\lim_n a^m_n = 1$. Then $a^m_{\pi(q)} > a^m_{\pi(q)-1}$, and since s_m is extreme, there exists $p < \Pi(q) - 1$ such that

$$a^m_{\pi(q)} > a^m_{\pi(q)-1} =\ldots= a^m_{p+1} < a^m_p.$$

But $q > n > m$. Therefore $s_m - ts_q$ is also extreme if $t > 1$. This implies

$$a^m_p > a^m_{p+1} =\ldots= a^m_{\pi(q)-1} \leq a^m_{\pi(q)} - ta^q_{\pi(q)}$$

which yields a contradiction for large t.

Hence for $n \geq K$, $m < n$,

(1) $$a^m_{\pi(n)} = 1.$$

By Lemma 2.e.10, if $K \leq m < n$, $\Pi(m) < \Pi(n)$. But $0 < a^m_{\pi(m)}$ and since s_m is extreme, it is not possible that $0 < a^m_{\pi(m)} < a^m_{\pi(n)} = 1$. Hence $a^m_{\pi(m)} = 1$ and therefore $T^{**} s_m = s_{\pi(m)}$ for $m \geq K$.

Applying Lemmata 2.e.9, 2.e.10 and the results obtained so far to T^{-1}, we get Π' and K' such that for $i \geq K'$ we have $\Pi'(i) \leq i$,

$$(T^{-1})^{**} s_i = s_{\pi'(i)}$$

and Π' is unbounded. Therefore, if $i \geq K'$ and $\Pi'(i) \geq K$, then

$$s_i = T^{**} s_{\pi'(i)} = s_{\pi(\pi'(i))}$$

and hence, since Π is increasing for $i \geq K$, $i = \Pi(\Pi'(i)) \leq \Pi(i)$ and this, together with Lemma 2.e.10, gives us $i = \Pi(i)$ and $T^{**} s_i = s_i$ for i large enough.

Let $m = \max\{n : T^{**}s_n \neq s_n\}$. Then for $i > m$

(2) $$T^{**}s_i = s_i.$$

Now we will see that

$$T^{**}s_n = s_n \quad \text{for all } n,$$

and therefore T = identity.

Suppose $m \geq 1$. Then we may assume $K = m + 1$, and since $\Pi(n) = n$ for $n \geq K$, by (1) we get

(3) $$a_j^k = 1 \quad \text{if} \quad k \leq m < j.$$

Therefore

(4) $$T^{**}s_k = \sum_{n=1}^{m} a_n^k e_n + \sum_{n=m+1}^{\infty} e_n,$$

and since $T^{**}s_k \neq T^{**}s_{m+1}$ we obtain

(5) $$\Pi(k) < m + 1.$$

If $\Pi(m) = m$, then

$$T^{**}(s_{m+1} - s_m) = T^{**}e_m = a_m^m e_m$$

and since $a_m^m > 0$ and $1 = \|T^{**}e_m\| = a_m^m$, we get $a_m^m = 1$ and thus

$$T^{**}s_m = s_m.$$

So we may assume $\Pi(m) < m$ and $m > 1$.

If $1 \leq k \leq m$, by (5), $\Pi(k) \leq m$. This implies that $a_m^k > 0$ because otherwise from (4) we see that $\|T^{**}s_k\|^2 > 1$. Since s_k is extreme, if we suppose $0 < a_m^k < 1 = a_{m+1}^k$, then there exists $n < m$ such that

$$a_n^k > a_{n+1}^k = \ldots = a_m^k < a_{m+1}^k = 1.$$

But $a_m^k > a_{m+1}^k - 2a_{m+1}^{m+1} = -1$, and hence we have $a_n^k > a_m^k > a_{m+1}^k - 2a_{m+1}^{m+1}$, which is a contradiction to $s_k - 2s_{m+1}$ being extreme. Hence for $1 \leq k \leq m$

(6) $$a_m^k = 1.$$

Also for $1 \leq j < k \leq m$

(7) $$a_{m-1}^j \geq a_{m-1}^k.$$

Otherwise we would have by (6)

$$a_{m-1}^j - a_{m-1}^k < 0 = a_m^j - a_m^k < a_{m+1}^j - a_{m+1}^k + a_{m+1}^{m+1} = 1,$$

which contradicts the fact that $s_j - s_k + s_{m+1}$ is extreme.

Let b_j be defined by $T^{**}(\sum_{j=1}^{\infty} b_j e_j) = s_m$. By the same reasoning as for the a_i^j, we have $0 \le b_j \le 1$. By (3), and since $e_j = s_j - s_{j+1}$ and for $j > m$, $T^{**} s_j = s_j$, we get by (4)

$$\sum_{j=m}^{\infty} e_j = T^{**}(\sum_{j=1}^{\infty} b_j e_j) = \sum_{j=1}^{\infty} b_j T^{**}(s_j - s_{j+1}) =$$

$$= \sum_{j=1}^{m} b_j \sum_{n=1}^{m} (a_n^j - a_n^{j+1}) e_n + \sum_{j=m+1}^{\infty} b_j e_j.$$

Therefore, if $n < m$

(8) $$\sum_{j=1}^{m} (a_n^j - a_n^{j+1}) b_j = 0$$

and if $n = m$, since by (6) if $j \le m$, $a_m^j = 1$ and by (2) $a_m^{m+1} = 0$, it follows that

$$1 = \sum_{j=1}^{m} (a_m^j - a_m^{j+1}) b_j = b_m.$$

Since by (7) if $j < m$, $a_{m-1}^j - a_{m-1}^{j+1} \ge 0$, since by (2) $a_{m-1}^{m+1} = 0$ and since $b_j \ge 0$, we have by, (8) applied to $n = m - 1$, that $0 = b_j(a_{m-1}^j - a_{m-1}^{j+1})$ for $j = 1,\dots,m$. In particular from $b_m = 1$ we get

(9) $$0 = b_m (a_{m-1}^m - a_{m-1}^{m+1}) = a_{m-1}^m.$$

Since by assumption $\Pi(m) \le m - 1$ and $a_{\pi(m)}^m > 0$, (9) implies that $\Pi(m) < m - 1$. Hence by (4), (6) and (9)

$$1 = \|T^{**} s_m\|^2 = \|\sum_{j=1}^{\infty} a_j^m e_m\|^2 = \|\sum_{j=1}^{m-2} a_j^m e_m + \sum_{j=m}^{\infty} e_j\|^2 \ge (a_{\pi(m)}^m)^2 + 1 > 1$$

which is a contradiction. Thus $m = 0$ and $T^{**} s_n = s_n$ for all n.

Remark: Bellenot in [4] proved that the condition given in Lemma 2.e.3 is not only sufficient but also necessary for $x \in J$ to be extreme.

2.f. J as a conjugate space

Another important aspect in the theory of Banach spaces is the knowledge of whether a given non-reflexive Banach space may be viewed as the dual of another Banach space. In this section we discuss this question for J and will give James' construction of an isometric predual space I of J which is unique up to isometries. We will also show that I is isomorphic to J^* and in Section 2.i we will see that J^* and J (and therefore I and J) are incomparable, that is, no non-reflexive subspace of J^* is isomorphic to a subspace of J and vice versa, no non-reflexive subspace of J^* is isomorphic to a subspace of J.

Definition 2.f.1. A Banach space X is called an isomorphic (isometric) predual of a Banach space Y, if the dual X^* of X is (isometrically) isomorphic to Y. We say that Y has a unique isomorphic (isometric) predual, if Y has an isomorphic (isometric) predual and all isomorphic (isometric) preduals are mutually (isometrically) isomorphic.

To begin with, we show using general results for Banach spaces that J is a conjugate space.

Proposition 2.f.2. J is a conjugate space, and any predual of J is quasi-reflexive of order one.

Proof: Since $\{\xi_n\}$ is a monotone boundedly complete basis (Lemma 2.d.9), it follows immediately from Theorem 1.6 that J is a conjugate space, and by Theorem 1.14 we get that any predual of J is quasi-reflexive of order one.

Moreover, the isomorphic predual of J is unique. This is shown in the following theorem due to Civin and Yood [1].

Theorem 2.f.3. Let X_1 and X_2 be Banach spaces where X_1 is quasi-reflexive of order one. If X_1^* is isomorphic to X_2^* then X_1 is isomorphic to X_2.

Proof: By Theorem 1.14 X_1^* is quasi-reflexive of order one and so are X_2^* and X_2. Hence there exist $y \in X_1^{**}$ and $z \in X_2^{**}$ such that

$$X_1^{**} = j_{X_1}(X_1) \oplus [y]$$

and

$$X_2^{**} = j_{X_2}(X_2) \oplus [z],$$

where j denotes the canonical embedding of a Banach space in its double dual. Let T be the isomorphism between X_1^* and X_2^* and $T^*: X_2^{**} \longrightarrow X_1^{**}$ be its adjoint. Then there exists a subspace W of X_1^{**} such that

$$X_1^{**} = [y, T^*(z)] \oplus W.$$

Define $S : X_1^{**} \longrightarrow X_1^{**}$ as follows:

$$S(\lambda y + \mu T^*(z) + w) = \mu y + \lambda T^*(z) + w$$

for every $\lambda, \mu \in \mathbb{R}$ and $w \in W$. Then S is an automorphism of X_1^{**} such that

$$ST^* z = y.$$

Let P be the projection from X_1^{**} onto $j_{X_1} X_1$. Observe that if $PST^* j_{X_2} x_2 = 0$, then $ST^* j_{X_2} x_2 = \mu y$ for some μ and therefore $T^* j_{X_2} x_2 = \mu T^* z$, and $j_{X_2} x_2 = 0$. Also, from $PST^* z = Py = 0$, we get that for every $x_2^{**} \in X_2^{**}$, if $x_2^{**} = j_{X_2} x_2 + \mu z$, then $PST^* x_2^{**} = PST^* j_{X_2} x_2$. Hence $PST^* |_{j_{X_2}(X_2)}$ is an isomorphism onto $j_{X_1}(X_1)$, and $j_{X_1}^{-1} PST^* j_{X_2}$ is the promised isomorphism between X_2 and X_1.

Corollary 2.f.4. If X^* is isomorphic to J, X is isomorphic to J^*.

Proof: Since by Theorem 2.a.2 J is isomorphic to J^{**}, we have that X^* is isomorphic to J^{**} and therefore, using Theorem 2.f.3 we obtain the desired result.

Not only does J have a predual, but any isomorphic predual of J is primary.

Theorem 2.f.5. Any isomorphic predual X of J is primary.

Proof: By Proposition 2.f.2, X is quasi-reflexive of order one. Suppose $X = M \oplus N$. Then $J \approx M^* \oplus N^*$, and since J is primary (Theorem 2.d.22), either M^* or N^* is isomorphic to J. Hence by Theorem 2.f.3 either M or N is isomorphic to X.

Having solved the existence of an isomorphic predual, we now proceed to construct an isometric predual I of J as in James [4]. In what follows we will use the norm $\||\ |\|$ for J. First we start with some terminology:

Definition 2.f.6. Let $A \subseteq \mathbb{N} \cup \{0\}$ be a bounded set. We will denote by χ_A the function given by $\chi_A(n) = 1$ if $n \in A$ and $\chi_A(n) = 0$ otherwise and call it the characteristic function of A. If $m, n \in \mathbb{N}$, we will denote by $[m, n]$ the interval $\{x \in \mathbb{N} : m \leq x \leq n\}$ and by (m, n) the interval $\{x \in \mathbb{N} : m < x < n\}$. We will say that A and B are strongly disjoint intervals, if they are disjoint intervals of natural numbers which are separated by at least one integer.

Definition 2.f.7. Let $x = \sum_{i=1}^{n} a_i \chi_{A_i}$ with $a_i \neq 0$ for $i = 1, 2, \ldots, n$ and A_i, A_j strongly disjoint intervals if $i \neq j$. We will call x a strongly disjoint (S.D.) step-function.

If x is an S.D. step-function define

$$\langle\!\langle x \rangle\!\rangle = \left(\sum_{i=1}^{n} a_i^2\right)^{1/2}.$$

The space I is then defined as the completion of the normed linear space of real sequences x with finite support with

$$\|x\| = \inf \left\{\sum_{k=1}^{n} \langle\!\langle x^{(k)} \rangle\!\rangle : x = \sum_{k=1}^{n} x^{(k)}\right\}$$

where the inf is taken over all the representations of x as a finite sum of S.D. step-functions.

It is easy to verify that $\|\ \|$ is indeed a norm and that the sequence $\{\zeta_n\}_{n=1}^{\infty}$ of unit vectors is a monotone basis in I.

Lemma 2.f.8. If x is an S.D. step-function, then $\langle\!\langle x \rangle\!\rangle = \|x\|$.

Proof: Suppose that

$$x = \sum_{i=1}^{n} a_i \chi_{A_i}$$

with $A_1, \ldots, A_n$ strongly disjoint intervals contained in some bounded interval $[0, N]$. Estimating $\|x\|$ by use of Definition 2.f.7, we see that we can restrict each x^k to have support in $[0, N]$.

If $x = \sum_{k=1}^{m} x^{(k)}$ and for some k, $x^{(k)} = \sum_{i=1}^{\mu} b_i \chi_{B_i}$, for purposes of calculating $\|x\|$ we may assume that $x^{(k)}$ is the only step-function in this representation of x with support $\bigcup_{i=1}^{\mu} B_i$. Indeed, if $y = \sum_{i=1}^{\mu} c_i \chi_{B_i}$, since

$$(\sum_{i=1}^{\mu}(b_i + c_i)^2)^{1/2} \le (\sum_{i=1}^{\mu} b_i^2)^{1/2} + (\sum_{i=1}^{\mu} c_i^2)^{1/2},$$

we have

(1) $$«x^{(k)} + y» \le «x^{(k)}» + «y».$$

Therefore, if $C_1, C_2, \ldots, C_j$ is a combination of different subsets of [0,N] such that each C_k is a union of strongly disjoint intervals, there exists at most one representation of x of the form $x = \sum_{k=1}^{j} x^{(k)}$ such that $x^{(k)}$ has support in C_k for k = 1,...,j. Hence, since there are only finitely many sets of intervals in [0, N] such that any two intervals are separated by at least one integer, there exists a sequence of S.D. step-functions $\{x^{(k)}\}_{k=1}^{m}$ with $x = \sum_{k=1}^{m} x^{(k)}$ and

(2) $$\|x\| = \sum_{k=1}^{m} «x^{(k)}».$$

We may assume that max A_i < min A_{i+1} for i = 1,...,n.

(i) If $x^{(i)} = b\chi_B + \sum_{r=1}^{\mu} b_r \chi_{B_r}$ and $x^{(j)} = c\chi_C + \sum_{r=1}^{\nu} c_r \chi_{C_r}$ where $|c| = \theta|b|$ and $0 < \theta \le 1$, then $x^{(i)}$ may be replaced in (2) by the two summands $\theta x^{(i)}$ and $(1 - \theta)x^{(i)}$ since $«x^{(i)}» = «\theta x^{(i)}» + «(1 - \theta)x^{(i)}»$.
In this new representation, if sgn$b$ denotes the sign of b, there appear the summands

$$\theta x^{(i)} = (\mathrm{sgn} b)(\mathrm{sgn} c)c\chi_B + \sum_{r=1}^{\mu} \theta b_r \chi_{B_r}$$

and

$$x^{(j)} = c\chi_C + \sum_{r=1}^{\nu} c_r \chi_{C_r}.$$

(ii) Let u = min A_1, s = max A_1 and t = min A_2. Replace each $x^{(k)}$ by

$$x^{(k)}(s)\chi_{A_1} + x^{(k)}\chi_{[0,N]\setminus[0,s]}.$$

This does not change $\sum_{k=1}^{m} x^{(k)}$. Also the sum in (2) is not increased, since $«x^{(k)}»$ is not increased for any k. Therefore the new sum $\sum_{k=1}^{m} «x^{(k)}»$ is not changed.
Observe that in this new representation $x^{(k)}(n) = 0$ if $n \le u$ and $x^{(k)}(n) = x^{(k)}(s)$ for $n \in A_1$.

Let

$$\mathfrak{A} = \{\chi_A : [s, t] \subseteq A\},$$

$$\mathfrak{B} = \{\chi_A : s \in A,\ t \notin A\},$$

$$\mathfrak{C} = \{\chi_A : s \notin A,\ s+1 \in A\}$$

and $x^{(k)} = \sum_{i=1}^{\mu_k} b_i^{(k)} \chi_{B_{i,k}}$ for $k = 1,\dots,m$.

(iii) All $b_i^{(k)}$ such that $\chi_{B_{i,k}} \in \mathfrak{A} \cup \mathfrak{B}$ have the same sign. Otherwise by (i) we may suppose that there are $x^{(i)} = (-1)c\chi_B + \sum_{r=1}^{\mu} b_r \chi_{B_r}$ and $x^{(j)} = c\chi_C + \sum_{r=1}^{\nu} c_r \chi_{C_r}$ with $s \in B \cap C$. Then by (ii), $A_1 \subset B \cap C$ and $u = \min B = \min C$, and therefore we may assume that the interval B is contained in C and hence $C \backslash B$ is an interval. Replacing $x^{(i)}$ by $x^{(i')} = \sum_{r=1}^{\mu} b_r \chi_{B_r}$ and $x^{(j)}$ by

$$x^{(j')} = (-1)c\chi_B + c\chi_C + \sum_{r=1}^{\nu} c_r \chi_{C_r} = c\chi_{C\backslash B} + \sum_{r=1}^{\nu} c_r \chi_{C_r}$$

we get $«x^{(i)}» + «x^{(j)}» > «x^{(i')}» + «x^{(j')}»$, and this is impossible.

(iv) There is no loss of generality if we suppose that all $b_i^{(k)}$ such that $\chi_{B_{i,k}} \in \mathfrak{A} \cup \mathfrak{C}$ have the same sign. For if not, by (i) we may suppose that there are $x^{(i)} = (-1)c\chi_B + \sum_{r=1}^{\mu} b_r \chi_{B_r}$ and $x^{(j)} = c\chi_C + \sum_{r=1}^{\nu} c_r \chi_{C_r}$ with $\chi_B \in \mathfrak{A}$, $\chi_C \in \mathfrak{C}$, or $\chi_B, \chi_C \in \mathfrak{C}$.

If $\chi_B \in \mathfrak{A}$, $\chi_C \in \mathfrak{C}$, then $u = \min B$ and there are three cases:

(a) $\max B = \max C$. Then $C \subset B$ and $B \backslash C$ is an interval. Hence replacing $x^{(i)}$ by

$$x^{(i')} = c\chi_C + (-1)c\chi_B + \sum_{r=1}^{\mu} b_r \chi_{B_r} = (-1)c\chi_{B\backslash C} + \sum_{r=1}^{\mu} b_r \chi_{B_r}$$

and $x^{(j)}$ by

$$x^{(j')} = \sum_{r=1}^{\nu} c_r \chi_{C_r}$$

we get that $«x^{(i')}» = «x^{(i)}»$ and $«x^{(j')}» < «x^{(j)}»$ and this is impossible.

(b) $\max C > \max B$. Then $-c\chi_B + c\chi_C = -c\chi_{B\backslash C} + c\chi_{C\backslash B}$, where $B\backslash C = A_1$ and $C\backslash B \subset [t+1, \max C]$ are strongly disjoint intervals which are not members of $\mathfrak{A} \cup \mathfrak{C}$. Letting

$$x^{(i')} = (-1)c\chi_{B\backslash C} + \sum_{r=1}^{\mu} b_r \chi_{B_r} \quad \text{and} \quad x^{(j')} = c\chi_{C\backslash B} + \sum_{r=1}^{\nu} c_r \chi_{C_r}$$

we have $«x^{(i')}» + «x^{(j')}» = «x^{(i)}» + «x^{(j)}»$.

(c) $\max B > \max C$. Then $-c\chi_B + c\chi_C = -c\chi_{A_1} - c\chi_{[1+\max C, \max B]}$. Letting

$$x^{(i')} = -c\chi_{[1+\max C,\max B]} + \sum_{r=1}^{\mu} b_r \chi_{B_r} \quad \text{and} \quad x^{(j')} = -c\chi_{A_1} + \sum_{r=1}^{\nu} c_r \chi_{C_r}$$

we obtain $«x^{(i')}» = «x^{(i)}»$ and $«x^{(j')}» = «x^{(j)}»$. Also we have that neither A_1 nor $[1 + \max C, \max B]$ belongs to $\mathfrak{A} \cup \mathfrak{C}$.

If $\chi_B \in \mathfrak{C}$ and $\chi_C \in \mathfrak{C}$, then $\min B = \min C = s + 1$ and we may suppose that $C \subset B$ and the result follows as in (a).

Successive applications of this process lead to a representation of x such that (2) holds but so that all $b_i^{(k)}$ for which $\chi_{B_{i,k}} \in \mathfrak{A} \cup \mathfrak{C}$ have the same sign.

Since $x(s + 1) = 0$, and $b_i^{(k)}\chi_{B_{i,k}}(s + 1) \neq 0$ implies $\chi_{B_{i,k}} \in \mathfrak{A} \cup \mathfrak{B} \cup \mathfrak{C}$, (iii) and (iv) imply that $\mathfrak{A} = \emptyset$ and $\mathfrak{C} = \emptyset$.

Hence if $x^{(k)} = \sum_{i=1}^{\mu_k} b_i^{(k)}\chi_{B_{i,k}}$, then either $s \notin B_{i,k}$ or $t \notin B_{i,k}$, and we can replace $x^{(k)}$ by $x^{(k')}$ such that $x^{(k')}(r) = 0$ for $r \in (s, t)$ and $x^{(k')}(r) = x^{(k)}(r)$ for $r \notin (s, t)$. Clearly $x = \sum_{k'=1}^{n} x^{(k')}$, and since $«x^{(k)}» \geq «x^{(k')}»$, by the definition of $\|x\|$ and (2), we conclude that $«x^{(k)}» = «x^{(k')}»$. In this new representation, if we assume that $\max B_{i,k} < \min B_{i+1,k}$, we have that $B_{1,k} = A_1$.

Carrying out this procedure inductively for every i we arrive at a representation $\sum x^{(k)}$ which satisfies (2) and for which all $B_{i,k}$'s are A_i's and therefore $\|x\| = «x»$.

Now we prove that I is in fact an isometric predual of $(J, \||\ |\|)$.

Theorem 2.f.9. I^* is isometric to $(J, \||\ |\|)$ and the unit vector basis of I is shrinking.

Proof: Let $\{\zeta_n\}_{n=1}^{\infty}$ be the unit vector basis for I and $\{\zeta_n^*\}_{n=1}^{\infty}$ be the corresponding coefficient functionals. Define the following correspondence between I^* and J:

To $f \in I^*$ associate $\sum_{i=1}^{\infty} f(\zeta_i)\xi_i \in J$, where $\{\xi_n\}_{n=1}^{\infty}$ is the summing basis of J. Observe that to ζ_i^* we associate ξ_i.

Let

$$\||f|\|_J = \||\sum_{i=1}^{\infty} f(\zeta_i)\xi_i|\| = \sup \left(\sum_{k=1}^{n}\left(\sum_{i=p_{2k-1}}^{p_{2k}-1} f(\zeta_i)\right)^2\right)^{1/2},$$

where the sup is taken over all choices of n and $p_1 < p_2 < \ldots < p_{2n}$.
Let $p_1 < p_2 < \ldots < p_{2n}$ and $x \in I$ be the S.D. step-function

$$x = \sum_{k=1}^{n}\Big(\sum_{i=p_{2k-1}}^{p_{2k}-1} f(\zeta_i)\Big)\chi_{A_k},$$

where $A_k = [p_{2k-1}, p_{2k} - 1]$. Then

$$\|x\| = \Big(\sum_{k=1}^{n}\Big(\sum_{i=p_{2k-1}}^{p_{2k}-1} f(\zeta_i)\Big)^2\Big)^{1/2}$$

and

$$\Big(\sum_{k=1}^{n}\Big(\sum_{i=p_{2k-1}}^{p_{2k}-1} f(\zeta_i)\Big)^2\Big)^{1/2}\|x\| = \sum_{k=1}^{n}\Big(\sum_{i=p_{2k-1}}^{p_{2k}-1} f(\zeta_i)\Big)^2 = |f(x)| \le \|f\|\,\|x\|.$$

Hence $\||f|\|_J \le \|f\|$.
On the other hand, if A_k is as before, and x is the S.D. step-function $x = \sum_{i=1}^{n} a_i\chi_{A_i}$, then by Lemma 2.f.8 $\|x\| = (\sum_{i=1}^{n} a_i^2)^{1/2}$ and by Hölder's inequality

$$|f(x)| = \Big|\sum_{k=1}^{n} a_k\Big(\sum_{i=p_{2k-1}}^{p_{2k}-1} f(\zeta_i)\Big)\Big| \le$$

$$\le \Big(\sum_{k=1}^{n}\Big(\sum_{i=p_{2k-1}}^{p_{2k}-1} f(\zeta_i)\Big)^2\Big)^{1/2}(\sum_{k=1}^{n} a_k^2)^{1/2} \le \||f|\|_J\|x\|.$$

Now suppose $x \in I$ has finite support. Let $\varepsilon > 0$ and $x^{(1)},\ldots,x^{(n)}$ be S.D. step-functions such that $x = \sum_{k=1}^{n} x^{(k)}$ and $\sum_{k=1}^{n}\|x^{(k)}\| < \varepsilon + \|x\|$. Then

$$|f(x)| = |f(\sum_{k=1}^{n} x^{(k)})| \le \||f|\|_J(\sum_{k=1}^{n}\|x^{(k)}\|) < \||f|\|_J(\varepsilon + \|x\|).$$

Since I is generated by the space of the x's with finite support,

$$\|f\| \le \||f|\|_J.$$

Hence $\|f\| = \||f|\|_J$ and I^* is isometric to $(J, \||\ |\|)$.
Since the coefficient functionals $\{\zeta_i^*\}$ of the natural basis for I are identified with the summing basis $\{\xi_i\}$ for J, we have that $\{\zeta_i\}$ is shrinking.

By Theorem 1.14 we know that I is quasi-reflexive of order one. This and the above commentary imply that the summing basis in J is 1-shrinking as defined in 2.a.3.

The next theorem gives an explicit description of the natural embedding of I into J^*.

Theorem 2.f.10. (a) The operator defined by $\zeta_n \longrightarrow e^*_n - e^*_{n+1}$ defines an isometry of I onto the closure of

$$\{\sum_{i=1}^{n} a_i e^*_i : n \geq 1 \text{ and } \sum_{i=1}^{n} a_i = 0\} \subset (J, \| \ \|)^*.$$

(b) $J^* = j(I) \oplus [e^*_1]$ where j is the canonical injection of I into J^*.

Proof: First observe that

$$\langle e^*_n - e^*_{n+1}, \xi_m \rangle = \langle e^*_n - e^*_{n+1}, e_1 + \ldots + e_m \rangle = \begin{cases} 0 \text{ if } m < n, \\ 1 \text{ if } n = m, \\ 0 \text{ if } n + 1 \leq m. \end{cases}$$

Hence $e^*_n - e^*_{n+1} = \xi^*_n$, and from the discussion above, we see that $\zeta_n \longleftrightarrow \xi^*_n$ is the natural embedding of I into $I^{**} = J^*$ since $\xi_n = \zeta^*_n$.
Now if $y = \sum_{i=1}^{n} a_i(e^*_i - e^*_{i+1})$, then $y = \sum_{i=1}^{n+1}(a_i - a_{i-1})e^*_i$ where $a_0 = 0$ and $a_{n+1} = 0$; therefore $\sum_{i=1}^{n+1} (a_i - a_{i-1}) = 0$.
Conversely, if $y = \sum_{i=1}^{n+1} b_i e^*_i$ with $\sum_{i=1}^{n+1} b_i = 0$, then

$$y = \sum_{i=1}^{n+1}(\sum_{j=1}^{i} b_j)(e^*_i - e^*_{i+1}).$$

This proves (a).
Since $\{e_n\}$ is shrinking,

$$I^{**} = [e^*_n] = [\{e^*_n - e^*_{n+1}\} \cup \{e^*_1\}].$$

We will prove that $e^*_1 \notin [e^*_n - e^*_{n+1}]$.
Suppose that $e^*_1 = \sum_{i=1}^{\infty} a_i(e^*_i - e^*_{i+1})$. Then for all n,

$$(a_1 - 1)e^*_1 + \sum_{i=2}^{n}(a_i - a_{i-1})e^*_i = a_n e^*_{n+1} - \sum_{i=n+1}^{\infty} a_i(e^*_i - e^*_{i+1}) =$$

$$= \sum_{i=n}^{\infty}(a_i - a_{i+1})e^*_{i+1}.$$

This implies that $a_i = 1$, $i = 1, 2, \ldots$, and $e^*_1 = \sum_{i=1}^{\infty}(e^*_i - e^*_{i+1})$. But it is easy to see that $\|\sum_{i=n}^{m}(e^*_i - e^*_{i+1})\| = \|\sum_{i=n}^{m}\zeta_i\| = 1$ for $n \leq m$, and hence $\sum_{i=1}^{\infty}(e^*_i - e^*_{i+1})$ cannot converge in norm and this proves (b).

Corollary 2.f.11. For $f \in J^*$, $\lim_{i\to\infty} f(\xi_i)$ exists and is 0 if and only if $f \in [\xi^*_i]_{i=1}^{\infty} = \tilde{J}$.

Proof: Since $J^* = \tilde{J} \oplus [e_1^*]$, for every $f \in J^*$, $f = \sum_{i=1}^{\infty} c_i \xi_i^* + de_1^*$, and hence $\lim_{i\to\infty} f(\xi_i) = d$ exists, and is 0 if and only if $f \in \tilde{J}$.

So far we have proved only that J has a unique isomorphic predual and that it has an isometric predual. Brown and Ito [1] proved in fact that the isometric predual is unique. The proof given here follows from the next two results by Godefroy [1].

The next lemma gives a sufficient condition for a space to have a unique isometric predual. Here $\sigma(X^{**}, X^*)$ will denote the weak star topology in X^{**}.

Lemma 2.f.12. Let X be a Banach space, $B_{X^{**}}$ be the unit ball in X^{**} and

$$R_X = \{f \in X^{***} : \ker f \cap B_{X^{**}} \text{ is } \sigma(X^{**}, X^*)\text{-dense in } B_{X^{**}}\}.$$

If R_X is a vector space, then X is the unique isometric predual of X^*.

Proof: Let $j_X : X \to X^{**}$ denote the canonical injection. Then $j_X(X)^{\perp} \subset R_X$ and by Lemma 1.12,

$$X^{***} = j_X^*(X^*) \oplus j_X(X)^{\perp}.$$

Let $f = j_X^*(x^*) \in j_X^*(X^*) \cap R_X$, then $\ker f = \{x^*\}^{\perp}$ which is $\sigma(X^{**}, X^*)$-closed in X^{**} (see e.g. Rudin [1]); therefore $\ker f = X^{**}$, and hence $j_X^*(X^*) \cap R_X = \{0\}$. It follows that if R_X is a vector space then $j_X(X)^{\perp} = R_X$.

Let $T : X^* \to Y^*$ be an isometric isomorphism between X^* and a dual space Y^*. Since $T^* : Y^{**} \to X^{**}$ is $\sigma(Y^{**}, Y^*)$-$\sigma(X^{**}, X^*)$ continuous (see e.g. Beauzamy [1]), it is easily seen that $(T^*(j_Y(Y)))^{\perp} \subset R_X = j_X(X)^{\perp}$. Therefore $j_X(X) \subset T^*(j_Y(Y))$. Similarly, since $T^{**} R_X = R_Y$, we get

$$j_Y(Y) \subset (T^{-1})^* j_X(X) = (T^*)^{-1} j_X(X),$$

and hence $T^*(j_Y(Y)) = j_X(X)$, and thus Y is isometrically isomorphic to X, since T^* is an isometry.

Theorem 2.f.13. Let X be a Banach space. If X^* is separable and does not contain a subspace isomorphic to ℓ_1 then X is the unique isometric predual of X^*.

Proof: One of the equivalences in Odell and Rosenthal's theorem (see e.g. Lindenstrauss and Tzafriri [1]) says that a Banach space Y does not contain ℓ_1 if and only if every element in Y^{**} is the weak star limit of a sequence of elements in the canonical image of Y in Y^{**}. On the other hand, it follows from Baire's category theorem (see e.g. Diestel [2]) that if $\{f_n\}$ is a sequence of continuous scalar valued functions converging pointwise on a compact set K, then the set of points where $\{f_n\}$ is equicontinuous is a dense $\mathfrak{G}_\delta$ in K. From this it follows that if X^* does not contain a subspace isomorphic to ℓ_1, then for every $f \in X^{***}$ the set of $\sigma(X^{**},X^*)$-continuity points in $B_{X^{**}}$ is a $\sigma(X^{**},X^*)$-dense $\mathfrak{G}_\delta$ in $B_{X^{**}}$.

Let $f \in R_X$. Then f is zero for every $\sigma(X^{**},X^*)$-continuity point and hence is zero on a $\sigma(X^{**},X^*)$-dense $\mathfrak{G}_\delta$ in $B_{X^{**}}$. If $g \in R_X$, $\ker(\lambda f + \mu g) \supset \ker f \cap \ker g$. Since the intersection of two dense $\mathfrak{G}_\delta$-sets is again a $\mathfrak{G}_\delta$, we get $\lambda f + \mu g \in R_X$. The result follows from Lemma 2.f.12.

Corollary 2.f.14. $(J, \|\ \|)$, J^*, J^{**},... each have a unique isometric predual.

Proof: This is an immediate consequence of Theorems 2.a.2, 2.f.13 and 2.f.9.

In his paper [4], James also proved a theorem stating that the space I is not isomorphic to any subspace of J. We will not use James' proof since it is very complicated, even longer than some we have given in the text and in James' own words "gruesome". However, we will prove this theorem in Section 2.i using arguments involving the notions of type and cotype.

2.g. The dual of the James space

We now turn our attention to J^*. The results we will describe in this section are the corresponding ones for J^* to some statements proved for J in Section 2.d. The main theorems are Theorem 2.g.7 where it is shown that every non-reflexive subspace of J^* contains a subspace isomorphic to J^* and complemented in J^* and Theorem 2.g.9, giving a characterization of the reflexive subspaces of J and J^*. For this we study the basis $\{e_i^*\}$ in J^* of the biorthogonal functionals associated to the unit vector basis in J, whose behavior is similar to that of the summing basis $\{\xi_i\}$ in J. All the results of this section are due to Andrew [1].

We start by establishing two elementary properties of the basis $\{e_i^*\}$.

Proposition 2.g.1. Let $x^* = \sum_{i=1}^{\infty} a_i e_i^* \in J^*$.

(a) If $a_i \geq 0$ for all i, then $\|x^*\| = \sum_{i=1}^{\infty} a_i$.

(b) $\|x^*\| \geq (1/\sqrt{2})(\sum_{i=1}^{\infty} a_i^2)^{1/2}$.

Proof: (a) Let $a_i \geq 0$. Since $\|e_i^*\| = 1$ for every i, we have

$$\|x^*\| \leq \sum_{i=1}^{\infty} |a_i| = \sum_{i=1}^{\infty} a_i.$$

Let $\mathbb{1}$ denote the sequence $(1, 1, 1,\ldots) \in J^{**}$. Then by Theorem 1.5 $\|\mathbb{1}\|_{J^{**}} = 1$ and $\sum_{i=1}^{\infty} a_i = \mathbb{1}(x^*) \leq \|x^*\|$.

(b) From Proposition 2.d.11(5) it follows that

$$\sqrt{2}(\sum_{i=1}^{n} a_i^2)^{1/2}\|x^*\| \geq \|\sum_{i=1}^{n} a_i e_i\|\|x^*\| \geq x^*(\sum_{i=1}^{n} a_i e_i) = \sum_{i=1}^{n} a_i^2.$$

Hence $\|x^*\| \geq (1/\sqrt{2})(\sum_{i=1}^{n} a_i^2)^{1/2}$ for every n, and this proves (b).

The next propositions will show that the basis $\{e_n^*\}$ in J^* plays a similar role in this space to that of the summing basis in J.

Proposition 2.g.2. The unit vector basis $\{e_n^*\}$ for J^* is spreading.

Proof: Suppose $\{e_{p_n}^*\}$ is a subsequence of $\{e_n^*\}$. Let $x = \sum_{k=1}^{\infty} c_k e_k \in J$ and

$x' = \sum_{k=1}^{\infty} c_k \sum_{i=p_{k-1}+1}^{p_k} e_i$ where $p_0 = 0$. Then clearly $\|x\| = \|x'\|$ and if $\sum_{k=1}^{\infty} b_k e_{p_k}^* \in J^*$,

$$\langle \sum_{k=1}^{\infty} b_k e_k^*, x \rangle = \sum_{k=1}^{\infty} b_k c_k = \langle \sum_{k=1}^{\infty} b_k e_{p_k}^*, x' \rangle \le \|\sum_{k=1}^{\infty} b_k e_{p_k}^*\| \|x\|$$

and thus $\|\sum_{k=1}^{\infty} b_k e_k^*\| \le \|\sum_{k=1}^{\infty} b_k e_{p_k}^*\|$.

On the other hand, if $\sum_{k=1}^{\infty} b_k e_k^* \in J^*$ and $x = \sum_{k=1}^{\infty} c_k e_k \in J$, using Proposition 2.d.11(3),

$$\langle \sum_{k=1}^{\infty} b_k e_{p_k}^*, x \rangle = \sum_{k=1}^{\infty} b_k c_{p_k} = \langle \sum_{k=1}^{\infty} b_k e_k^*, \sum_{k=1}^{\infty} c_{p_k} e_k \rangle \le \|\sum_{k=1}^{\infty} b_k e_k^*\| \|x\|.$$

Hence $\|\sum_{k=1}^{\infty} b_k e_{p_k}^*\| \le \|\sum_{k=1}^{\infty} b_k e_k^*\|$ and this proves the proposition.

The following results about block basic sequences of $\{e_n^*\}$ are the counterpart to those for spreading sequences in J, seen in Section 2.d.

Proposition 2.g.3. Suppose $0 \le p_1 < p_2 < \dots$. Let $y_n = \sum_{i=p_n+1}^{p_{n+1}} a_i e_i^*$ be a block basic sequence in J^* with $\frac{1}{M} \le \|y_n\| \le M$, and suppose $\sum_{i=p_n+1}^{p_{n+1}} a_i = 0$ for all n. Then

$$(1/\sqrt{2}M)(\sum_{n=1}^{m} b_n^2)^{1/2} \le \|\sum_{n=1}^{m} b_n y_n\| \le 2M(\sum_{n=1}^{m} b_n^2)^{1/2}.$$

Proof: We may assume that $a_{p_n} = 0$ for all n. In fact if we prove this case then

$$y'_n = \sum_{i=q_n+1}^{q_{n+1}-1} a_{p_n - q_n + i} e_i^* = \sum_{i=p_n+1}^{p_{n+1}} a_i e_{i+n-1}^*,$$

where $q_n = p_n + n - 1$, is equivalent to the unit vector basis of ℓ_2; but since $\{e_n^*\}$ is spreading, $\{y'_n\}$ is also equivalent to $\{y_n\}$.
Thus assume $a_{p_n} = 0$ for all n and let

$$X = \left[\sum_{i=p_n+1}^{p_{n+1}} e_i\right]_{n=1}^{\infty} \subset J.$$

Then X is complemented in J by a projection P, where

$$P(\sum_{i=1}^{\infty} c_i e_i) = \sum_{n=0}^{\infty} c_{p_{n+1}} (\sum_{i=p_n+1}^{p_{n+1}} e_i),$$

and has complement

$$Y = QJ = (I - P)J = [\{e_j : j \neq p_n \ \forall n\}].$$

By (3) of Proposition 2.d.11, we have that $\|P\| \le 1$ and $\|Q\| \le 2$. For every $g \in Y^*$ and $x \in Y$,

$$|g(x)| = |g(Qx)| \le \|g \circ Q\|_{J^*} \|x\|.$$

Hence

(1) $$\|g\|_{Y^*} \le \|g \circ Q\|_{J^*} \le \|g\|_{Y^*} \|Q\| \le 2\|g\|_{Y^*}.$$

Let $x \in J$, $x = \sum_{i=1}^{\infty} c_i e_i$ and $g = \sum_{n=1}^{m} b_n y_n$. Then

$$Qx = \sum_{n=1}^{\infty} \sum_{i=p_n+1}^{p_{n+1}-1} (c_i - c_{p_{n+1}}) e_i$$

and

$$(g \circ Q)(x) = \langle \sum_{n=1}^{m} b_n y_n, Qx \rangle = \sum_{n=1}^{m} b_n \langle y_n, \sum_{i=p_n+1}^{p_{n+1}-1} (c_i - c_{p_{n+1}}) e_i \rangle =$$

$$= \sum_{n=1}^{m} b_n \sum_{i=p_n+1}^{p_{n+1}-1} a_i (c_i - c_{p_{n+1}}) = \sum_{n=1}^{m} b_n \sum_{i=p_n+1}^{p_{n+1}-1} a_i c_i =$$

$$= \langle \sum_{n=1}^{m} b_n y_n, x \rangle = g(x).$$

Therefore

(2) $$\|g \circ Q\|_{J^*} = \|g\|_{J^*}.$$

Let $x \in Y$, $x = \sum_{n=1}^{\infty} \sum_{i=p_n+1}^{p_{n+1}-1} c_i e_i$, $x_n = \sum_{i=p_n+1}^{p_{n+1}-1} c_i e_i$ and g be as above. Using Hölder's inequality and applying Lemma 2.c.2, we get

$$|g(x)| = |\sum_{n=1}^{m} b_n \langle y_n, x_n \rangle| \le (\sum_{n=1}^{m} b_n^2 \|y_n\|^2)^{1/2} (\sum_{n=1}^{m} \|x_n\|^2)^{1/2} \le$$

$$\le (\sum_{n=1}^{m} b_n^2 \|y_n\|^2)^{1/2} \|x\|.$$

Hence

(3) $$\|g\|_{Y^*} \le (\sum_{n=1}^{m} b_n^2 \|y_n\|^2)^{1/2}.$$

Now let $x_n = \sum_{i=p_n+1}^{p_{n+1}-1} c_i e_i$ be such that $\|x_n\| = 1$ and $y_n(x_n) = \|y_n\|$, and let $x = \sum_{n=1}^{m} b_n \|y_n\| x_n$. Then by Lemma 2.c.2,

$$\|x\| \le \sqrt{2} (\sum_{n=1}^{m} b_n^2 \|y_n\|^2)^{1/2}$$

and

$$|g(x)| = |\sum_{n=1}^{m} \langle b_n y_n, b_n \|y_n\| x_n \rangle| = \sum_{n=1}^{m} b_n^2 \|y_n\|^2 \ge (\tfrac{1}{2} \sum_{n=1}^{m} b_n^2 \|y_n\|^2)^{1/2} \|x\|.$$

Therefore

(4) $$\|g\|_{Y^*} \geq (1/\sqrt{2})(\sum_{n=1}^{m} b_n^2 \|y_n\|^2)^{1/2}.$$

Thus, from (1), (2), (3), (4) we get

(5) $$(1/\sqrt{2}M)(\sum_{n=1}^{m} b_n^2)^{1/2} \leq \|\sum_{n=1}^{m} b_n y_n\|_{J^*} \leq 2M(\sum_{n=1}^{m} b_n^2)^{1/2},$$

and this proves the proposition.

Corollary 2.g.4. Let $\{x_j^*\}$ be a sequence in J^* with $\frac{1}{M} \leq \|x_j^*\| \leq M$ for all j. If $\{x_j^*\}$ converges weakly to zero, then it has a subsequence equivalent to the unit vector basis of ℓ_2.

Proof: Since the sequence $\{x_j^*\}$ converges weakly to zero, by Corollary 1.9 it has a subsequence $\{x_{j_n}^*\}$ equivalent to a block basic sequence $\{y_n\}$ of $\{e_n^*\}$, where $y_n = \sum_{i=p_n+1}^{p_{n+1}} a_i e_i^*$ for n = 1,2,....

Therefore, if $1 \in J^{**}$ is given by $1(e_n^*) = 1$ for $n \in \mathbb{N}$ as in Theorem 2.a.2, we get

(1) $$0 = \lim_n \langle 1, y_n \rangle = \lim_n \sum_{i=p_n+1}^{p_{n+1}} a_i.$$

Let $u_n = \sum_{i=p_n+1}^{p_{n+1}} a_i e_i^* - (\sum_{i=p_n+1}^{p_{n+1}} a_i) e_{p_{n+1}}^* = \sum_{i=p_n+1}^{p_{n+1}} b_i e_i^*$. Then

$$\sum_{i=p_n+1}^{p_{n+1}} b_i = 0 \quad \text{and} \quad \|y_n - u_n\| = |\sum_{i=p_n+1}^{p_{n+1}} a_i|.$$

Hence by (1) and Proposition 1.8, there exists a subsequence $\{y_{n_j}\}$ of $\{y_n\}$ equivalent to $\{u_{n_j}\}$. But by Proposition 2.g.3, $\{u_{n_j}\}$ is equivalent to the unit vector basis of ℓ_2 and this concludes the proof of the corollary.

Theorem 2.g.5. Suppose $p_1 < p_2 < \ldots$. Let $\{w_n\}$ be a block basic sequence given by $w_n = \sum_{i=p_n+1}^{p_{n+1}} a_i e_i^*$ for $n \in \mathbb{N}$ and suppose $\sum_{i=p_n+1}^{p_{n+1}} a_i = K > 0$. Then for any real sequence $\{b_n\}$:

(a) $\|\sum_{i=1}^{\infty} b_i w_i\| \geq K\|\sum_{i=1}^{\infty} b_i e_i^*\|$.

(b) If $a_i \geq 0$ for all i, then $\|\sum_{i=1}^{\infty} b_i w_i\| \leq \sqrt{2}K\|\sum_{i=1}^{\infty} b_i e_i^*\|$.

(c) If $\{w_n\}$ is seminormalized, then it is equivalent to $\{e_n^*\}$, and $[w_n]$ is complemented in J^*.

Proof: Let $x = \sum_{i=1}^{\infty} c_i e_i \in J$, and observe that

(1) $$\text{if } x' = \sum_{n=1}^{\infty} c_n \sum_{i=p_n+1}^{p_{n+1}} e_i, \quad \text{then } \|x\| = \|x'\|.$$

Hence

$$\langle \sum_{k=1}^{\infty} b_k w_k, x' \rangle = \langle \sum_{k=1}^{\infty} b_k w_k, \sum_{n=1}^{\infty} c_n \sum_{i=p_n+1}^{p_{n+1}} e_i \rangle = \sum_{n=1}^{\infty} b_n c_n \sum_{i=p_n+1}^{p_{n+1}} a_i =$$

$$= K \sum_{n=1}^{\infty} b_n c_n = K \langle \sum_{n=1}^{\infty} b_n e_n^*, \sum_{n=1}^{\infty} c_n e_n \rangle = K \langle \sum_{n=1}^{\infty} b_n e_n^*, x \rangle,$$

so that if $\|x\| = \|x'\| = 1$,

$$|K \langle \sum_{n=1}^{\infty} b_n e_n^*, x \rangle| \le \|\sum_{k=1}^{\infty} b_k w_k\|,$$

and (a) follows.

Suppose now that $a_i \ge 0$ for every i, and define $\check{c}_n = (1/K) \sum_{i=p_n+1}^{p_{n+1}} a_i c_i$ for each n. Let $A_n = \{p_n + 1, p_n + 2, \ldots, p_{n+1}\}$ and $q_1 < q_2 < \ldots < q_k$. Then

$$\frac{1}{2} \sum_{i=1}^{k-1} (\check{c}_{q_i} - \check{c}_{q_{i+1}})^2 \le$$

$$\le \frac{1}{2} \sum_{i=1}^{k-1} \max \left((\max_{j \in A_{q_i}} c_j - \min_{j \in A_{q_{i+1}}} c_j)^2, (\min_{j \in A_{q_i}} c_j - \max_{j \in A_{q_{i+1}}} c_j)^2 \right) \le$$

$$\le 2 \|\sum_{i=1}^{\infty} c_i e_i\|^2.$$

It follows that

$$\|\sum_{i=1}^{\infty} \check{c}_i e_i\| \le \sqrt{2} \|\sum_{i=1}^{\infty} c_i e_i\|$$

and

$$|\langle \sum_{k=1}^{\infty} b_k w_k, \sum_{n=1}^{\infty} c_n e_n \rangle| = |\sum_{n=1}^{\infty} b_n \sum_{i=p_n+1}^{p_{n+1}} a_i c_i| = K |\sum_{i=1}^{\infty} b_i \check{c}_i| =$$

$$= K |\langle \sum_{i=1}^{\infty} b_i e_i^*, \sum_{k=1}^{\infty} \check{c}_k e_k \rangle| \le K \|\sum_{i=1}^{\infty} b_i e_i^*\| \|\sum_{k=1}^{\infty} \check{c}_k e_k\| \le$$

$$\le \sqrt{2} K \|\sum_{i=1}^{\infty} b_i e_i^*\| \|\sum_{k=1}^{\infty} c_k e_k\|.$$

Therefore (b) holds.

Now,

$$w_n = \sum_{i=p_n+1}^{p_{n+1}} a_i e_i^* = \frac{K}{p_{n+1} - p_n} \sum_{i=p_n+1}^{p_{n+1}} e_i^* + \sum_{i=p_n+1}^{p_{n+1}} \left(a_i - \frac{K}{p_{n+1} - p_n} \right) e_i^* =$$

$$= z_n + y_n.$$

By Proposition 2.g.1 we have $\|z_n\| = K$ for all n, and since $\{w_n\}$ is seminormalized, there exists a constant M such that $\|y_n\| \le M$ for all n. Hence from (b), Proposition 2.g.3 and Proposition 2.g.1, it follows that

(2) $$\|\sum_{n=1}^{\infty} b_n w_n\| \le \|\sum_{n=1}^{\infty} b_n z_n\| + \|\sum_{n=1}^{\infty} b_n y_n\| \le$$

$$\le \sqrt{2}K\|\sum_{n=1}^{\infty} b_n e_n^*\| + 2M(\sum_{n=1}^{\infty} b_n^2)^{1/2} \le \sqrt{2}(K + 2M)\|\sum_{n=1}^{\infty} b_n e_n^*\|.$$

Thus by (a) and (2), $\{w_n\}$ is equivalent to $\{e_n^*\}$.

Let $Q : J \longrightarrow J$ be the averaging projection defined by

$$Q(\sum_{i=1}^{\infty} c_i e_i) = \frac{1}{K}\sum_{n=1}^{\infty}(\sum_{i=p_n+1}^{p_{n+1}} a_i c_i)(\sum_{i=p_n+1}^{p_{n+1}} e_i).$$

To see that Q is bounded, since $\{w_n\}$ is equivalent to $\{e_n^*\}$, let L be such that

(3) $$\frac{1}{L}\|\sum_{n=1}^{\infty} b_n w_n\| \le \|\sum_{n=1}^{\infty} b_n e_n^*\| \le L\|\sum_{n=1}^{\infty} b_n w_n\|,$$

and let $x = \sum_{i=1}^{\infty} c_i e_i \in J$, $f = \sum_{n=1}^{\infty} b_n e_n^* \in J^*$. Then

(4) $$\|\sum_{n=1}^{\infty} w_n(x) e_n\| = \sup_{f\in J^*} \frac{\langle \sum_{n=1}^{\infty} b_n e_n^*, \sum_{n=1}^{\infty} w_n(x) e_n\rangle}{\|\sum_{n=1}^{\infty} b_n e_n^*\|} =$$

$$= \sup_{f\in J^*} \frac{|\sum_{n=1}^{\infty} w_n(x) b_n|}{\|\sum_{n=1}^{\infty} b_n e_n^*\|} = \sup_{f\in J^*} \frac{|\sum_{n=1}^{\infty} b_n \sum_{i=p_n+1}^{p_{n+1}} a_i c_i|}{\|\sum_{n=1}^{\infty} b_n e_n^*\|}.$$

On the other hand, by (3) and (4) if $\varepsilon > 0$ and $f = \sum_{n=1}^{\infty} b_n e_n^* \in J^*$ is such that $\|\sum_{n=1}^{\infty} w_n(x) e_n\| < \frac{1}{\|f\|}|f(\sum_{n=1}^{\infty} w_n(x) e_n)| + \varepsilon$ then

(5) $$\|x\| = \|\sum_{n=1}^{\infty} c_n e_n\| \ge \frac{\langle \sum_{n=1}^{\infty} b_n w_n, \sum_{n=1}^{\infty} c_n e_n\rangle}{\|\sum_{n=1}^{\infty} b_n w_n\|} = \frac{|\sum_{n=1}^{\infty} b_n \sum_{i=p_n+1}^{p_{n+1}} c_i a_i|}{\|\sum_{n=1}^{\infty} b_n w_n\|} \ge$$

$$\ge \frac{1}{L} \frac{|\sum_{n=1}^{\infty} b_n \sum_{i=p_n+1}^{p_{n+1}} a_i c_i|}{\|\sum_{n=1}^{\infty} b_n e_n^*\|} > \frac{1}{L}\|\sum_{n=1}^{\infty} w_n(x) e_n\| - \frac{\varepsilon}{L}.$$

Thus using (1) and (5),

$$\|Qx\| = \|Q(\sum_{i=1}^{\infty} c_i e_i)\| = \|\frac{1}{K}\sum_{n=1}^{\infty}(\sum_{i=p_n+1}^{p_{n+1}} a_i c_i)(\sum_{i=p_n+1}^{p_{n+1}} e_i)\| =$$

$$= \|\tfrac{1}{K}\sum_{n=1}^{\infty}(\sum_{i=p_n+1}^{p_{n+1}} a_i c_i)e_n\| = \tfrac{1}{K}\|\sum_{i=1}^{\infty} w_i(x)e_i\| \le \tfrac{L}{K}\|x\|.$$

A simple calculation shows that

$$Q^*(\sum_{n=1}^{\infty} d_n e_n^*) = \sum_{n=1}^{\infty}(\tfrac{1}{K}\sum_{i=p_n+1}^{p_{n+1}} d_i)w_n.$$

Hence $[w_n]$ is complemented and this proves (c).

Remark: Observe that if $\{w_{n_i}\}$ is a subsequence of $\{w_n\}$, then

$$w_{n_i} = \sum_{j=p_{n_i}+1}^{p_{n_i+1}} a_j e_j^* = \sum_{j=p_{n_i}+1}^{p_{n_{i+1}}} a'_j e_j^*$$

where $a'_j = a_j$ for $p_{n_i} + 1 \le j \le p_{n_i+1}$ and $a'_j = 0$ if $n_i + 1 < n_{i+1}$ and $p_{n_i+1} < j \le p_{n_{i+1}}$. Thus $\{w_{n_i}\}$ may be regarded as a sequence of the same kind as $\{w_n\}$ and $[w_{n_i}]$ is complemented in J^* and isomorphic to J^*.

Proposition 2.g.6. Let $\{z_n\} \subset J^*$ be a seminormalized sequence with no weak cluster points and suppose it w^*-converges to $z \in J^*$. Then there exists a subsequence $\{z_{n_i} - z\}_i$ of $\{z_n - z\}$ which is equivalent to the basis $\{e_n^*\}$ and such that $[z_{n_i} - z]_i$ is complemented in J^*.

Proof: Since $z_n - z$ does not converge to zero weakly, and for $n \in \mathbb{N}$ $\|z_n - z\| \le 1 + \|z\|$, it follows that one may assume, by passing to a subsequence, that $\{z_n - z\}$ is seminormalized.
Let $z'_n = z_n - z$. Since z'_n does not converge to zero weakly, there exists $x \in J^{**}$ such that $\langle x, z'_n\rangle$ does not converge to zero. By Theorem 2.a.2 there exist $x_0 \in J$ and $\lambda \in \mathbb{R}$ such that

$$x = jx_0 + \lambda\mathbb{1},$$

where $\mathbb{1} \in J^{**}$ denotes the functional given by $\langle \mathbb{1}, e_n^*\rangle = 1$ for every n, $x_0 \in J$, $\lambda \in \mathbb{R}$ and j is the canonical injection of J into J^{**}. It follows from the w^*-convergence of z'_n to zero that $\langle z'_n, x_0\rangle \to 0$, and therefore $\lambda \ne 0$ and $\langle \mathbb{1}, z'_n\rangle$ does not converge to zero. Since $\{\langle \mathbb{1}, z'_n\rangle\}$ is bounded, one can extract a subsequence converging to a non-zero limit. Using this, the w^*-convergence of $\{z'_n\}$ to zero and passing to an adequate subsequence, we obtain as in Corollary 1.9 a block basic sequence $\{z''_n\}$

with $z''_n = \sum_{i=p_n+1}^{p_{n+1}} b_i e_i^*$ for every n, equivalent to a subsequence $\{z'_{n_i}\}$ of $\{z'_n\}$, such that

(1) $$\lim_i \|z''_i - z'_{n_i}\| = 0$$

and

(2) $$\lim_n \langle 1, z''_n \rangle = \lim_n \sum_{i=p_n+1}^{p_{n+1}} b_i = K \neq 0.$$

Let

$$w_n = \sum_{i=p_n+1}^{p_{n+1}} b_i e_i^* + (K - \sum_{i=p_n+1}^{p_{n+1}} b_i) e_{p_n+1}^* = \sum_{i=p_n+1}^{p_{n+1}} a_i e_i^*.$$

Since $\langle 1, w_n \rangle = \sum_{i=p_n+1}^{p_{n+1}} a_i = K$, by Theorem 2.g.5 $[w_n]$ is complemented in J^* and $\{w_n\}$ is equivalent to $\{e_n^*\}$. But by (2), $\lim_n \|z''_n - w_n\| = 0$, therefore by passing to a subsequence such that $\sum_{n=1}^{\infty} \|w_n - z''_n\|$ is small enough and by the remark to Theorem 2.g.5, we may apply Proposition 1.8 in order to get that $\{z''_n\}$ is equivalent to $\{w_n\}$, and $[z''_n]$ is complemented in J^* and isomorphic to J^*. Applying the same arguments, again using (1), we get a subsequence of $\{z'_{n_i}\}$ such that the generated space $[z'_{n_i}]$ is complemented in J^* and isomorphic to J^*.

Now we are able to describe the non-reflexive subspaces of J^*.

Theorem 2.g.7. If Y is a non-reflexive subspace of J^*, then Y contains a subspace isomorphic to J^* and complemented in J^*.

Proof: Let $Y \subset J^*$ be non-reflexive. Then there exists a sequence of norm one vectors $\{w'_n\} \subset Y$ having no limit in the weak topology of J^* (see e.g. Beauzamy [1]). We may as above assume, by passing to a subsequence, that $\{w'_n\}$ has a w^*-limit $w \in J^*$ and that $\{w'_n - w\}$ is semi-normalized.

Consider first the case when $w \in Y$; in this case we may assume $w = 0$. By Proposition 2.g.6 we get a subsequence $\{w'_{n_i}\}$ of $\{w'_n\}$ such that $[w'_{n_i}]$ is complemented in J^* and $\{w'_{n_i}\}$ is isomorphic to $\{e_i^*\}$.

Now we consider the case where w does not belong to Y. Let $z_n = w_n - w$. Applying the first case to $\{z_n\} \subset Y \oplus [w]$, we get a subsequence $\{z'_n\}$ of

$\{z_n\}$ such that $\{z'_n\}$ is equivalent to $\{e^*_n\}$, and $[z'_n]$ is complemented in J^*. Then $z'_n - z'_{n+1} = w'_n - w'_{n+1} \in Y$, and $\{z'_n - z'_{n+1}\}$ is equivalent to $\{e^*_n - e^*_{n+1}\}$. But $e^*_n - e^*_{n+1} = \xi^*_n$ where $\{\xi_n\}$ is the boundedly complete summing basis of J. Hence, by Theorem 1.6, $[e^*_n - e^*_{n+1}]^*$ is isomorphic to J, and thus by Corollary 2.f.4, $[w'_n - w'_{n+1}]$ is isomorphic to J^*. Since

$$[w'_n - w'_{n+1}]_n \oplus [z'_1] = [z'_n]_n$$

with $[z'_n]$ complemented in J^*, $[w'_n - w'_{n+1}]$ is complemented in J^*.

Corollary 2.g.8. J^* is somewhat reflexive.

Proof: By Corollary 2.d.4, ℓ_2 is a complemented subspace of J and thus of J^*. Now, if Y is a non-reflexive subspace of J^*, by the previous theorem, Y contains a subspace isomorphic to J^* and thus contains ℓ_2.

Finally we will study the reflexive subspaces of J and J^*, using Theorem 2.g.7 in the case of J and Corollary 2.d.21 in the case of J^*. Furthermore, as a consequence we obtain a complete description of the reflexive subspaces of J.

Theorem 2.g.9. If $X \subset J$ $(X \subset J^*)$ is reflexive, then there exists a reflexive space $Y \subset J$ $(Y \subset J^*)$ such that Y is complemented in J (J^*) and $X \subset Y$.

Proof: Let $X \subset J$ and let I be the predual of J. Since X is reflexive, J/X is non-reflexive and hence

$$(J/X)^* = X^\perp = \{f \in J^* : f(X) = 0\}$$

is non-reflexive. Let j denote the canonical injection of I into J^* and let

$$X_\perp = \{z \in I : \langle x, z\rangle = 0 \ \forall x \in X\}.$$

Then $jX_\perp \subset X^\perp$ is of codimension at most one in $X^\perp$. In fact, since $X^\perp \cap jI = jX_\perp$, by Theorem 2.f.10 if $g_1, g_2 \in X^\perp \setminus jX_\perp$, there exist $z_1, z_2 \in I$ and $\lambda_1, \lambda_2 \in \mathbb{R}\setminus\{0\}$ such that $g_1 = jz_1 + \lambda_1 e^*_1$ and $g_2 = jz_2 + \lambda_2 e^*_1$; hence

$$\lambda_2 g_1 - \lambda_1 g_2 = \lambda_2 jz_1 - \lambda_1 jz_2 \in jX_\perp.$$

Now we will show that

$$jX_\perp \oplus [g_1] = jX_\perp \oplus [g_1, g_2],$$

which proves our claim about the codimension of $X_\perp$. Let $z \in jX_\perp$ and $a, b \in \mathbb{R}$. Then

$$z + ag_1 + bg_2 = z - \frac{b}{\lambda_1}(\lambda_2 g_1 - \lambda_1 g_2) + ag_1 + \frac{b}{\lambda_1}\lambda_2 g_1 = z' + (a + \frac{b}{\lambda_1}\lambda_2)g_1$$

with $z' \in jX_\perp$.

Hence $X_\perp$ is non-reflexive. Since by Corollary 2.f.4 I is isomorphic to J^*, by Theorem 2.g.7, there exists a subspace $Z \subset X_\perp$ with Z isomorphic to J^* and complemented in I by a projection P. If we view P as a function from I into I, then we see that X is contained in the complemented space $\ker P^*$. But P^*J is isomorphic to $(PI)^* = Z^* \approx J^{**} \approx J$ (see e. g. Beauzamy [1]), and hence using Corollary 2.d.23 we get that $(I - P)^*J = \ker P^*$ is reflexive.

As mentioned, the proof of the case $X \subset J^*$ is the same, using Corollary 2.d.21 in place of Theorem 2.g.7.

Corollary 2.g.10. If $X \subset J$ is reflexive, then X is isomorphic to a subspace of $\mathfrak{J}$.

Proof: By Theorem 2.g.9, X is contained in a complemented reflexive subspace Y of J and by Proposition 2.c.18 Y is isomorphic to a complemented subspace of $\mathfrak{J}$.

2.h. The Banach-Saks properties and the spreading models of J and J^*

In their classical paper, Banach and Saks [1], while investigating the convergence in L^p, $1 < p < \infty$, proved that these spaces have the following property, which now bears their name:

Definition 2.h.1. A Banach space X has the Banach-Saks (B.S.) property if every bounded sequence $\{x_n\}$ in X has a subsequence $\{x_{n_i}\}$ such that its arithmetic means $\frac{1}{k}\sum_{i=1}^{k} x_{n_i}$ converge in norm when $k \longrightarrow \infty$, that is, $\{x_{n_i}\}$ is Cesàro summable.

This property has been extensively studied since Banach and Saks proved their result. Among other things, it is known that any space possessing the Banach-Saks property is reflexive, so that in particular J and J^* don't have it. However, we will see that both spaces have a weakened form of this property, the so-called alternate Banach-Saks property, and thus also the weak Banach-Saks property.

Definition 2.h.2. A Banach space X has the weak Banach-Saks (W.B.S.) property, if every sequence weakly converging to zero has a subsequence Cesàro summable to zero.

Definition 2.h.3. A Banach space X has the alternate Banach-Saks (A.B.S.) property if every bounded sequence $\{x_n\}$ in X has a subsequence $\{x_{n_i}\}$ such that its alternating arithmetic means $\frac{1}{k}\sum_{i=1}^{k}(-1)^i x_{n_i}$ converge in norm when $k \longrightarrow \infty$.

As mentioned, it is true in general that the alternate Banach-Saks property implies the weak Banach-Saks property. However, since for the James spaces J and J^* it is not difficult to see that they possess both properties, we will give both proofs.

Proposition 2.h.4. J and J^* have the weak Banach-Saks property.

Proof: Let $\{x_n\}$ be a sequence in J which converges weakly to zero. If it has a subsequence converging in norm to zero, the result is obvious. Otherwise $\{x_n\}$ is seminormalized, and by Proposition 2.d.16 it has a subsequence equivalent to the unit vector basis of ℓ_2. This subsequence obviously is Cesàro summable to zero.
The proof for the space J^* is the same using Corollary 2.g.4.

Proposition 2.h.5. J and J^* have the alternate Banach-Saks property.

Proof: We will prove the proposition for J first. Let $\{x_n\}$ be a bounded sequence in J. As above, we may suppose that $\{x_n\}$ is seminormalized. Then by Proposition 2.d.15 either $\{x_n\}$ has a weakly convergent subsequence or it has a subsequence equivalent to the summing basis

$\{\xi_n\}$; we will again call the subsequence $\{x_n\}$.

In the first case let x_0 be the weak limit of $\{x_n\}$. Then by Proposition 2.d.16, $\{x_n - x_0\}$ has a subsequence $\{x_{m_n} - x_0\}$ equivalent to the unit vector basis of ℓ_2; thus the alternate arithmetic means of this subsequence tend to zero in norm, and hence the alternate arithmetic means of $\{x_{m_n}\}$ also tend to zero in norm.

In the second case

$$\frac{1}{k}\sum_{i=1}^{k}(-1)^i\xi_i = \begin{cases} \frac{1}{k}(-e_1 - e_3 - \ldots - e_k) & \text{if k is odd,} \\ \frac{1}{k}(e_2 + e_4 + \ldots + e_k) & \text{if k is even.} \end{cases}$$

and clearly $\frac{1}{k}\sum_{i=1}^{k}(-1)^i\xi_i$ tends to zero in norm.

For J^* the proof is similar using Corollary 2.g.4 and Proposition 2.g.6. It only remains to prove that the alternate arithmetic means of $\{e_n^*\}$ converge to zero in norm. Indeed if $x = \sum_{i=1}^{\infty} a_i e_i \in J$

$$\left|\frac{1}{k}\sum_{i=1}^{k}(-1)^i e_i^*(x)\right| = \begin{cases} \frac{1}{k}\left|\sum_{i=1}^{(k-1)/2}(a_{2i} - a_{2i-1}) - a_k\right| & \text{if k is odd,} \\ \frac{1}{k}\left|\sum_{i=1}^{k/2}(a_{2i} - a_{2i-1})\right| & \text{if k is even.} \end{cases}$$

Thus $\left|\frac{1}{k}\sum_{i=1}^{k}(-1)^i e_i^*(x)\right| \le \frac{1}{k}\left(\sqrt{k}\|x\| + \|x\|\right)$, which implies that as k tends to infinity, $\left\|\frac{1}{k}\sum_{i=1}^{k}(-1)^i e_i^*\right\|$ tends to zero.

An important consequence of this proposition is its role in the classification of the spreading models of J and J^*. The concept of a spreading model of a Banach space X based on a sequence $\{x_n\}$ in X was introduced by Brunel and Sucheston in [1] to study the summability of this sequence, and several other applications of this notion have since arisen, for example in the study of stable Banach spaces.

These results and those that follow can be found in Beauzamy and Lapresté [1].

First we give the formal definition of a spreading model.

Definition 2.h.6. Let X be a Banach space and $\{x_n\} \subset X$ be a bounded sequence with no convergent subsequences. A Banach space Y is called a spreading model of X based on $\{x_n\}$ if the following hold:

(a) There exists a sequence $\{e_n\}$ such that Y is the closed linear span of $\{e_n\}$.

(b) For all $k \in \mathbb{N}$ and every real sequence $a_1, a_2, \dots, a_k$ we have for every $x \in X$ that $\lim \|x + a_1 x_{n_1} + \dots + a_k x_{n_k}\|_X$ as $n_1 \longrightarrow \infty$ exists whenever $n_1 < n_2 < \dots < n_k$ and the limit depends only on x and $a_1, a_2, \dots, a_k$.

(c) $\|a_1 e_1 + \dots + a_k e_k\|_Y = \lim \|a_1 x_{n_1} + \dots + a_k x_{n_k}\|_X$ as $n_1 \longrightarrow \infty$.

The sequence $\{e_n\}$ is called the fundamental sequence of Y.

Our next task is to investigate conditions on a given sequence $\{x_n\}$ in X, guaranteeing the existence of spreading models based on it. For this, the following terminology will be useful.

Definition 2.h.7. Let X be a Banach space. A bounded sequence $\{x_n\}$ in X will be called a good sequence, if for every $x \in X$, every $k \in \mathbb{N}$ and every real sequence $a_1, a_2, \dots, a_k$

$$\lim \|x + a_1 x_{n_1} + \dots + a_k x_{n_k}\|$$

as $n_1 \longrightarrow \infty$ exists, whenever $n_1 < n_2 < \dots < n_k$, and this limit depends only on x and $a_1, a_2, \dots, a_k$.

Lemma 2.h.8. Every subsequence of a good sequence $\{x_n\}$ is also a good sequence, and if in addition $\{x_n\}$ has a convergent subsequence, then the sequence $\{x_n\}$ itself converges.

Proof: The first part is clear. Now suppose $x_{n_i} \longrightarrow x$ as $i \longrightarrow \infty$; then taking $a = -1$ we get

$$\lim_{i\to\infty} \|x - x_{n_i}\| = \lim_{n\to\infty} \|x - x_n\| = 0.$$

From here on, we will denote by $\mathbb{R}^{(\mathbb{N})}$ the space of finite real sequences.

Definition 2.h.9. Let X be a separable Banach space, $\{x_n\}$ a bounded sequence in X and $\{x'_n\}$ a good subsequence of $\{x_n\}$. For $x \in X$ and $a = (a_1, \dots, a_k)$ belonging to $\mathbb{R}^{(\mathbb{N})}$ let

$$L(x,a) = \lim_{n_1<n_2<\dots<n_k} \|x + a_1 x'_{n_1} + \dots + a_k x'_{n_k}\| \text{ as } n_1 \longrightarrow \infty.$$

Theorem 2.h.10. Let X be a separable Banach space, $\{x_n\}$ a bounded sequence in X, $\{x'_n\}$ a good subsequence of $\{x_n\}$ and $\{e_n\}$ the unit vector basis of $\mathbb{R}^{(\mathbb{N})}$. For $x \in X$ and $a = (a_1,\dots,a_k) \in \mathbb{R}^{(\mathbb{N})}$ define

$$\|x + a_1e_1 + \dots + a_ke_k\| = L(x,a).$$

Then L is a seminorm in $X \times \mathbb{R}^{(\mathbb{N})}$ and is a norm if and only if $\{x'_n\}$ doesn't converge in X.

Proof: It is obvious that L is a seminorm.
Now suppose that $\{x'_n\}$ does not converge in X, and let $x \in X$ and $a_1,\dots,a_k \in \mathbb{R}$ be such that $\|x + a_1e_1 + \dots + a_ke_k\| = 0$. This implies that for every $\varepsilon > 0$ there exists N such that for every $N < n_1 < \dots < n_k$

$$\|x + a_1x'_{n_1} + \dots + a_kx'_{n_k}\| < \varepsilon.$$

Hence if m and n are such that $N < n < m < n_2 < \dots < n_k$ we have that

$$\|x + a_1x'_n + a_2x'_{n_2} + \dots + a_kx'_{n_k}\| < \varepsilon$$

and

$$\|x + a_1x'_m + a_2x'_{n_2} + \dots + a_kx'_{n_k}\| < \varepsilon.$$

Therefore for n, m > N,

(1) $$\|a_1(x'_n - x'_m)\| < 2\varepsilon,$$

and this means that the sequence $\{a_1x'_n\}$ converges. But since $\{x'_n\}$ is not convergent, this is only possible if $a_1 = 0$. Similarly we prove that $a_i = 0$ for every $1 \le i \le k$. Hence also $x = 0$ and L is a norm.
Conversely, if $\{x'_n\}$ converges to $x \in X$, we obtain that

$$\|x - e_1\| = L(x,-1) = \lim_{n\to\infty}\|x - x'_n\| = 0,$$

and L is not a norm.

To prove the existence of good sequences, we will use the well known theorem in combinatorics of Ramsey which we prove next.

Theorem 2.h.11. Let S be an infinite set, $k \ge 1$ an integer, P_kS the family of all subsets of S with k elements. Suppose P_kS is partitioned into two sets A and B. Then there exists an infinite subset M of S such that P_kM is contained either in A or in B.

Proof: The theorem is trivial if k = 1. We will proceed by induction; thus assume that the assertion is true for any infinite set and for i = k - 1.

Suppose that there are $x_1 \in S$ and an infinite subset $T_1 \subset S \setminus \{x_1\}$ such that

$$\{x_1, a_2,\dots,a_k\} \in A \text{ for all } \{a_2,\dots,a_k\} \subset T_1;$$

now suppose that there are $x_2 \in T_1$ and an infinite subset $T_2 \subset T_1 \setminus \{x_2\}$ such that

$$\{x_2, a_2,\dots,a_k\} \in A \text{ for all } \{a_2,\dots,a_k\} \subset T_2;$$

and so forth: suppose that there are $x_n \in T_{n-1}$ and an infinite subset $T_n \subset T_{n-1} \setminus \{x_n\}$ such that $\{x_n, a_2,\dots,a_k\} \in A$ for all $\{a_2,\dots,a_k\} \subset T_n$.

If this process is infinite, let $M = \{x_1, x_2,\dots,x_n,\dots\}$. Since $x_i \neq x_j$ for $i \neq j$, M is an infinite subset of S. Now let $\{x_{r_1},\dots,x_{r_k}\} \in P_k M$, and suppose that $r_1 < r_j$ for $j = 2,\dots,k$; then by construction $\{x_{r_2},\dots,x_{r_k}\} \subset T_{r_1}$ and $\{x_{r_1},\dots,x_{r_k}\} \in A$. Hence $P_k M \subset A$.

If the process stops at stage n, let $y_1 \in T_{n-1}$ (where $T_0 = S$). Then $P_{k-1}(T_{n-1} \setminus \{y_1\}) = A_1 \cup B_1$, where

$$A_1 = \{\{a_2,\dots,a_k\} \subset T_{n-1} \setminus \{y_1\} : \{y_1, a_2,\dots,a_k\} \in A\}$$

and

$$B_1 = \{\{a_2,\dots,a_k\} \subset T_{n-1} \setminus \{y_1\} : \{y_1, a_2,\dots,a_k\} \in B\}.$$

By the induction hypothesis there exists an infinite subset R_1 of $T_{n-1} \setminus \{y_1\}$ such that $P_{k-1} R_1 \subset A_1$ or $P_{k-1} R_1 \subset B_1$, or equivalently, either

(1) $$\{\{y_1, a_2,\dots,a_k\} : \{a_2,\dots,a_k\} \subset R_1\} \subset A$$

or

(2) $$\{\{y_1, a_2,\dots,a_k\} : \{a_2,\dots,a_k\} \subset R_1\} \subset B.$$

But (1) cannot happen, otherwise we could take $x_n = y_1$ and $T_n = R_1$ in the previous construction; hence (2) holds. Let $y_2 \in R_1$, then

$$P_{k-1}(R_1 \setminus \{y_2\}) = A_2 \cup B_2,$$

where

$$A_2 = \{\{a_2,\dots,a_k\} \subset R_1 \setminus \{y_2\} : \{y_2, a_2,\dots,a_k\} \in A\}$$

and

$$B_2 = \{\{a_2,\ldots,a_k\} \subset R_1 \setminus \{y_2\} : \{y_2, a_2,\ldots,a_k\} \in B\}.$$

By the induction hypothesis there exists an infinite subset R_2 of $R_1 \setminus \{y_2\}$ such that $P_{k-1}R_2 \subset A_2$ or $P_{k-1}R_2 \subset B_2$, or equivalently either

(3) $$\{\{y_2, a_2,\ldots,a_k\} : \{a_2,\ldots,a_k\} \subset R_2\} \subset A$$

or

(4) $$\{\{y_2, a_2,\ldots,a_k\} : \{a_2,\ldots,a_k\} \subset R_2\} \subset B.$$

But (3) cannot happen, otherwise we could take $x_n = y_2$ and $T_n = R_2$ in the previous construction; hence (4) holds.

Proceeding in this fashion we construct a sequence $\{y_1, y_2,\ldots,y_n,\ldots\}$ and a family of infinite sets $\{R_1, R_2,\ldots,R_n,\ldots\}$ such that $y_n \in R_{n-1}$, $R_n \subset R_{n-1} \setminus \{y_n\}$ and

$$\{\{y_n, a_2,\ldots,a_k\} : \{a_2,\ldots,a_k\} \subset R_n\} \subset B.$$

Let $M = \{y_1, y_2,\ldots,y_n,\ldots\}$, then clearly $P_kM \subset B$ and this proves the theorem.

The next theorem, about the existence of good subsequences, was proved in a slightly weaker form in Brunel and Sucheston [1].

Theorem 2.h.12. Let X be a separable Banach space and $\{x_n\} \subset X$ be a bounded sequence. Then $\{x_n\}$ contains a good subsequence.

Proof: Fix $x \in X$, $k \in \mathbb{N}$, $k \geq 1$ and $a_1,\ldots,a_k \in \mathbb{R}$. For $n \in \mathbb{N}^k$ let $\bar{n} = (n_1,\ldots,n_k)$ where $n_1 \leq n_2 \leq\ldots\leq n_k$ are the components of n re-arranged in increasing order. Define $\psi : \mathbb{N}^k \longrightarrow \mathbb{R}$, by

$$\psi(n) = \psi(\bar{n}) = \|x + a_1x_{n_1} +\ldots+ a_kx_{n_k}\|.$$

This function is bounded since for every $n \in \mathbb{N}^k$

$$\psi(n) \leq \|x\| + \sup_n\|x_n\|(\textstyle\sum_{i=1}^k|a_i|) = C.$$

Let M be an infinite subset of $\mathbb{N}$ and P_k be as in Ramsey's theorem. Then P_kM is a subset of $\mathbb{N}^k$ such that if $n \in P_kM$ all its components are different. Let

$$A_1^0 = \{n \in P_kM : 0 \leq \psi(n) \leq \tfrac{C}{2}\}, \quad A_2^0 = \{n \in P_kM : \tfrac{C}{2} < \psi(n) \leq C\}.$$

By Ramsey's theorem there is an infinite subset $\mathcal{M}_1$ of M such that either

every $n \in P_k\mathcal{M}_1$ is in A_1^0 or every $n \in P_k\mathcal{M}_1$ is in A_2^0, call this subset S_1.
Suppose for example that $S_1 = A_1^0$ and let

$$A_1^1 = \{n \in P_kS_1 : 0 \le \psi(n) \le \frac{C}{4}\}, \quad A_2^1 = \{n \in P_kS_1 : \frac{C}{4} < \psi(n) \le \frac{C}{2}\}.$$

Applying Ramsey's theorem again, we obtain an infinite subset $\mathcal{M}_2$ of $\mathcal{M}_1$ such that either every $n \in P_k\mathcal{M}_2$ is in A_1^1 or every $n \in P_k\mathcal{M}_2$ is in A_2^1, call this subset S_2. Continue in this fashion, obtaining $\mathcal{M}_i = \{s_r^i\}_{r=1}^{\infty}$ such that $\mathcal{M}_{i+1} \subset \mathcal{M}_i \subset \mathbb{N}$, $s_r^i < s_{r+1}^i$ and for $n, m \in P_k\mathcal{M}_i$,

(1) $$|\psi(n) - \psi(m)| \le C/2^i.$$

Consider the sequence $\{p_r\}_{r=1}^{\infty} = \{s_r^r\}_{r=1}^{\infty}$. Then from (1) it follows that

(2) $$\lim_{n_1\to\infty} \|x + a_1x_{n_1} + \ldots + a_kx_{n_k}\|$$

exists, whenever $n_1 < \ldots < n_k$ and $n_i \in \{p_r\}_{r=1}^{\infty}$ for $i = 1,\ldots,k$.

Observe that the sequence $\{p_r\}_{r=1}^{\infty} \subset M$ constructed here depends only on M, x, k and $\mathcal{A} = (a_1,\ldots,a_k)$.

Since X and $\mathbb{R}^{(\mathbb{N})}$ equipped with the norm ℓ_1 are separable, $X \times \mathbb{R}^{(\mathbb{N})}$ is separable also, hence let $\{(y_j, \mathcal{A}_j)\}_{j=1}^{\infty}$ be a dense subset.

Let $\{p_r^1\}_{r=1}^{\infty}$ be the sequence constructed above for $x = y_1$, $M = \mathbb{N}$ and $\mathcal{A} = \mathcal{A}_1$. Let k_1 be the cardinality of $\mathcal{A}_1$.

Repeat the construction for $x = y_2$, $M = \{p_r^1\}_{r=1}^{\infty}$ and $\mathcal{A} = \mathcal{A}_2$.

Proceeding inductively we get a sequence $\{p_r^j\}_{r=1}^{\infty}$ such that

(i) $\{p_r^{j+1}\}_{r=1}^{\infty}$ is a subsequence of $\{p_r^j\}_{r=1}^{\infty}$,

(ii) $\lim_{n_1\to\infty} \|y_j + a_1^{(j)}x_{n_1} + \ldots + a_{k_j}^{(j)}x_{n_{k_j}}\|$ exists, whenever $n_1 < \ldots < n_{k_j}$ and $n_i \in \{p_r^j\}_{r=1}^{\infty}$ for $i = 1, \ldots, k_j$, where $(a_1^{(j)},\ldots,a_{k_j}^{(j)}) = \mathcal{A}_j$.

Hence for the diagonal sequence $\{q_r\}_{r=1}^{\infty}$

$$\lim_{n_1\to\infty} \|x + a_1x_{n_1} + \ldots + a_kx_{n_k}\|$$

exists, whenever $n_1 < \ldots < n_k$, $n_i \in \{q_r\}_{r=1}^{\infty}$ for $i = 1,\ldots,k$ and for every $x \in X$ and $\mathcal{A} = (a_1,\ldots,a_k) \in \mathbb{R}^{(\mathbb{N})}$ such that $(x, \mathcal{A}) \in \{(y_j, \mathcal{A}_j)\}_{j=1}^{\infty}$.

By a density argument the limit above exists for every $(x,\mathcal{A}) \in X \times \mathbb{R}^{(\mathbb{N})}$, hence $\{x_{q_r}\}_{r=1}^{\infty}$ is a good subsequence of $\{x_n\}$.

Theorem 2.h.13. Let X be a Banach space and $\{x_n\}$ be a bounded sequence in X having a good non-convergent subsequence $\{x'_n\}$. Then there exists a spreading model of X based on $\{x'_n\}$.

Proof: Suppose first that X is separable and let $\mathcal{X}$ be the completion of $X \times \mathbb{R}^{(\mathbb{N})}$ with respect to the norm L. Then the space Y which is the closure of the linear span of the $\{e_n\}$ in $\mathcal{X}$ is a spreading model of X based on $\{x'_n\}$, and the sequence $\{e_n\}$ is the fundamental sequence of Y. If X is not separable and $\{x_n\}$ is a bounded sequence in X having a good non-convergent subsequence $\{x'_n\}$, then the spreading model for the separable space $X_0 = [x_n]$ based on $\{x'_n\}$ is the spreading model for X based on $\{x'_n\}$.

Before continuing, we point out that the spreading models may also be constructed in a different way, using ultraproducts, see Beauzamy and Lapresté [1] for this approach.

Now we proceed to the classification of the spreading models of J, a result due to Andrew [2]. For this we need the next two theorems which we will state without proof, since they would require concepts not directly related to this monograph; both theorems and their proofs can be found in Beauzamy and Lapresté [1]. The first one, characterizing the Banach spaces not containing ℓ_1 and not having ℓ_1 as spreading model, is due to Beauzamy [2] and the second one is by Guerre and Lapresté [1].

Theorem 2.h.14. Let X be a Banach space not containing ℓ_1. Then X doesn't have ℓ_1 as spreading model if and only if X has the alternate Banach-Saks property.

Theorem 2.h.15. Let $\{x_n\}$ be a good non-convergent sequence in a Banach space X and $\{e_n\}$ be the fundamental sequence of the spreading model Y based on $\{x_n\}$. If $\{e_n\}$ is not equivalent to the unit vector basis of ℓ_1, then $\{x_n\}$ is a weak Cauchy sequence if and only if $\{e_n\}$ is weakly Cauchy in Y, and $\{x_n\}$ converges weakly in X if and only if $\{e_n\}$ converges weakly in Y, and the limit, viewing both X and Y as subspaces of $\mathcal{X}$, is the same.

The fundamental sequence $\{e_n\}$ of a spreading model Y is not necessarily a basis of Y; however, Guerre and Lapresté [1] proved that every spreading model of X is isomorphic to a spreading model of X with a fundamental basic sequence. This allows us to consider only spreading models with fundamental basic sequences.

We can now establish the main theorem of this section, classifying the spreading models of J and J^*:

Theorem 2.h.16. (i) The only spreading models of J are ℓ_2 and J.
(ii) The only spreading models of J^* are ℓ_2 and J^*.

Proof: (i) Let Y be a spreading model of J based on a good non-convergent sequence $\{x_n\}$ and with fundamental basic sequence $\{e_n\}$. Since by Theorem 2.a.2 J does not contain ℓ_1, and by Proposition 2.h.5 it has the alternate Banach-Saks property, by Theorem 2.h.14 $\{e_n\}$ is not equivalent to the unit vector basis of ℓ_1. Hence by the dichotomy theorem of Rosenthal, which states that every bounded sequence in a Banach space not containing ℓ_1 has a weak Cauchy subsequence (see e.g. Lindenstrauss and Tzafriri [1]), it follows that $\{e_n\}$ has a weak Cauchy subsequence, and since $\{e_n\}$ is spreading, it is itself weakly Cauchy. There are two different cases:

(a) $\{e_n\}$ converges weakly to some y in Y. Since $\{e_n\}$ is basic, $y = 0$ and by Theorem 2.h.15 $\{x_n\}$ also converges weakly to 0.
Using Proposition 2.d.16, $\{x_n\}$ has a subsequence equivalent to the unit vector basis of ℓ_2 and therefore Y is isomorphic to ℓ_2.

(b) $\{e_n\}$ is weakly Cauchy but not weakly convergent in Y. By Theorem 2.h.15 $\{x_n\}$ is also weakly Cauchy but not weakly convergent. It follows from Proposition 2.d.15 that $\{x_n\}$ has a subsequence equivalent to the summing basis $\{\xi_n\}$ of J. This implies the equivalence of $\{e_n\}$ and $\{\xi_n\}$

and therefore Y is isomorphic to J.

(ii) The proof for J^* is the same using Corollary 2.g.4 and Proposition 2.g.6.

2.i. J^* has cotype 2

The notions of type and cotype were introduced by Hoffmann-Jørgensen in [1] while studying the convergence of sums of independent Banach space valued random variables. Since then, these concepts have proved to be very useful in the study of the geometry of Banach spaces, due to their close relation with the factorization of operators through L_p spaces.

In this section we prove that J has no cotype other than infinity and J^* has cotype two, which is perhaps the only tangible difference between J and J^* that is proved in this monograph.

There are several general topics needed for our purposes, but since they don't deal directly with J^*, we have placed them in Appendix 2.k for the interested reader to see.

To start with, we will define a set of very special functions in $L_p[0,1]$, $1 \le p \le \infty$, called the Rademacher functions, which are the basis for the concepts of type and cotype.

Definition 2.i.1. The set of Rademacher functions $\{r_n\}_{n=0}^{\infty}$ is given by

$$r_0(t) = 1 \text{ for all } t \in [0,1],$$
$$r_n(t) = \operatorname{sgn}(\sin(2^n \pi t)) \text{ for all } t \in [0,1],\ n = 1, 2, \ldots,$$

where sgn(0) is defined as +1.

Let X be a Banach space and $x_1,\ldots,x_n$ a finite sequence of points in X. Then

$$\left(\int_0^1 \left\|\sum_{i=1}^n r_i(t)x_i\right\|^p dt\right)^{1/p} = \left(2^{-n}\sum_{\varepsilon_i=\pm1}\left\|\sum_{i=1}^n \varepsilon_i x_i\right\|^p\right)^{1/p},$$

where the second sum is taken over all possible choices of signs $\varepsilon_j = \pm1$

for $j = 1,\dots,n$. The type and cotype indicate how this average is situated with respect to $(\sum_{i=1}^{n}\|x_i\|^p)^{1/p}$; if the convergence in $L_1(X)$ of $\sum_{i=1}^{\infty} r_i(t)x_i$ implies the convergence of $\sum_{i=1}^{\infty}\|x_i\|^q$, we will say that X has cotype q, and vice versa, if the convergence of $\sum_{i=1}^{\infty}\|x_i\|^p$ implies the convergence in $L_1(X)$ of $\sum_{i=1}^{\infty} r_i(t)x_i$, we say that X has type p.

Definition 2.i.2. A space X is of cotype q for some $2 \le q \le \infty$, respectively of type p for some $1 \le p \le 2$, if there exists a constant $L < \infty$, respectively $M < \infty$, such that for every finite set of vectors $\{x_j\}_{j=1}^{n} \subset X$ we have

$$(\sum_{i=1}^{n}\|x_i\|^q)^{1/q} \le L\int_0^1 \|\sum_{i=1}^{n} r_i(t)x_i\|dt,$$

respectively

$$\int_0^1 \|\sum_{i=1}^{n} r_i(t)x_i\|dt \le M(\sum_{i=1}^{n}\|x_i\|^p)^{1/p}.$$

Observe that it follows immediately from the definitions, that if two spaces are isomorphic, they have the same type and cotype.

The notions of type and cotype can also be defined using a sequence $\{g_n\}_n$ of independent identically distributed gaussian random variables with the standard normal distribution, instead of the Rademacher functions. However, it is known that the two definitions are equivalent (see e.g. Maurey and Pisier [1]).

Using the well known Kahane inequality (see Beauzamy [1]) stating that for every $q \ge 1$ there exists K_q such that, for every Banach space X and for every $x_1,\dots,x_n \in X$,

$$\int_0^1 \|\sum_{i=1}^{n} r_i(t)x_i\|dt \le \left(\int_0^1 \|\sum_{i=1}^{n} r_i(t)x_i\|^q dt\right)^{1/q} \le K_q \int_0^1 \|\sum_{i=1}^{n} r_i(t)x_i\|dt,$$

one gets that in the definitions of cotype and type one may take the average $\left(\int_0^1 \|\sum_{i=1}^{n} r_i(t)x_i\|^q dt\right)^{1/q}$ in L_q for any $1 \le q < \infty$, instead of $\int_0^1 \|\sum_{i=1}^{n} r_i(t)x_i\|dt$.

It is clear that every Banach space is of type 1 and cotype ∞, and it is not too hard to see, using Khintchine's inequalities (see Beauzamy [1]), that a Banach space X cannot have type $p > 2$ or cotype $q < 2$.

The notions of type and cotype are intimately related to the concept of finite representability (see Section 2.b), as the following theorem shows. Part of it was proved by Maurey and Pisier in [1] and part by Pisier in [3].

Theorem 2.i.3. Let X be a Banach space, then

$$\mathfrak{p}(X) = \sup\{r : X \text{ has type } r\} = \inf\{r : \ell_r \text{ is f. r. in } X\}$$

and

$$\mathfrak{q}(X) = \inf\{r : X \text{ has cotype } r\} = \sup\{r : \ell_r \text{ is f. r. in } X\}.$$

It also holds that $\ell_{\mathfrak{p}(X)}$ and $\ell_{\mathfrak{q}(X)}$ are both finitely representable in X and, if $\mathfrak{p}(X) < 2$, then ℓ_r is finitely representable in X for every r in the interval $[\mathfrak{p}(X), 2]$.

From the result that every Banach space is finitely representable in J (Corollary 2.b.10) and the above, the next corollary follows immediately.

Corollary 2.i.4. J has no type other than 1 and no cotype other than ∞.

Now we turn our attention to J^*; first we will show that J^* has only type 1, which is not very difficult using Theorem 2.d.6. The core of this section is based on Pisier's paper [1], proving that J^* has cotype 2, which is a fundamental difference between J and J^*. James [4] proved that J and J^* are not isomorphic, by showing explicitly that the existence of an isomorphism from J^* into J is impossible. However, this proof is even more complicated than the one we are going to present here. Although J and J^* are not isomorphic, James in the same paper proved that there exist quasi-reflexive Banach spaces which in fact are isomorphic to their dual.

Theorem 2.i.5. J^* has no type other than 1.

Proof: By Theorem 2.d.6, $(\sum_{n=1}^{\infty}\ell_{\infty}^{n})_{\ell_2}$ is isomorphic to a complemented subspace X of J. Hence $(\sum_{n=1}^{\infty}\ell_{1}^{n})_{\ell_2}$ is isomorphic to the complemented subspace X^* of J^*. But by Lemma 2.b.6 and Corollary 2.b.7 ℓ_1 is f. r. in $(\sum_{n=1}^{\infty}\ell_{1}^{n})_{\ell_2}$ and thus ℓ_1 is f. r. in J^*. Hence by Theorem 2.i.3 $p(J^*) = 1$ and this proves the theorem.

In his paper [1], Pisier proved that a whole family of Banach spaces, a member of which is J^*, have cotype 2. The proof we give here is the same but applied only to J^*. We start with one of Pisier's generalizations of the James space.

Definition 2.i.6. Let X be a Banach space and $x = \{x_n\}_{n=0}^{\infty}$ be a sequence in X with $x_0 = 0$. Let

$$\|x\|_{J(X)} = \sup\left(\sum_{i=1}^{n}\|x_{p_i} - x_{p_{i-1}}\|_X^2\right)^{1/2},$$

where the supremum is taken over all choices of n and all positive integers $0 = p_0 < p_1 <\ldots< p_n$. J(X) is then defined as the $\|\ \|_{J(X)}$ completion of the space of sequences $\{x_n\}_{n=0}^{\infty}$ with finite support and $x_0 = 0$.

Similarly we define $\ell_p(X)$ for $1 \le p < \infty$ as the space of sequences $x = \{x_n\}_{n=0}^{\infty}$ in X, where $\|x\|_{\ell_p} = (\sum_{n=0}^{\infty}\|x_n\|_X^p)^{1/p} < \infty$ and $\ell_\infty(X)$ as the space of sequences $x = \{x_n\}_{n=0}^{\infty}$ in X, where $\|x\|_{\ell_\infty} = \sup_n\|x_n\|_X < \infty$.

Observe that if $X = \mathbb{R}$, then J(X) is isomorphic to J^{**}, the bidual of the James space.

It will be convenient for our purposes to consider in the remainder of this section only the sequences $\{x_n\}_{n=0}^{\infty}$ with $x_0 = 0$ and the norm in J given by $\|\ \|_J = \sqrt{2}\|\ \|$.

Lemma 2.i.7. Let X be a Banach space and $x = \{x_n\}_{n=0}^{\infty}$ a sequence in X. Then the following are equivalent:

(i) $\|x\|_{J(X)} \leq 1$.

(ii) There exist a non-decreasing function $\phi : \mathbb{N} \to [0,1]$ with $\phi(0) = 0$, $\phi(n) \to 1$ as $n \to \infty$, and a function $f : [0,1] \to X$ with $f\phi(n) = x_n$ and such that for every t, $s \in [0,1]$

(1) $$\|f(t) - f(s)\|_X \leq |t - s|^{1/2}.$$

Proof: (i) $\Rightarrow$ (ii) Suppose $\|x\|_{J(X)} \leq 1$, and for $n = 0, 1, \ldots$ let

$$x^{(n)} = (x_0, x_1, x_2, \ldots, x_n, x_n, x_n, \ldots).$$

Define

$$\phi(n) = \|x^{(n)}\|^2_{J(X)} / \|x\|^2_{J(X)}.$$

Then ϕ is non-decreasing, $\phi(0) = 0$, $\phi(n) \to 1$ as $n \to \infty$, and if $m < n$, then $\|x_n - x_m\|^2_X / \|x\|^2_{J(X)} + \phi(m) \leq \phi(n)$. Hence

(2) $$\|x_n - x_m\|_X \leq |\phi(n) - \phi(m)|^{1/2} \|x\|_{J(X)} \leq |\phi(n) - \phi(m)|^{1/2}.$$

For $t \in \phi(\mathbb{N})$ let n_t be such that $\phi(n_t) = t$. Define $f(t) = x_{n_t}$; this is well defined, for if m is such that $\phi(m) = t$, by (2) $x_{n_t} = x_m$. Hence for s, $t \in \phi(\mathbb{N})$

(3) $$\|f(s) - f(t)\|_X = \|x_{n_s} - x_{n_t}\|_X \leq |\phi(n_s) - \phi(n_t)|^{1/2} = |t - s|^{1/2}.$$

We may extend f to the closure of $\phi(\mathbb{N})$ so that (3) still holds and then taking the affine extension to $[0,1]$ we get (ii).

(ii) $\Rightarrow$ (i) This is obvious, since by (1), if $n > m$,

$$\|x_n - x_m\|^2_X / \|x\|^2_{J(X)} \leq \phi(n) - \phi(m)$$

and thus

$$\sum_{i=1}^{n} \|x_{p_i} - x_{p_{i-1}}\|^2_X \leq \|x\|^2_{J(X)} \sum_{i=1}^{n} (\phi(p_i) - \phi(p_{i-1})) = \|x\|^2_{J(X)} \phi(p_n) \leq 1.$$

Observe that if f and ϕ are as in (ii), but instead of (1) we pose the condition $\|f(t) - f(s)\|_X \leq K|t - s|^{1/2}$, then $\|x\|_{J(X)} \leq K$ and f belongs to $\mathrm{Lip}_{1/2}(X)$ with $\|f\|_{1/2} \leq K$, where this space is defined below.

Definition 2.i.8. Let $0 < \alpha < 1$; define the space of Lipschitz functions in $[0,1]$ with values in X, denoted by $\mathrm{Lip}_\alpha([0,1], X)$, as the space of all functions $f : [0,1] \to X$ such that

$$\sup_{\{s,t\in[0,1]:s\neq t\}} \frac{\|f(s) - f(t)\|}{|s - t|^{\alpha}} < \infty,$$

equipped with the norm

$$\|f\|_{\alpha,X} = \|f(0)\|_X + \sup_{s\neq t} |s - t|^{-\alpha}\|f(s) - f(t)\|_X.$$

Since we will only use Lipschitz functions in [0,1], we will denote $Lip_{\alpha}([0,1], X)$ simply by $Lip_{\alpha}(X)$.

The next two lemmata are very simple, but we will need them to prove that J(X/Y) is isomorphic to the quotient of J(X) and J(Y) for every closed subspace Y of X.

Lemma 2.i.9. Let X and Y be Banach spaces and $S : X \longrightarrow Y$ a bounded operator. Then, if $x = \{x_n\} \in J(X)$ (respectively $\ell_\infty(X)$), the operator $\tilde{S} : J(X) \longrightarrow J(Y)$ (respectively $\tilde{S} : \ell_\infty(X) \longrightarrow \ell_\infty(Y)$) given by $(\tilde{S}x)_n = Sx_n$ for n = 0, 1,... is also bounded and $\|\tilde{S}\| = \|S\|$.

Proof: By the definition of $\| \ \|_{J(X)}$ and $\| \ \|_{J(Y)}$ we get for every $x \in J(X)$

$$\|\tilde{S}x\|_{J(Y)} = \|\{Sx_n\}_{n=1}^{\infty}\|_{J(Y)} \le \|S\|\,\|x\|_{J(X)}.$$

Hence $\|\tilde{S}\| \le \|S\|$.

On the other hand, let $x = (0, x_1, x_1, x_1, \ldots)$ with $x_1 \in X$; then

$$\|Sx_1\|_Y = \|\tilde{S}x\|_{J(Y)} \le \|\tilde{S}\|\,\|x\|_{J(X)} = \|\tilde{S}\|\,\|x_1\|_X,$$

and thus $\|S\| \le \|\tilde{S}\|$.

The proof for ℓ_∞ instead of J is exactly the same.

Lemma 2.i.10. Let Y be a closed subspace of a Banach space X and let $q : X \longrightarrow X/Y$ be the quotient map. Then for every $\varepsilon > 0$, if $x = \{x_n\} \in \ell_\infty(X/Y)$, there exists $\tilde{x} = \{\tilde{x}_n\} \in \ell_\infty(X)$ such that $q(\tilde{x}_n) = x_n$ and $\|\tilde{x}\|_{\ell_\infty(X)} \le \|x\|_{\ell_\infty(X/Y)} + \varepsilon$.

Proof: If $x = \{x_n\} \in \ell_\infty(X/Y)$ then $\|x_n\|_{X/Y} \le \|x\|_{\ell_\infty(X/Y)}$ for every n; therefore, by the definition of the quotient norm, for every $\varepsilon > 0$ there exists $\tilde{x}_n \in X$ with $q(\tilde{x}_n) = x_n$ and $\|\tilde{x}_n\| \le \|x_n\|_{X/Y} + \varepsilon \le \|x\|_{\ell_\infty(X/Y)} + \varepsilon$. Then $\tilde{x} = \{\tilde{x}_n\}$ has the required properties.

Using the results and definitions in Appendix 2.k, mainly the Theorem 2.k.3 stating that $\mathrm{Lip}_\alpha(X)$ and $\ell_\infty(X)$ are isomorphic for $0 < \alpha < 1$, we will see that with J instead of ℓ_∞, the same lifting property as above is satisfied.

In the proof of the following proposition we will need the functions $\varphi_n^{(\alpha)} : [0,1] \longrightarrow \mathbb{R}$ defined in the appendix as

$$\varphi_{2^n+k}^{(\alpha)}(t) = 2^{(n+1)(1-\alpha)-(n/2)}\int_0^t h_{2^n+k}(\tau)d\tau,$$

where $n = 0, 1, \dots,\ k = 1, \dots, 2^n$ and $\{h_n\}$ is the set of Haar functions in [0,1].

Proposition 2.i.11. Let Y be a closed subspace of a Banach space X and $q : X \longrightarrow X/Y$ be the quotient map. Then there exists K and for every $x = \{x_n\} \in J(X/Y)$ there exists $\tilde{x} = \{\tilde{x}_n\} \in J(X)$, such that $q(\tilde{x}_n) = x_n$ and $\|\tilde{x}\|_{J(X)} \le K\|x\|_{J(X/Y)}$, where K does not depend on X and Y.

Proof: Let $x = \{x_n\} \in J(X/Y)$ with $\|x\|_{J(X/Y)} = 1$. By Lemma 2.i.7 there exist $f \in \mathrm{Lip}_{1/2}(X/Y)$ and $\phi : \mathbb{N} \longrightarrow [0,1]$ such that

(1) $$x_n = f\phi(n)$$

and

(2) $$\|f\|_{1/2,X/Y} \le 1.$$

Let

$$T_{X/Y} : \mathrm{Lip}_{1/2}(X/Y) \longrightarrow \ell_\infty(X/Y) \quad\text{and}\quad T_X : \mathrm{Lip}_{1/2}(X) \longrightarrow \ell_\infty(X)$$

be the isomorphisms given by Theorem 2.k.3 with $\|T_{X/Y}\| \le 1$ and $\|T_X^{-1}\| \le K_{1/2}$. Let $\tilde{q} : \ell_\infty(X) \longrightarrow \ell_\infty(X/Y)$ be defined by $\tilde{q}(y) = \{q(y_n)\}_{n=1}^\infty$ for every $y = \{y_n\}_{n=1}^\infty \in \ell_\infty(X)$; then by Lemma 2.i.9, $\|\tilde{q}\| = \|q\|$.

Now let $\varepsilon > 0$; by Lemma 2.i.10 there exists $g \in \ell_\infty(X)$ such that $\tilde{q}(g) = T_{X/Y}f$ and

(3) $$\|g\|_{\ell_\infty(X)} \le (1+\varepsilon)\|T_{X/Y}f\|_{\ell_\infty(X/Y)}.$$

If $F = T_X^{-1}g$, then

(4) $$\tilde{q}T_XF = T_{X/Y}f.$$

Let $T_XF = \{a_n\} \in \ell_\infty(X)$ and $T_{X/Y}f = \{b_n\} \in \ell_\infty(X/Y)$; then (4) says that for every n = 0, 1,... $q(a_n) = b_n$. Since by Proposition 2.k.2 and Theorem 2.k.3

$$F(t) = \sum_{n=1}^{\infty} a_n \varphi_n^{(1/2)}(t) \quad \text{and} \quad f(t) = \sum_{n=1}^{\infty} b_n \varphi_n^{(1/2)}(t),$$

where these representations are unique and besides q is continuous, we get for every $t \in [0,1]$

(5) $$q(F(t)) = f(t).$$

Now let $K = (1 + \varepsilon)K_{1/2}$, then by (2) and (3)

(6) $$\|F\|_{1/2,X} \leq \|T_X^{-1}\| \|g\|_{\ell_\infty(X)} \leq (1 + \varepsilon)\|T_X^{-1}\| \|T_{X/Y}\| \leq K.$$

Finally let $\tilde{x} = \{F(\phi(n))\}$. Applying Lemma 2.i.7 we get $\tilde{x} \in J(X)$ and

$$\|\tilde{x}\|_{J(X)} \leq K,$$

and by (5) and (1) $q(\tilde{x}_n) = x_n$, and this finishes the proof.

Corollary 2.i.12. Let X be a Banach space and Y a subspace of X. Then J(X/Y) is K-isomorphic to J(X)/J(Y), where K which does not depend on X, and Y is as in Proposition 2.i.11.

Proof: Let $q : X \longrightarrow X/Y$ be the quotient map. Then if $\tilde{q} : J(X) \longrightarrow J(X/Y)$ is given by $\tilde{q}(y) = \{q(y_n)\}$ for every $y = \{y_n\} \in J(X)$, by Lemma 2.i.9, $\tilde{q}(y) \in J(X/Y)$ and $\|\tilde{q}\| = \|q\| \leq 1$.

On the other hand, let $y = \{y_n\} \in J(X)$. Then $\tilde{q}(y) = 0$ if and only if $q(y_n) = 0$ for n = 1, 2,..., and this holds if and only if $y_n \in Y$ for n = 1, 2,... and thus, if and only if $y \in J(Y)$. Hence $\ker\tilde{q} = J(Y)$ and then, if $[\![y]\!]$ denotes the equivalence class of y in J(X)/J(Y), $Q : J(X)/J(Y) \longrightarrow J(X/Y)$ given by $Q([\![y]\!]) = \tilde{q}(y)$ is obviously injective, and by Proposition 2.i.11, Q is surjective. Also for every $z \in [\![y]\!]$,

$$\|Q([\![y]\!])\| = \|\tilde{q}(y)\| = \|\tilde{q}(z)\| \leq \|z\|.$$

Hence $\|Q\| \leq 1$. Now let $x \in J(X/Y)$ and $\tilde{x} \in J(X)$ be as in Proposition 2.i.11. Then, since $\tilde{q}(\tilde{x}) = x$ and

$$\|[\![\tilde{x}]\!]\|_{J(X)/J(Y)} \leq \|\tilde{x}\|_{J(X)} \leq K\|x\|_{J(X/Y)},$$

we have that $Q^{-1}(x) = [\![\tilde{x}]\!]$ and $\|Q^{-1}\| \leq K$. Thus Q is a well defined isomorphism.

This is one of the two main properties of the spaces J(X) that will be used in the proof of the principal theorem of this section (Theorem 2.i.19). The other one is given in the next lemma and shows the relation between $L_2(J)$ and $J(L_2)$, and $L_\infty(J)$ and $J(L_\infty)$.

For notations used and not explained in this section, we refer the reader to Appendix 2.k.

Definition 2.i.13. Let X be a Banach space and $N = 2^n$ for some $n \in \mathbb{N}$, define

$$\mathcal{K}_n(X) = \left\{\phi \in L_\infty^N(X) : \int_\Omega \phi(\omega) r_i(\omega) d\mu = 0,\ i = 1,\dots,n\right\},$$

where $L_\infty^N(X) = \{\sum_{i=0}^{N-1} x_i \chi_{[i/N,(i+1)/N)} : x_i \in X\} \subset L_\infty(X)$ and $\Omega = [0,1]$.

It is easy to see that both $L_\infty^N(X)$ and $\mathcal{K}_n(X)$ are closed subspaces of $L_\infty(X)$.

Lemma 2.i.14. Let $f \in X(J)$, where $X = L_2$ or L_∞, and define $\mathcal{V}$ as follows: for $f : \Omega \longrightarrow J$, $f(\omega) = \{f_n(\omega)\}_{n=0}^\infty \in J$, let $\mathcal{V}(f) = \{f_n\}_{n=0}^\infty$. Then

(a) $\|\mathcal{V}(f)\|_{J(L_2)} \le \|f\|_{L_2(J)}$ if $f \in L_2(J)$.

(b) $\|f\|_{L_\infty(J)} \le \|\mathcal{V}(f)\|_{J(L_\infty)}$ if $\mathcal{V}(f) \in J(L_\infty)$.

(c) $\Phi \in \mathcal{K}_n(J)$ if and only if $\mathcal{V}(\Phi) \in J(\mathcal{K}_n(\mathbb{R}))$.

Proof: (a) Let $f \in L_2(J)$; since $f_0 = 0$ we have that for $n = 0, 1,\dots$

$$\int_\Omega |f_n(\omega)|^2 d\mu \le \int_\Omega \|f(\omega)\|_J^2\, d\mu < \infty;$$

hence $f_n \in L_2(\mathbb{R})$. Also

$$\|\mathcal{V}(f)\|_{J(L_2)} = \sup\left(\sum_{j=1}^m \int_\Omega |f_{p_j}(\omega) - f_{p_{j-1}}(\omega)|^2 d\mu\right)^{1/2} \le$$

$$\le \left(\int_\Omega \sup \sum_{j=1}^m |f_{p_j}(\omega) - f_{p_{j-1}}(\omega)|^2 d\mu\right)^{1/2} = \left(\int_\Omega \|f(\omega)\|_J^2 d\mu\right)^{1/2} = \|f\|_{L_2(J)},$$

where the sup is taken over all finite sequences of positive integers $0 = p_1 < p_2 <\dots< p_m$.

(b) Suppose $\mathcal{V}(f) \in J(L_\infty)$, then μ-a.e.

$$\sup(\textstyle\sum_{j=1}^{m}|f_{p_j}(\omega) - f_{p_{j-1}}(\omega)|^2)^{1/2} \le \sup(\sum_{j=1}^{m}\|f_{p_j} - f_{p_{j-1}}\|_{L_\infty}^2)^{1/2} = \|\mathcal{V}(f)\|_{J(L_\infty)},$$

where the sup is taken over all finite sequences of positive integers $0 = p_1 < p_2 <\ldots< p_m$. Therefore $\|f\|_{L_\infty(J)} \le \|\mathcal{V}(f)\|_{J(L_\infty)}$.

(c) Let $\Phi \in \mathcal{K}_n(J)$. Then $\int_\Omega \phi(\omega)r_i(\omega)d\mu = 0$ for $i = 1,\ldots,n$, and for every $\omega \in \Omega$, since $\Phi \in L_\infty^N(J)$, $\Phi(\omega) = \{\phi_m(\omega)\}_{m=0}^\infty \in J$, with $\phi_m \in L_\infty^N(\mathbb{R})$ for every $m > 0$ and $\phi_0 = 0$. Let $\{e_m^*\}_{m=1}^\infty$ be the canonical basis in J^*. Then, using Theorem 2.k.5 of the appendix, we get for $m = 1, 2,\ldots$

$$(1) \qquad 0 = e_m^*\left(\int_\Omega \Phi(\omega)r_i(\omega)d\mu\right) = \int_\Omega e_m^*\circ\Phi(\omega)r_i(\omega)d\mu = \int_\Omega \phi_m(\omega)r_i(\omega)d\mu.$$

Now suppose $\Phi = \sum_{i=0}^{2^n-1}x_i\chi_{[i/2^n,(i+1)/2^n)}$ with $x_i = \{x_{im}\}_{m=0}^\infty \in J$; then

$$\phi_m = \textstyle\sum_{i=0}^{2^n-1}x_{im}\chi_{[i/2^n,(i+1)/2^n)}.$$

For $0 = p_0 < p_1 <\ldots< p_r$ we have

$$\textstyle\sum_{j=1}^{r}\|\phi_{p_j} - \phi_{p_{j-1}}\|_{L_\infty}^2 =$$

$$= \textstyle\sum_{j=1}^{r}(\text{ess sup}|\sum_{i=0}^{2^n-1}(x_{ip_j} - x_{ip_{j-1}})\chi_{[i/2^n,(i+1)/2^n)}(\omega)|)^2$$

$$\le \textstyle\sum_{j=1}^{r}(\sum_{i=0}^{2^n-1}|x_{ip_j} - x_{ip_{j-1}}|)^2 \le 2^n\sum_{j=1}^{r}\sum_{i=0}^{2^n-1}|x_{ip_j} - x_{ip_{j-1}}|^2 \le 2^n\sum_{i=0}^{2^n-1}\|x_i\|_J^2.$$

Therefore $\mathcal{V}(\Phi) = \{\phi_m\} \in J(\mathcal{K}_n(\mathbb{R}))$.

Now suppose $\Psi \in J(\mathcal{K}_n(\mathbb{R}))$; then $\Psi = \{\phi_m\}$ with $\phi_m \in \mathcal{K}_n(\mathbb{R})$. It is easy to see that $\|\{\phi_m(\omega)\}\|_J \le \|\Psi\|_{J(\mathcal{K}_n)}$ μ-a.e. and hence we may assume that $\{\phi_m(\omega)\} \in J$. So we can define $\Phi : \Omega \longrightarrow J$ by $\Phi(\omega) = \{\phi_m(\omega)\}$. As above, but inverting the steps in (1), we get that for $m = 1,2,\ldots$, $e_m^*\left(\int_\Omega \Phi(\omega)r_i(\omega)d\mu\right) = 0$; therefore $\int_\Omega \Phi(\omega)r_i(\omega)d\mu = 0$. This, together with the fact that $\phi_m \in \mathcal{K}_n(\mathbb{R})$, implies $\Phi \in \mathcal{K}_n(J)$; it is obvious that $\mathcal{V}(\Phi) = \Psi$.

Definition 2.i.15. Let $J^n = \{x = (x_1,\ldots,x_n) : x_i \in J,\ i = 1,\ldots,n\}$. Let $\mathcal{U} : J^n \longrightarrow J(\ell_2^n)$ be defined as follows: for $x \in J^n$ with $x_i = \{x_{ij}\}_{j=0}^\infty$, $i = 1,\ldots,n$, let $x^{(j)} = (x_{1j},\ldots,x_{nj})$ and $\mathcal{U}(x) = \{x^{(j)}\}_{j=0}^\infty$.

It is not difficult to show that $\mathcal{U}$ is bijective and

$$(\tfrac{1}{n}\sum_{i=1}^{n}\|x_i\|_J^2)^{1/2} \le \|\mathcal{U}(x)\|_{J(\ell_2^n)} \le (\sum_{i=1}^{n}\|x_i\|_J^2)^{1/2},$$

that is, $\mathcal{U}$ is an isomorphism between $\ell_2^n(J)$ and $J(\ell_2^n)$.

The key to the proof of the main theorem in this section is Theorem 2.i.17, which shows that we can equip J^n with a norm in such a way that it is L-isomorphic to $J(\ell_n^2)$, where L is independent of n.

Definition 2.i.16. Let X be a Banach space, n a fixed positive integer and $N = 2^n$. For $x = (x_1,\dots,x_n) \in X^n$ we define

$$[x] = \inf\left\{\|\sum_{i=1}^{n} r_i x_i + \phi\|_{L_\infty(X)} : \phi \in \mathcal{K}_n(X)\right\},$$

where $\{r_i\}$ is the sequence of Rademacher functions defined in 2.i.1. Obviously [ ] defines a norm in X^n.

Theorem 2.i.17. If $\mathcal{U}$ is as in Definition 2.i.15, there exists a positive constant L such that for all $n \in \mathbb{N}$ and every $x \in J^n$

$$\|\mathcal{U}(x)\|_{J(\ell_2^n)} \le [x] \le L\|\mathcal{U}(x)\|_{J(\ell_2^n)}.$$

Proof: Define $T : L_2 \longrightarrow \ell_2^n$ by $Tf = (\langle f, r_1\rangle,\dots,\langle f, r_n\rangle)$, where $\langle f, g\rangle$ denotes the inner product in L_2. Since $\{r_i\}$ is an orthonormal sequence in L_2,

$$\|Tf\|_{\ell_2^n}^2 = \sum_{i=1}^{n}|\langle f, r_i\rangle|^2 = \|\sum_{i=1}^{n}\langle f, r_i\rangle r_i\|_{L_2}^2 \le \|f\|_{L_2}^2.$$

By Lemma 2.i.9, the operator $\tilde{T} : J(L_2) \longrightarrow J(\ell_2^n)$ given by $(\tilde{T}\mathfrak{f})_m = Tf_m$ for every $\mathfrak{f} = \{f_m\} \in J(L_2)$ satisfies

(1) $$\|\tilde{T}\| \le 1.$$

If $x=(x_1,\dots,x_n) \in J^n$ and $\Phi\in\mathcal{K}_n(J)$, by Lemma 2.i.14(c), $\mathcal{V}(\Phi)\in J(\mathcal{K}_n(\mathbb{R}))$, and by (1) of the proof of the said lemma, $\mathcal{V}(\Phi) = \{\phi_m\}$ with $\int_\Omega \phi_m(\omega)r_i(\omega)d\mu = 0$ for $i = 1,\dots,n$. Hence $T\phi_m = 0$ for every m and therefore

(2) $$\tilde{T}(\mathcal{V}(\Phi)) = 0.$$

Suppose $x_i = \{x_{im}\}_{m=0}^{\infty} \in J$; we get

(3) $$\tilde{T}(\mathcal{V}(\sum_{i=1}^{n} r_i x_i)) = \{T(\sum_{i=1}^{n} r_i x_{im})\}_{m=0}^{\infty} =$$

$$= \left\{\left(\langle\sum_{i=1}^{n} r_i x_{im}, r_1\rangle,\dots,\langle\sum_{i=1}^{n} r_i x_{im}, r_n\rangle\right)\right\}_{m=0}^{\infty} =$$

$$= \{(x_{1m},\dots,x_{nm})\}_{m=0}^{\infty} = \mathcal{U}(x).$$

By (2) and (3), $\tilde{T}(\mathcal{V}(\sum_{i=1}^{n} r_i x_i + \Phi)) = \mathcal{U}(x)$. Since by Lemma 2.i.14(a)

$$\|\mathcal{V}(\sum_{i=1}^{n} r_i x_i + \Phi)\|_{J(L_2)} \le \|\sum_{i=1}^{n} r_i x_i + \Phi\|_{L_2(J)} \le \|\sum_{i=1}^{n} r_i x_i + \Phi\|_{L_\infty(J)},$$

we obtain using (1)

$$\|\mathcal{U}(x)\|_{J(\ell_2^n)} \le \|\mathcal{V}(\sum_{i=1}^{n} r_i x_i + \Phi)\|_{J(L_2)} \le \|\sum_{i=1}^{n} r_i x_i + \Phi\|_{L_\infty(J)},$$

that is

(4) $$\|\mathcal{U}(x)\|_{J(\ell_2^n)} \le [x].$$

To prove the reverse inequality, we need the famous Khintchine inequality (see for example Beauzamy [1]) for p = 1, which states that there exist A and B such that for every $n \in \mathbb{N}$ and every $\alpha_1,\dots,\alpha_n \in \mathbb{R}$

$$A(\sum_{i=1}^{n} \alpha_i^2)^{1/2} \le \|\sum_{i=1}^{n} \alpha_i r_i\|_{L_1} \le B(\sum_{i=1}^{n} \alpha_i^2)^{1/2}.$$

Therefore if $\mathcal{R}_n(\mathbb{R})$ is the subspace of L_1 generated by the functions $\sum_{i=1}^{n} \alpha_i r_i$ with $\alpha_i \in \mathbb{R}$ and if $\mathfrak{I} : \mathcal{R}_n(\mathbb{R}) \longrightarrow \ell_2^n$ is given by

$$\mathfrak{I}\sum_{i=1}^{n} \alpha_i r_i = (\alpha_1,\dots,\alpha_n),$$

Khintchine's inequality yields $\|\mathfrak{I}\| \le A^{-1}$.

By Proposition 2.k.6 $\mathcal{R}_n(\mathbb{R}) = (\mathbb{R}^n,[\ ]_{\mathbb{R}})^*$ and thus $\mathcal{R}_n(\mathbb{R})^* = (\mathbb{R}^n,[\ ]_{\mathbb{R}})$. It is easy to see that for every $\beta = (\beta_1,\dots,\beta_n) \in \ell_2^n = (\ell_2^n)^*$, $\mathfrak{I}^*\beta = \beta$; hence

(5) $$[\beta]_{\mathbb{R}} = [\mathfrak{I}^*\beta]_{\mathbb{R}} \le \|\mathfrak{I}^*\|(\sum_{i=1}^{n} \beta_i^2)^{1/2} \le A^{-1}(\sum_{i=1}^{n} \beta_i^2)^{1/2}.$$

Next we define $S : \ell_2^n \longrightarrow L_\infty^N/\mathcal{K}_n(\mathbb{R})$ as follows: for $\alpha = (\alpha_1,\dots,\alpha_n) \in \ell_2^n$ let $S\alpha = [\![\sum_{i=1}^{n} \alpha_i r_i]\!]$, where $[\![\sum_{i=1}^{n} \alpha_i r_i]\!]$ denotes the equivalence class of this element in the quotient; then by (5) and the definition of the quotient norm in $L_\infty^N/\mathcal{K}_n(\mathbb{R})$, we get

$$\|S\alpha\|_{L_\infty^N/\mathcal{K}_n} = [\alpha]_{\mathbb{R}} \le A^{-1}(\sum_{i=1}^{n} \alpha_i^2)^{1/2}.$$

Thus $\|S\| \le A^{-1}$.

Applying Lemma 2.i.9, the operator $\tilde{S} : J(\ell_2^n) \to J(L_\infty^N/\mathcal{K}_n(\mathbb{R}))$ given by $(\tilde{S}x)_n = Sx_n$ for every $x \in J(\ell_2^n)$ satisfies $\|\tilde{S}\| = \|S\|$ and therefore

(6) $$\|\tilde{S}x\|_{J(L_\infty^N/\mathcal{K}_n)} \le A^{-1}\|x\|_{J(\ell_2^n)}.$$

By Corollary 2.i.12, if $q : L_\infty^N \to L_\infty^N/\mathcal{K}_n(\mathbb{R})$ is the quotient map and for $y \in J(L_\infty^N)$, $\tilde{q}(y) = \{q(y_n)\} \in J(L_\infty^N/\mathcal{K}_n(\mathbb{R}))$, the isomorphism

$$Q : J(L_\infty^N)/J(\mathcal{K}_n(\mathbb{R})) \to J(L_\infty^N/\mathcal{K}_n(\mathbb{R}))$$

given by $Q([\![y]\!]) = \tilde{q}(y)$ is such that $\|Q^{-1}\| \le K$, where K does not depend on n.

Using Lemma 2.i.14(b) and (c), (6) and the easily checked fact that for every $x = (x_1,\dots,x_n)$ with $x_i \in J$, $\tilde{q}V(\sum_{i=1}^n r_i x_i) = \tilde{S}\mathcal{U}(x)$, where V is defined in Lemma 2.i.14, we obtain

$$[x] = \inf\left\{\|\textstyle\sum_{i=1}^n r_i x_i + \phi\|_{L_\infty^N(J)} : \phi \in \mathcal{K}_n(J)\right\} \le$$

$$\le \inf\left\{\|V(\textstyle\sum_{i=1}^n r_i x_i + \phi)\|_{J(L_\infty^N)} : V(\phi) \in J(\mathcal{K}_n)\right\} =$$

$$= \|[\![V(\textstyle\sum_{i=1}^n x_i r_i)]\!]\|_{J(L_\infty^N)/J(\mathcal{K}_n)} \le \|Q^{-1}\|\|\tilde{q}V(\textstyle\sum_{i=1}^n r_i x_i)\|_{J(L_\infty^N/\mathcal{K}_n)} \le$$

$$\le K\|\tilde{S}\mathcal{U}(x)\|_{J(L_\infty^N/\mathcal{K}_n)} \le KA^{-1}\|\mathcal{U}(x)\|_{J(\ell_2^n)}.$$

Taking $L = KA^{-1}$ we arrive at the desired result.

By duality, using Proposition 2.k.6 which characterizes the dual norm of [ ], we obtain

Corollary 2.i.18. For every $x^* = (x_1^*,\dots,x_n^*) \in (J^*)^n$, $\mathcal{U}$ and L as above,

$$L^{-1}\|(\mathcal{U}^{-1})^* x^*\|_{(J(\ell_2^n))^*} \le \|\textstyle\sum_{i=1}^n r_i x_i\|_{L_1(J^*)} \le \|(\mathcal{U}^{-1})^* x^*\|_{(J(\ell_2^n))^*}.$$

Proof: Using Proposition 2.k.6 and Theorem 2.i.17 we have for $x^* = (x_1^*,\dots,x_n^*)$,

$$\|\textstyle\sum_{i=1}^n r_i x_i^*\|_{L_1(J^*)} = \sup\left\{\langle x^*, x\rangle : x = (x_1,\dots,x_n) \in J^n \text{ and } [x] \le 1\right\} =$$

$$= \sup\left\{\langle(\mathcal{U}^{-1})^* x^*, \mathcal{U}x\rangle : x = (x_1,\dots,x_n) \in J^n \text{ and } [x] \le 1\right\} \le$$

$$\leq \sup\left\{\langle(\mathcal{U}^{-1})^* x^*, \mathcal{U}x\rangle : x = (x_1,\dots,x_n) \in J^n \text{ and } \|\mathcal{U}x\|_{J(\ell_2^n)} \leq 1\right\} =$$

$$= \|(\mathcal{U}^{-1})^* x^*\|_{(J(\ell_2^n))^*}.$$

The other side of the inequality is obtained similarly, using the fact that $[x] \leq L\|\mathcal{U}(x)\|_{J(\ell_2^n)}$.

Finally, combining all the inequalities proved in the course of this section we get

Theorem 2.i.19. J^* has cotype 2.

Proof: Let $x \in \ell_2^n(J)$; then as was observed in Definition 2.i.15,

$$\|\mathcal{U}(x)\|_{J(\ell_2^n)} \leq (\textstyle\sum_{i=1}^n \|x_i\|_J^2)^{1/2} = \|x\|_{\ell_2^n(J)}.$$

Since for $x^* = (x_1^*,\dots,x_n^*) \in \ell_2^n(J)^* = \ell_2^n(J^*)$ we have

$$\langle x^*, x\rangle = \langle(\mathcal{U}^{-1})^* x^*, \mathcal{U}x\rangle,$$

from the above we get

$$\|x^*\|_{\ell_2^n(J^*)} \leq \|(\mathcal{U}^{-1})^*(x^*)\|_{(J(\ell_2^n))^*}$$

and from Corollary 2.i.18,

$$L^{-1}(\textstyle\sum_{i=1}^n \|x_i^*\|_{J^*}^2)^{1/2} = L^{-1}\|x^*\|_{\ell_2^n(J^*)} \leq \|\textstyle\sum_{i=1}^n r_i x_i\|_{L_1(J^*)}.$$

This proves the statement.

The above theorem has several deep consequences:

First of all, by Theorems 2.i.3 and 2.i.5, we get that ℓ_r is finitely representable in J^* if and only if $1 \leq r \leq 2$.

Secondly, by known results in the theory of type and cotype, the above theorem also implies that every operator from J into J^* factors through a Hilbert space and thus is weakly compact.

And last but not least, it enables us to show that J and J^* are different.

Lemma 2.i.20. The space J is not isomorphic to any subspace of J^*.

Proof: By Corollary 2.i.4 J has only cotype ∞. Then, if J were isomorphic to a subspace X of J^*, X would have also only cotype ∞ but this contradicts the fact that J^* has cotype 2, since from the definition of cotype it follows that any subspace of a space with cotype q also has cotype q.

Furthermore, J and J^* are non-comparable:

Theorem 2.i.21. If $X \subset J$ and $Y \subset J^*$ are non-reflexive, then X and Y are not isomorphic.

Proof: Suppose there is a non-reflexive subspace $X \subset J$ isomorphic to a subspace $Y \subset J^*$ via an isomorphism $T : X \longrightarrow Y$. Then, by Corollary 2.d.21, X contains a subspace W isomorphic to J. Thus let $S : J \longrightarrow W$ be an isomorphism. Then TS is an isomorphism from J into J^*, contradicting Lemma 2.i.20.

2.j. Appendix: π_λ-spaces, bounded approximation property and finite dimensional decompositions

This section doesn't deal with the James spaces directly and is mainly to be thought of as a complement to the text; the results are necessary to prove some of the main theorems in Section c of this chapter. Most of this material can be found in Johnson, Rosenthal and Zippin [1].

In general a Banach space X does not possess a Schauder basis, but in many instances it is possible to find a family $\{X_\alpha\}_{\alpha \in A}$ of finite dimensional subspaces which span the space; depending on the conditions these spaces satisfy, the Banach space X may possess some very interesting properties as we will show below. Specifically we will treat the notions of π_λ-spaces, spaces with the bounded approximation property and spaces which admit a finite dimensional decomposition, and we will see that there is a close relation between these notions.

It is known that for every subspace F of dimension k of a Banach space X, there exists a projection from X onto F with norm less than or equal to k; the spaces that have the property that there exist uniformly bounded projections onto the finite dimensional subspaces are called π_λ-spaces.

Definition 2.j.1. Let X be a Banach space and let $1 \le \lambda < \infty$. Then X is called a π_λ-space if $X = \overline{\cup_{\alpha\in A} E_\alpha}$ where $\{E_\alpha\}_{\alpha\in A}$ is a family of finite dimensional subspaces of X directed by inclusion, and for every $\alpha \in A$ there is a projection T_α from X onto E_α with $\|T_\alpha\| \le \lambda$.

An obvious example of a π_λ-space is a Banach space X with a basis $\{x_n\}_{n=1}^{\infty}$, since the projections $P_n : X \longrightarrow [x_i]_{i=1}^{n}$ are of norm less than or equal to the basis constant.

We will give an equivalent characterization of π_λ-spaces which will be more easy to handle in the applications, and in order to do so we will use the following lemma.

In what follows I_X will denote the identity operator in X.

Lemma 2.j.2. Let T be an operator from a Banach space X onto an n-dimensional subspace E of X. Let $k \le n$ and let F be a k-dimensional subspace of X such that $\|T|_F - I_F\| < \varepsilon < 1$, where $(1 - \varepsilon)^{-1}\varepsilon k < 1$. Then there is an operator S from X onto an n-dimensional subspace of X such that

$$S|_F = I_F \quad \text{and} \quad \|S - T\| < \frac{k\varepsilon\|T\|}{1 - \varepsilon} < \|T\|.$$

If in addition T is a projection, then S can be chosen to be a projection.

Proof: Let $U = T|_F$. Then U is an invertible operator from F onto TF with $\|U\| < 1 + \varepsilon$. Also if $y \in TF$ then

$$\|y - U^{-1}y\| = \|(T - I_F)\, U^{-1}y\| < \varepsilon\|U^{-1}y\|.$$

Therefore

$$\|U^{-1}\| - 1 \le \|I_{T(F)} - U^{-1}\| < \varepsilon\|U^{-1}\|,$$

$$\|U^{-1}\| < (1 - \varepsilon)^{-1}$$

and

$$\|U^{-1} - I_{T(F)}\| < \frac{\varepsilon}{1 - \varepsilon}.$$

As we mentioned before, it is known (see e.g. Beauzamy [1]), that there exists a projection from a Banach space X onto its k-dimensional subspaces of norm less than or equal to k. Thus let P be a projection from $E = TX$ onto TF with $\|P\| \le k$, put $V = U^{-1}P + I_E - P$ and $S = VT$. Then

$$\|V - I_E\| = \|U^{-1}P - P\| = \|(U^{-1} - I_{T(F)})P\| < k\varepsilon(1 - \varepsilon)^{-1} < 1.$$

Hence $V : E \longrightarrow X$ is a one to one operator, $S|_F = I_F$,

$$\|S - T\| < \frac{k\varepsilon\|T\|}{1 - \varepsilon},$$

and if T is a projection, then it is easy to see that S is a projection.

Lemma 2.j.3. Let X be a Banach space. Then the following are equivalent:

(i) There exists $\lambda \ge 1$ such that X is a π_λ-space.

(ii) There is a $\lambda \ge 1$ such that for every finite dimensional subspace $E \subset X$ there exists a projection S on X with finite dimensional range such that $Sx = x$ for every $x \in E$ and $\|S\| \le \lambda$.

Proof: That (ii) implies (i) follows easily on observing that $X = \overline{\cup_F T_F X}$, where the union is taken over all finite dimensional subspaces F of X and T_F is the projection given by (ii) such that $T_F x = x$ for $x \in F$.
Now suppose X is a π_λ-space. Let $E \subset X$ be a finite dimensional space. Suppose $\{f_1,\dots,f_n\}$ is a basis for E. Then for every $\varepsilon > 0$, if δ is such that for every real sequence $a_1,\dots,a_n$, $\sum_{i=1}^n |a_i| \le \delta\|\sum_{i=1}^n a_i f_i\|$, we have by (i) that there exist $\alpha \in A$ and $g_1,\dots,g_n \in E_\alpha$ with $\|g_i - f_i\| < \varepsilon/\delta$ for $i = 1,\dots,n$ and a projection T_α from X onto E_α with $\|T_\alpha\| \le \lambda$. Clearly if ε is small enough $\dim E_\alpha = n$. Since $T_\alpha g_i = g_i$ we have for every $x = \sum_{i=1}^n a_i f_i \in E$ that

$$\|T_\alpha x - I_E x\| = \|\sum_{i=1}^n a_i(T_\alpha f_i - f_i)\| \le \sum_{i=1}^n |a_i| \left(\|T_\alpha f_i - T_\alpha g_i\| + 0 + \|g_i - f_i\|\right) \le$$

$$\le \sum_{i=1}^n |a_i| (\tfrac{\varepsilon}{\delta})(\lambda + 1) \le \varepsilon(\lambda + 1)\|x\|.$$

Choose ε and ε' such that $\varepsilon(\lambda + 1) < \varepsilon'$ and $\frac{\varepsilon' n}{1 - \varepsilon'} < 1$. By Lemma 2.j.2 there is a projection S_α on X with finite dimensional range such that $S_\alpha|_E = I_E$ and $\|S_\alpha\| < (1 + \frac{\varepsilon' n}{1 - \varepsilon'})\lambda < 2\lambda$.

For every Banach space X it is true that the space of compact operators from X to X is norm closed; however, it is not always true that every compact operator is the limit of a sequence of finite rank operators, as was first shown in an example by Enflo [1]. Here we will study a special class of Banach spaces in which this property is satisfied.

Definition 2.j.4. Let X be a Banach space and let $1 \le \mu < \infty$. We say that X has the μ-approximation property (μ-A.P.) if for every $\varepsilon > 0$ and every compact subset K of X there is a finite rank operator T in X, with $\|T\| \le \mu$, such that

$$\|Tx - x\| \le \varepsilon \text{ for every } x \in K.$$

X is said to have the bounded approximation property (B.A.P.), if it has the μ-A.P. for some μ.

The next lemma, due to Grothendieck, shows that in fact if X has the bounded approximation property, then every compact operator from X into X is a limit of finite rank operators.

Lemma 2.j.5. Let X be a Banach space having the bounded approximation property, then for every compact operator $T : X \longrightarrow X$ and every $\varepsilon > 0$, there is a finite rank operator $S : X \longrightarrow X$ such that $\|T - S\| < \varepsilon$.

Proof: Let $T : X \longrightarrow X$ be a compact operator, then the closure K of the set $T(B_X)$ is a compact set, and hence there is a finite rank operator $S_1 : X \longrightarrow X$ such that $\|S_1 x - x\| < \varepsilon$ for $x \in K$. Then $\|S_1 Ty - Ty\| < \varepsilon$ for every $y \in B_X$, and thus $S = S_1 T$ satisfies the conclusion of the lemma.

A more general structure than that of a Schauder basis is that of a Schauder decomposition, which decomposes a Banach space into subspaces of dimension larger than 1. We will be interested in the case in which those subspaces are all of finite dimension.

Definition 2.j.6. Let X be a Banach space. A sequence $\{X_n\}_{n=1}^{\infty}$ of closed subspaces of X is called a Schauder decomposition of X if every $x \in X$ has a unique representation of the form $x = \sum_{n=1}^{\infty} x_n$ with $x_n \in X_n$ for every n.

Let $P_n : X \longrightarrow X$ be the projection given by $P_n \sum_{i=1}^{\infty} x_i = \sum_{i=1}^{n} x_i$ for every n. The number $C = \sup_n \|P_n\|$ is called the decomposition constant.
Conversely, it is easy to see that every bounded sequence of projections $\{P_n\}$ on X such that if $n < m$ then $P_n P_m = P_m P_n = P_n$ and $\lim_{n\to\infty} P_n x = x$ for every $x \in X$, determines a unique Schauder decomposition by setting $X_1 = P_1 X$ and $X_n = (P_n - P_{n-1})X$ for $n > 1$.

If for every n, $\dim X_n < \infty$, the decomposition is called a finite dimensional decomposition or F.D.D. for short.

The decomposition is called shrinking, if for every $x^* \in X^*$ we have

$$\|P_n^* x^* - x^*\| \longrightarrow 0 \text{ as } n \longrightarrow \infty,$$

where P_n^* denotes the transpose of P_n.

A decomposition $\{X_n\}_{n=1}^{\infty}$ of X is boundedly complete, if for every sequence $\{x_n\}_{n=1}^{\infty}$ with $x_n \in X_n$ for $n = 1, 2, \ldots$ for which

$$\sup_n \|\sum_{i=1}^{n} x_i\| < \infty,$$

the series $\sum_{i=1}^{\infty} x_i$ converges.

Let $\{X_n\}_{n=1}^{\infty}$ be a Schauder decomposition of X. Let $1 = k_1 < k_2 < \ldots$ be an increasing sequence of integers, and for $i = 1, 2, \ldots$ set

$$Y_i = X_{k_i} \oplus X_{k_i+1} \oplus \ldots \oplus X_{k_{i+1}-1}.$$

Then the decomposition $\{Y_i\}_{i=1}^{\infty}$ of X is called a blocking of the decomposition $\{X_n\}_{n=1}^{\infty}$.

A sequence $\{X_n\}_{n=1}^{\infty}$ of closed subspaces of X is called a boundedly complete skipped blocking decomposition of X if:

(i) $X = [X_n]_{n=1}^{\infty}$.

(ii) $X_n \cap [X_m]_{m \neq n} = \{0\}$ for $m, n \in \mathbb{N}$.

(iii) For every pair of sequences $\{m_k\}_{k=1}^{\infty}$ and $\{n_k\}_{k=1}^{\infty}$ in $\mathbb{N}$ such that $m_k < n_k + 1 < m_{k+1}$ for all k, the sequence $Y_k = X_{m_k} \oplus X_{m_k+1} \oplus \ldots \oplus X_{n_k}$ is a boundedly complete Schauder decomposition of its closed linear span.

Next we mention the principle of local reflexivity, which together with the next two lemmata will help us to construct a shrinking F.D.D. in separable π_λ-spaces whose duals are also separable and have the bounded approximation property. This theorem, which we state in its most general version, says in particular that X^{**} is finitely representable in X; (i) and (ii) are due to Lindenstrauss and Rosenthal and (iii) to Johnson, Rosenthal and Zippin [1]. The proof of it can be found in Lacey [1].

Theorem 2.j.7. Let X be a Banach space, j_X the canonical embedding of X into X^{**}, $\varepsilon > 0$ and F a finite dimensional subspace of X^*. If E is a finite dimensional subspace of X^{**}, there is an isomorphism $V : E \longrightarrow X$ such that:

(i) $\max(\|V\|, \|V^{-1}\|) < 1 + \varepsilon$,

(ii) $Vj_X(x) = x$ for $x \in j_X^{-1}E$,

(iii) $\langle f, Ve\rangle = \langle e, f\rangle$ for every $e \in E$ and $f \in F$.

Lemma 2.j.8. Let X and Y be Banach spaces with Y contained in X^* and $\dim Y < \infty$. Let F be a finite dimensional subspace of X^*, let T be an operator from X^* into Y and let $\varepsilon > 0$. Then there is an operator S from X to X such that:

(a) S^* is an operator from X^* into Y,

(b) $S^*|_F = T|_F$,

(c) $\|S^*\| \le \|T\|(1 + \varepsilon)$.

Proof: By the principle of local reflexivity, there is an invertible operator V from $E = T^*(Y^*)$ into X such that $\langle f, Ve\rangle = \langle e, f\rangle$ for every $e \in E$ and $f \in F$ and $\max(\|V\|, \|V^{-1}\|) < 1 + \varepsilon$. Define $U : Y^* \longrightarrow X$ as VT^*. Let j_Y be the natural embedding from Y onto Y^{**}. Since U^* is an adjoint operator, $j_Y^{-1}U^* : X^* \longrightarrow Y \subset X^*$ is w^*-continuous and therefore there exists a continuous operator $S : X \longrightarrow X$ such that $j_Y^{-1}U^* = S^*$ (see e.g. Kelley and Namioka [1]). For $f \in F$ and $y^* \in Y^*$ we have

$$\langle y^*, S^*f\rangle = \langle y^*, j_Y^{-1}U^*f\rangle = \langle U^*f, y^*\rangle = \langle f, VT^*y^*\rangle = \langle T^*y^*, f\rangle = \langle y^*, Tf\rangle.$$

Also

$$\|S^*\| \le \|T^{**}\|\|V^*\| \le (1+\varepsilon)\|T\|.$$

This completes the proof.

Lemma 2.j.9. Let X be a π_λ-space such that X^* has the μ-A.P. Let $E \subset X$, $F \subset X^*$ be finite dimensional subspaces. Then there exists a projection Q with finite dimensional range on X such that:

(a) $Qe = e$ for all $e \in E$,

(b) $Q^*f = f$ for all $f \in F$,

(c) $\|Q\| \le \lambda + 2\mu + 2\lambda\mu$.

Proof: Let $\delta > 0$ be such that $\delta \dim F < \frac{1}{4}$. Since X^* has the μ-A.P. and B_F is a compact set, there exists an operator R with finite dimensional range on X^* such that for all $f \in F$

$$\|Rf - f\| \le \delta\|f\| \text{ and } \|R\| \le \mu.$$

By Lemma 2.j.2 there exists a finite rank operator $T : X^* \longrightarrow X^*$ such that $Tf = f$ for every $f \in F$ and

$$\|T - R\| < \left(\frac{\delta}{1-\delta}\dim F\right)\|R\| < \frac{\mu}{4(1-\delta)}.$$

Hence, since $\delta < \frac{1}{4}$, $\|T\| < \frac{4}{3}\mu$. By Lemma 2.j.8 we obtain an operator S on X with

$$\|S\| < \frac{3}{2}\|T\| < 2\mu$$

such that $S^*f = Tf = f$ for every $f \in F$ and $S^*(X^*) \subset TX^*$. Since the rank of S^* is finite, the same holds for S.

Now let $G = [E \cup S(X)]$ which is clearly of finite dimension. Since X is a π_λ-space, by Lemma 2.j.3 there is a finite rank projection P from X into X such that $\|P\| \le \lambda$ and $Pg = g$ for every $g \in G$. Let Q be defined on X as $Q = S + P - SP$. Then

$$\|Q\| \le 2\mu + \lambda + 2\mu\lambda,$$

and since S and P have finite dimensional ranges, so does Q. From $I - Q = (I - S)(I - P)$ we obtain $I - Q^* = (I - P^*)(I - S^*)$.

Since $\text{range}S \subset G \subset \text{range}P$, $\ker S^* = (\text{range}S)^\perp \supset (\text{range}P)^\perp = \ker P^*$. Thus

$$(I - S^*)|_{\ker P^*} = I_{\ker P^*}$$

and for $z \in \ker P^*$, $(I - Q^*)z = (I - P^*)z = z$.

Also, from $QP = SP + P - SP = P$, we get $(I - Q)P = 0$ and hence

$$P^*(I - Q^*) = 0.$$

Therefore $I - Q^*$ is a projection onto $\ker P^*$. Consequently Q^* and thus Q are projections. Now let $g \in G$, then

$$Qg = Sg + Pg - SPg = Sg + g - Sg = g.$$

Since $E \subset G$, (a) holds. On the other hand, if $f \in F$, recalling that $S^*f = f$, we get

$$(I - Q^*)f = (I - P^*)(I - S^*)f = 0.$$

Hence (b) holds.

The following theorem is the main result of this appendix and establishes a connection among the structures of a Banach space X and its dual X^*, if the latter is separable.

Theorem 2.j.10. Let X be a separable π_λ-space such that X^* is a separable space with the μ-approximation property. Then X has a shrinking finite dimensional decomposition.

Proof: Let $\{x_n\}$ and $\{y_n\}$ be dense sequences in X and X^* respectively with $x_1 = 0$ and $y_1 = 0$. We will construct a sequence $\{Q_n\}$ of finite rank projections on X such that for every i, j $\in \mathbb{N}$

(i) $Q_iQ_j = Q_jQ_i = Q_{\min(i,j)}$,

(ii) $Q_i(X) \supset \{x_1, x_2, \ldots, x_i\}$,

(iii) $Q_i^*(X^*) \supset \{y_1, y_2, \ldots, y_i\}$,

(iv) $\|Q_i\| \le \lambda + 2\mu + 2\lambda\mu = \nu$.

In fact, if (i)-(iv) are true, then if $\varepsilon > 0$, $x \in X$ and x_i is such that $\|x - x_i\| \le \dfrac{\varepsilon}{1 + \nu}$, for $j \ge i$ we have

$$\|x - Q_jx\| \le \|x - x_i\| + \|x_i - Q_jx_i\| + \|Q_jx_i - Q_jx\| \le$$
$$\le (1 + \|Q_j\|)\|x - x_i\| \le (1 + \nu)\|x - x_i\| < \varepsilon.$$

It follows that $\{Q_n\}$ determines an F.D.D. of X (see remark in Definition

2.j.6). Similarly $\{Q_n^*\}$ determines an F.D.D. of X^* and this proves the theorem.

So we proceed to construct the sequence $\{Q_n\}$ by induction:

Let $Q_1 \equiv 0$ and suppose $Q_1,\dots,Q_k$ have been constructed so that (i)-(iv) hold. By Lemma 2.j.9 there exists a finite rank projection Q_{k+1} such that

(a) $Q_{k+1}(X) \supset \{x_1,\dots,x_{k+1}\} \cup Q_k(X)$,

(b) $Q_{k+1}^*(X^*) \supset \{y_1,\dots,y_{k+1}\} \cup Q_k^*(X^*)$,

(c) $\|Q_{k+1}\| \le \nu$.

It is clear from (a) and the induction hypothesis, that for $i \le k$, $Q_{k+1}Q_i = Q_i$ and from (b) that $(Q_iQ_{k+1})^* = Q_{k+1}^*Q_i^* = Q_i^*$, thus $Q_iQ_{k+1} = Q_i$. This finishes the proof.

The next two lemmata give further properties of finite dimensional decompositions and are needed for the classification of the complemented reflexive subspaces of J, therefore we put them in this section. Lemma 2.j.11 can be found in Casazza [1] and is a minor modification of a result by Johnson [1]; Lemma 2.j.12, also by Johnson, can be found in Lindenstrauss and Tzafriri [1].

Lemma 2.j.11. If $\{E_n\}$ is an F.D.D. for a reflexive Banach space X, then for each natural number m and every $\varepsilon > 0$ there is an $n > m$ such that if $x \in X$, $\|x\| \le 1$ and $x = \sum_{i=1}^\infty x_i$ with $x_i \in E_i$, then there is some j with $m \le j < n$ such that $\max\{\|x_j\|, \|x_{j+1}\|\} < \varepsilon$.

Proof: Suppose the statement is false. Then there exist m, ε and a sequence $\{x^k\}$ of unit vectors in X, with $x^k = \sum_{i=1}^\infty x_i^k$ and $x_i^k \in E_i$, such that for $m \le i \le m + k$ either $\|x_i^k\| \ge \varepsilon$ or $\|x_{i+1}^k\| \ge \varepsilon$. Since $\dim E_i < \infty$ and for $k = 1, 2,\dots$, $\|x_i^k\| \le 2C$, where C is the decomposition constant, we may assume, by passing to a subsequence, that there is $x_i \in E_i$ such that $\lim_k x_i^k = x_i$ for $i = 1, 2,\dots$. But then, for every $i \ge m$, either $\|x_i\| \ge \varepsilon$ or $\|x_{i+1}\| \ge \varepsilon$. Now, for each n there is k such that

$$\|\textstyle\sum_{i=m}^n x_i\| < 1 + \|\sum_{i=m}^n x_i^k\| = 1 + \|(P_n - P_{m-1})x^k\| \le 1 + 2C,$$

where $P_0 = 0$ and P_n are the projections given in Definition 2.j.6.

Therefore $\sup_{m<n}\|\sum_{i=m}^{n} x_i\| < \infty$.
Since X is reflexive, $\{E_n\}$ is a boundedly complete decomposition (see e.g. Lindenstrauss and Tzafriri [1]). Therefore $\sum_{i=m}^{\infty} x_i$ converges in X but $\lim_i\|x_i\| \neq 0$ which yields a contradiction.

The following lemma concerning blockings of finite dimensional decompositions is very useful in the study of the structure of subspaces of L_p, but more interesting for this monograph is its application to the classification of complemented reflexive subspaces of J (see Theorem 2.c.17).

Lemma 2.j.12. Let X, Y be Banach spaces, $T : X \longrightarrow Y$ a bounded operator. Let $\{B_n\}_{n=1}^{\infty}$ be a shrinking finite dimensional decomposition of X and let $\{C_n\}_{n=1}^{\infty}$ be a finite dimensional decomposition of Y. Let $\{\varepsilon_n\}_{n=1}^{\infty}$ be a sequence of positive real numbers tending to 0. Then there are blockings $\{B'_i\}_{i=1}^{\infty}$ of $\{B_n\}_{n=1}^{\infty}$ and $\{C'_i\}_{i=1}^{\infty}$ of $\{C_n\}_{n=1}^{\infty}$, such that for every $x \in B'_i$ there is a $y \in C'_i \oplus C'_{i+1}$ such that $\|Tx - y\| \le \varepsilon_i\|x\|$.

Proof: Let $\{P_n\}_{n=1}^{\infty}$ and $\{Q_n\}_{n=1}^{\infty}$ be the projections associated to the given decompositions of X and Y respectively. Then

(1) For every $\varepsilon > 0$ and every $n \in \mathbb{N}$ there exists $m \in \mathbb{N}$ such that if $x \in X$ and $P_m x = 0$ then $\|Q_n Tx\| < \varepsilon\|x\|$.

If this were not so, there would be an n, an $\varepsilon > 0$ and a sequence $\{x_m\}_{m=1}^{\infty} \subset X$ with $\|x_m\| = 1$ and

(2) $$P_m x_m = 0$$

such that $\|Q_n Tx_m\| \ge \varepsilon$ for all m. Since $Q_n Y$ is finite dimensional, we may assume, by passing to a subsequence if necessary, that there exists $y \in Q_n Y$ with $\lim_{m\to\infty} Q_n Tx_m = y$. Hence $\|y\| \ge \varepsilon$ and there exists $z^* \in Y^*$ with $\|z^*\| = 1$ and $\langle z^*, y\rangle > \frac{3}{4}\varepsilon$. Let $y^* = Q_n^* z^*$; then there is K such that for $m \ge K$

(3) $$\langle y^*, Tx_m\rangle = \langle Q_n^* z^*, Tx_m\rangle = \langle z^*, Q_n Tx_m\rangle > \frac{\varepsilon}{2}.$$

By (2) and (3) for every $m \ge K$,

$$\|P_m^* T^* y^* - T^* y^*\| \geq |\langle P_m^* T^* y^*, x_m\rangle - \langle T^* y^*, x_m\rangle| =$$

$$= |\langle T^* y^*, P_m x_m\rangle - \langle T^* y^*, x_m\rangle| = |\langle T^* y^*, x_m\rangle| > \frac{\varepsilon}{2},$$

contradicting the fact that $\{B_n\}_{n=1}^{\infty}$ is shrinking.

Having established (1), we construct two sequences of integers $m_1 < m_2 < \ldots$ and $1 = k_1 < k_2 < \ldots$ as follows:

Choose m_1 so that if $P_{m_1} x = 0$, then $\|Q_{k_1} Tx\| < \varepsilon_2\|x\|/2$. Next pick $k_2 > k_1$ such that for every $x \in P_{m_1} X$, $\|Tx - Q_{k_2} Tx\| < \varepsilon_1\|x\|/2$, which is possible since $\dim P_{m_1} X < \infty$. If $m_1,\ldots,m_{i-1}$ and $k_1,\ldots,k_{i-1},k_i$ have been chosen, select $m_i > m_{i-1}$ so that if $P_{m_i} x = 0$, then $\|Q_{k_i} Tx\| < \varepsilon_{i+1}\|x\|/2$. Next pick $k_{i+1} > k_i$ such that for every $x \in P_{m_i} X$, $\|Tx - Q_{k_{i+1}} Tx\| < \varepsilon_i\|x\|/2$.

For i = 1, 2,... let

$$B_i' = B_{m_{i-1}+1} \oplus B_{m_{i-1}+2} \oplus \ldots \oplus B_{m_i}$$

and

$$C_i' = C_{k_{i-1}+1} \oplus C_{k_{i-1}+2} \oplus \ldots \oplus C_{k_i}.$$

Then, if $x \in B_i'$ and $y = Q_{k_{i+1}} Tx - Q_{k_{i-1}} Tx$, the conclusion of the lemma is satisfied.

2.k. Appendix: $\mathrm{Lip}_\alpha([0,1], X)$ and $L_p(X)$

This appendix is intended as a complement to Section 2.i, developing some results that do not directly involve the space J^*, but are necessary in the proof of J^* having cotype 2.

(i) An isomorphism between $\ell_\infty(X)$ and $\mathrm{Lip}_\alpha([0,1], X)$.

The existence of this isomorphism is a very beautiful result, due originally to Ciesielski ([1], [2]), who demonstrated it for the particular case in which $X = \mathbb{R}$. The definitions of the spaces $\mathrm{Lip}_\alpha([0,1],X)$ and $\ell_\infty(X)$ can be found in Section 2.i.

Definition 2.k.1. The orthonormal Haar system $\{h_m\}_{m=1}^{\infty}$ in $L_2([0,1])$ is given as follows:
For $n = 0, 1,\dots$ and $k = 1,\dots,2^n$,

$$h_1(t) = 1, \quad h_{2^n+k}(t) = \begin{cases} \sqrt{2^n} & \text{if } t \in [2^{-(n+1)}(2k-2),\ 2^{-(n+1)}(2k-1)], \\ -\sqrt{2^n} & \text{if } t \in (2^{-(n+1)}(2k-1),\ 2^{-(n+1)}2k], \\ 0 & \text{otherwise.} \end{cases}$$

Now we define the auxiliary sequences $\{\varphi_n\}_n$ and $\{\varphi_n^{(\alpha)}\}_n$:

For $n = 1,2,\dots$ and $t \in [0,1]$, let $\varphi_n(t) = \int_0^t h_n(\tau)d\tau$ and for $0 < \alpha < 1$, define $\varphi_1^{(\alpha)} = \varphi_1$, $\varphi_{2^n+k}^{(\alpha)} = \left(2^{(n+1)(1-\alpha)-(n/2)}\right)\varphi_{2^n+k}$.

Observe that

$$(\ddagger) \qquad \varphi_{2^n+k}^{(\alpha)}(t) = \begin{cases} 2^{(n+1)(1-\alpha)}\left(t - \dfrac{2k-2}{2^{n+1}}\right) & \text{if } t \in \left[\dfrac{2k-2}{2^{n+1}}, \dfrac{2k-1}{2^{n+1}}\right], \\ 2^{(n+1)(1-\alpha)}\left(\dfrac{2k}{2^{n+1}} - t\right) & \text{if } t \in \left(\dfrac{2k-1}{2^{n+1}}, \dfrac{2k}{2^{n+1}}\right], \\ 0 & \text{otherwise.} \end{cases}$$

Proposition 2.k.2. Let X be a Banach space, $0 < \alpha < 1$ and $f : [0,1] \longrightarrow X$ be a continuous function. Then the series $f(0) + \sum_{n=1}^{\infty} a_n \varphi_n^{(\alpha)}(t)$, where

$$a_1 = f(1) - f(0),$$

$$a_{2^n+k} = \frac{2^{(n+1)\alpha}}{2}\left(f((2k-1)/2^{n+1}) - f((2k-2)/2^{n+1}) - f((2k)/2^{n+1}) + f((2k-1)/2^{n+1})\right)$$

for $n = 0, 1,\dots,$ and $k = 1,\dots,2^n$, converges uniformly in $[0,1]$ to $f(t)$. Moreover, the representation of f in terms of the family $\{\varphi_n^{(\alpha)}\}_{n=1}^{\infty}$ is unique.

Proof: We order the dyadic rationals in $[0,1]$ as follows: let $s_0 = 0$, $s_1 = 1$ and for $m > 1$, $m = 2^n + k$ with $1 \le k \le 2^n$, $s_m = (2k-1)/2^{n+1}$. Denote by p_0 the function defined as $p_0(t) = f(0)$ for $t \in [0,1]$ and by $p_m : [0,1] \longrightarrow X$ the piecewise linear function given by $p_m(s_i) = f(s_i)$ for $i = 1,\dots,m$, and linear between two consecutive s_i's. (s_i and s_j are consecutive if there is no s_k between them). It is obvious that $p_m(t)$

converges uniformly to $f(t)$ as $m \longrightarrow \infty$; therefore the series

$$p_0(t) + \sum_{m=1}^{\infty}(p_m(t) - p_{m-1}(t))$$

converges uniformly to $f(t)$ in $[0,1]$.

Since for $m = 2^n + k$, $(2k-2)/2^{n+1} < s_m < (2k)/2^{n+1}$, it is a straightforward calculation to show that if $t \notin [(2k-2)/2^{n+1}, (2k)/2^{n+1}]$, then $p_m(t) = p_{m-1}(t)$, and also that for every $t \in [0,1]$

$$p_m(t) - p_{m-1}(t) = a_m \varphi_m^{(\alpha)}(t).$$

This proves the convergence of the series $f(0) + \sum_{n=1}^{\infty} a_n \varphi_n^{(\alpha)}$ to f.

To show the uniqueness, suppose

$$\sum_{n=1}^{\infty} a_n \varphi_n^{(\alpha)}(t) = 0$$

for every $t \in [0,1]$. By (‡), $\varphi_1^{(\alpha)}(1) = 1$, for $n > 1$ $\varphi_n^{(\alpha)}(1) = 0$ and

$$\varphi_{2^n+r}^{(\alpha)}\left(\frac{2k-1}{2^{n+1}}\right) = 0 \text{ for } 1 \le r < k, \quad \varphi_{2^n+k}^{(\alpha)}\left(\frac{2k-1}{2^{n+1}}\right) = \frac{1}{2^{(n+1)\alpha}},$$

and

$$\varphi_m^{(\alpha)}\left(\frac{2k-1}{2^{n+1}}\right) = 0 \text{ for } m > 2^n + k.$$

Therefore $\sum_{n=1}^{\infty} a_n \varphi_n^{(\alpha)}(1) = 0$ implies $a_1 = 0$. Suppose we know already that $a_1 = \ldots = a_{m-1} = 0$. Then if $m = 2^n + k$, $\sum_{n=m}^{\infty} a_n \varphi_n^{(\alpha)}\left(\frac{2k-1}{2^{n+1}}\right) = 0$ implies $a_m = 0$ and we are done.

Theorem 2.k.3. Let $0 < \alpha < 1$, then $\mathrm{Lip}_\alpha([0,1],X)$ is isomorphic to $\ell_\infty(X)$.

Proof: Let $T_\alpha : \mathrm{Lip}_\alpha([0,1],X) \longrightarrow \ell_\infty(X)$ be given by $T_\alpha f = \{a_n\}_{n=0}^{\infty}$, where $a_0 = f(0)$, $a_1 = f(1) - f(0)$ and for $n = 0, 1, \ldots$ and $k = 1, \ldots, 2^n$,

$$a_{2^n+k} = \frac{2^{(n+1)\alpha}}{2}\Big(f((2k-1)/2^{n+1}) - f((2k-2)/2^{n+1}) - f((2k)/2^{n+1}) + f((2k-1)/2^{n+1})\Big)$$

for every $f \in \mathrm{Lip}_\alpha([0,1], X)$.

Then $|a_{2^n+k}| \le \|f\|_\alpha$ for every $n = 0, 1, \ldots$ and $k = 1, \ldots, 2^n$. Hence $\|T_\alpha\| \le 1$. Using induction it is easy to see that $T_\alpha f = 0$ implies $f((2k-1)/2^{n+1}) = 0$ for all dyadic rationals, and by the continuity of f it follows that $f = 0$. It remains to show that T_α is onto $\ell_\infty(X)$. Thus let $\{c_m\}_{m=0}^{\infty} \in \ell_\infty(X)$ and for $t \in [0,1]$, let

$$f(t) = c_0 + \sum_{m=1}^{\infty} c_m \varphi_m^{(\alpha)}(t).$$

We will see that $f \in \mathrm{Lip}_\alpha([0,1], X)$: we have for $s, t \in [0, 1]$

$$\text{(1)} \quad \|f(t) - f(s)\|_X \le \|\{c_m\}\|_\infty \Big(|t - s| + \sum_{n=0}^{\infty} \sum_{k=1}^{2^n} |\varphi_{2^n+k}^{(\alpha)}(t) - \varphi_{2^n+k}^{(\alpha)}(s)| \Big).$$

Suppose N denotes the non-negative integer such that

$$\text{(2)} \qquad \frac{1}{2^{N+1}} < |t - s| \le \frac{1}{2^N}.$$

Using (‡) it is not very hard to establish that

$$\text{(3)} \qquad \sum_{k=1}^{2^n} |\varphi_{2^n+k}^{(\alpha)}(t) - \varphi_{2^n+k}^{(\alpha)}(s)| \le \begin{cases} 2^{(1-\alpha)(n+1)} |t - s| & \text{if } 0 \le n \le N, \\ 2^{1-\alpha}\, 2^{-n\alpha} & \text{if } n > N. \end{cases}$$

In order to do this observe that for every pair $s, t \in [0,1]$ and for every n there are either one or two summands in (3) different from 0. Hence, using (1), (2) and (3),

$$\frac{\|f(t) - f(s)\|_X}{|t - s|^\alpha} + \|f(0)\|_X \le$$

$$\le \|\{c_m\}\|_\infty \Big(1 + |t-s|^{1-\alpha} + |t-s|^{1-\alpha} \sum_{n=0}^{N} 2^{(1-\alpha)(n+1)} + |t-s|^{-\alpha} 2^{1-\alpha} \sum_{n=N+1}^{\infty} 2^{-n\alpha} \Big)$$

$$\le \|\{c_m\}\|_\infty \Big(2 + 2^{-N(1-\alpha)} \sum_{n=0}^{N} 2^{(1-\alpha)(n+1)} + 2^{1-\alpha} \sum_{n=0}^{\infty} 2^{-n\alpha} \Big) \le K_\alpha \|\{c_m\}\|_\infty.$$

Hence $\|T_\alpha^{-1}\| \le K_\alpha$, where K_α depends only on α, and since obviously $T_\alpha f = \{c_n\}_{n=0}^{\infty}$, this finishes the proof.

(ii) About the space $L_p(X)$ of Bochner integrable functions.

In this part of the appendix we give the definitions of Bochner integrable functions and some related results needed in Section 2.i. The reader interested in these matters is referred to Diestel and Uhl [1] for further details.

First recall the definitions of μ-measurable functions and of Bochner integrability.

Definition 2.k.4. Let (Ω,Σ,μ) be a probability space and X a Banach space. A function $f : \Omega \longrightarrow X$ is called simple if there exist $x_1,\dots,x_n \in X$ and $E_1,\dots,E_n \in \Sigma$ such that $f(\omega) = \sum_{i=1}^n x_i \chi_{E_i}(\omega)$, where χ_E denotes the characteristic function of the set E. In this case for every $E \in \Sigma$

$$\int_E f(\omega)d\mu = \sum_{i=1}^n x_i \mu(E_i \cap E) \in X.$$

The function f is called μ-measurable if there exists a sequence $\{f_n\}$ of simple functions with $\lim_{n\to\infty} \|f(\omega) - f_n(\omega)\|_X = 0$ μ-almost everywhere (μ-a.e.).

A μ-measurable function $f : \Omega \longrightarrow X$ is called Bochner integrable if there exists a sequence $\{f_n\}$ of simple functions such that

$$\lim_{n\to\infty} \int_\Omega \|f(\omega) - f_n(\omega)\|_X d\mu = 0.$$

In this case $\int_E f(\omega)d\mu$ is defined for every $E \in \Sigma$ as

$$\int_E f(\omega)d\mu = \lim_{n\to\infty} \int_E f_n(\omega)d\mu.$$

It is not difficult to see that f is Bochner integrable if and only if

$$\int_\Omega \|f(\omega)\|_X d\mu < \infty.$$

We are interested in the spaces of Bochner p-integrable functions, where $\Omega = [0,1]$, μ is the Lebesgue measure in $[0,1]$ and Σ the Borel σ-algebra in $[0,1]$.

We define the spaces $L_p(\Omega,\Sigma,\mu,X)$, $1 \le p < \infty$, as the spaces of all equivalence classes of Bochner integrable functions $f : \Omega \longrightarrow X$ such that $\int_\Omega \|f(\omega)\|_X^p d\mu < \infty$ with the norm

$$\|f\|_{L_p(X)} = \left(\int_\Omega \|f(\omega)\|_X^p d\mu\right)^{1/p}.$$

$L_\infty(\Omega,\Sigma,\mu,X)$ will denote the space of all the equivalence classes of essentially bounded Bochner integrable functions with the norm

$$\|f\|_{L_\infty(X)} = \text{ess sup}\|f(\omega)\|_X.$$

Since we are only dealing with $\Omega = [0,1]$ and μ the Lebesgue measure, we are going to denote $L_p(\Omega,\Sigma,\mu,X)$ simply by $L_p(X)$ and $L_p(\mathbb{R})$ by L_p.

If n is a fixed positive integer and $N = 2^n$, then for $1 \le p \le \infty$, $L_p^N(X)$ will denote the subspace of $L_p(X)$ given by

$$L_p^N(X) = \{\textstyle\sum_{i=0}^{N-1} x_i \chi_{[i/N,(i+1)/N)} : x_i \in X\}.$$

The next theorem tells us conditions under which we may interchange a bounded operator with the integral. In its general version it is due to Hille, but we will only state and prove its simple version for bounded operators.

Theorem 2.k.5. Let X, Y be Banach spaces and $T : X \longrightarrow Y$ a bounded linear operator. If $f : \Omega \longrightarrow X$ and $T \circ f : \Omega \longrightarrow Y$ are both Bochner integrable with respect to μ, then for every $E \in \Sigma$

$$T\left(\int_E f(\omega)d\mu\right) = \int_E T \circ f(\omega)d\mu.$$

Proof: The result is obvious for simple functions; now let $\{f_n\}$ be a sequence of simple functions such that $\lim_{n\to\infty} \int_E \|f(\omega) - f_n(\omega)\|_X d\mu = 0$. Then

$$\int_E \|T\circ f(\omega) - T\circ f_n(\omega)\|_X d\mu \le \|T\| \int_E \|f(\omega) - f_n(\omega)\|_X d\mu.$$

Therefore

$$\int_E T\circ f(\omega)d\mu = \lim_{n\to\infty} \int_E T\circ f_n(\omega)d\mu = \lim_{n\to\infty} T\left(\int_E f_n(\omega)d\mu\right) = T\left(\int_E f(\omega)d\mu\right).$$

The last result of this appendix, due to Pisier [2], plays a very important role in the proof in Section 2.i of J^* having cotype 2; it characterizes the dual space of $(X^n,[\ ])$ where

$$[x] = \inf\left\{\|\textstyle\sum_{i=1}^n r_i x_i + \phi\|_{L_\infty(X)} : \phi \in L_\infty^N(X) \text{ and } \displaystyle\int_\Omega \phi(\omega) r_i(\omega)d\mu = 0,\ i = 1,\dots,n\right\},$$

and $\{r_i\}$ is the set of Rademacher functions.

Proposition 2.k.6. Let X be a Banach space, then we have for every $x_1^*,\dots,x_n^* \in X^*$

$$\|\textstyle\sum_{i=1}^n r_i x_i^*\|_{L_1(X^*)} = \sup\left\{\sum_{i=1}^n \langle x_i^*, x_i\rangle : x = (x_1,\dots,x_n) \in X^n \text{ and } [x] \le 1\right\}.$$

In particular, if $\mathcal{R}_n(X^*)$ is the closed subspace of $L_1(X^*)$ generated by the functions of the form $\sum_{i=1}^n r_i x_i^*$, then $\mathcal{R}_n(X^*)$ is isometric to the dual of the space X^n equipped with the norm [ ].

Proof: Let $N = 2^n$, $\psi \in L_\infty^N(X)$ and define

$$x_i = \int_\Omega r_i(\omega)\psi(\omega)d\mu,$$

$x = (x_1,\ldots,x_n)$ and $\phi = \psi - \sum_{i=1}^n r_i x_i$. Then clearly $\phi \in L_\infty^N(X)$ since $\sum_{i=1}^n r_i x_i \in L_p^N(X)$ for every $1 \le p \le \infty$ and for every $i = 1,\ldots,n$

$$\int_\Omega r_i(\omega)\phi(\omega)d\mu = \int_\Omega r_i(\omega)\psi(\omega)d\mu - \sum_{j=1}^n x_j \int_\Omega r_i(\omega)r_j(\omega)d\mu = x_i - x_i = 0$$

and

(1) $$[x] \le \|\psi\|_{L_\infty(X)}.$$

Also if $x_1^*,\ldots,x_n^* \in X^*$, by Theorem 2.k.5 $\int_\Omega r_i(\omega)x_i^*\phi(\omega)d\mu = 0$ and

(2) $$\int_\Omega \langle \sum_{i=1}^n r_i(\omega)x_i^*,\ \psi(\omega)\rangle d\mu = \int_\Omega \langle \sum_{i=1}^n r_i(\omega)x_i^*,\ \phi(\omega) + \sum_{i=1}^n r_i(\omega)x_i\rangle d\mu =$$

$$= \int_\Omega \langle \sum_{i=1}^n r_i(\omega)x_i^*,\ \sum_{i=1}^n r_i(\omega)x_i\rangle d\mu = \sum_{i=1}^n \langle x_i^*,\ x_i\rangle.$$

By the definition of $L_1^N(X)$ it is easy to see that one may identify $L_1^N(X^*)$ with $(X^*)^N$ equipped with the following norm: if $x^* = (x_1^*,\ldots,x_N^*)$ belongs to $(X^*)^N$ then

$$|\|x^*\|| = \frac{1}{N}\sum_{i=1}^N \|x_i^*\|_{X^*}.$$

Similarly $L_\infty^N(X)$ can be identified with X^N endowed with the norm

$$|\|x\||_\infty = \sup_{1\le i\le N}\|x_i\|_X.$$

Thus $(X^N, |\|\ \||_\infty)^* = ((X^*)^N, |\|\ \||)$, where to every $(y_1^*,\ldots,y_N^*) \in (X^*)^N$ we associate the functional $y^* \in (X^N, |\|\ \||_\infty)^*$ given by

$$\langle y^*,\ x\rangle = \frac{1}{N}\sum_{i=1}^N \langle y_i^*,\ x_i\rangle$$

for every $x = (x_1,\ldots,x_N) \in X^N$. Clearly this is an onto application from $((X^*)^N, |\|\ \||)$ into $(X^N, |\|\ \||_\infty)^*$. Furthermore

$$\|y^*\| = \sup\{\langle y^*,\ x\rangle : |\|x\||_\infty \le 1\} =$$

$$= \sup\left\{\frac{1}{N}\sum_{i=1}^N \langle y_i^*, x_i\rangle : \sup_{1\le i\le N}\|x_i\|_X \le 1\right\} \le \frac{1}{N}\sum_{i=1}^N \|y_i^*\|_{X^*}.$$

On the other hand, let $\varepsilon > 0$ and $x_i \in X$ with $\|x_i\|_X = 1$ be such that

$$\|y_i^*\|_{X^*} < \langle y_i^*, x_i\rangle(1 + \varepsilon).$$

Then if $x = (x_1,\dots,x_N)$, $|\|x\||_\infty = 1$ and

$$\|y^*\| \ge \langle y^*, x\rangle = \frac{1}{N}\sum_{i=1}^N \langle y_i^*, x_i\rangle > (1+\varepsilon)^{-1}\frac{1}{N}\sum_{i=1}^N \|y_i^*\|_{X^*}.$$

From the above it follows that $L_\infty^N(X)^* = L_1^N(X^*)$. Hence by (1) and (2)

$$\left\|\sum_{i=1}^n r_i x_i^*\right\|_{L_1^N(X^*)} = \left\|\sum_{i=1}^n r_i x_i^*\right\|_{L_\infty^N(X)^*} =$$

$$= \sup\left\{\int_\Omega \left\langle \sum_{i=1}^n r_i(\omega)x_i^*, \psi(\omega)\right\rangle d\mu : \psi \in L_\infty^N(X), \|\psi\|_{L_\infty(X)} \le 1\right\} \le$$

$$\le \sup\left\{\sum_{i=1}^n \langle x_i^*, x_i\rangle : [x] \le 1\right\}.$$

On the other hand, if $[x] = 1$, then for every $\varepsilon > 0$ there exists $\phi_\varepsilon \in L_\infty^N(X)$ such that for $i = 1,\dots,n$

$$\int_\Omega \phi_\varepsilon(\omega) r_i(\omega)d\mu = 0$$

and

$$\left\|\sum_{i=1}^n r_i x_i + \phi_\varepsilon\right\|_{L_\infty(X)} \le 1 + \varepsilon.$$

Let $\psi_\varepsilon = \sum_{i=1}^n r_i x_i + \phi_\varepsilon$. Then $x_i = \int_\Omega r_i(\omega)\psi_\varepsilon(\omega)d\mu$, and by (2)

$$\frac{1}{1+\varepsilon}\sum_{i=1}^n \langle x_i^*, x_i\rangle = \int_\Omega \left\langle \sum_{i=1}^n r_i(\omega)x_i^*, \psi_\varepsilon(\omega)/(1+\varepsilon)\right\rangle d\mu \le$$

$$\le \left\|\sum_{i=1}^n r_i x_i^*\right\|_{L_\infty^N(X)^*} = \left\|\sum_{i=1}^n r_i x_i^*\right\|_{L_1^N(X^*)}$$

and this proves the proposition.

Mit dem Wissen wächst der Zweifel

Johann Wolfgang von Goethe

CHAPTER 3. THE JAMES TREE SPACE JT

Creció en mi frente un árbol.
Creció hacia dentro.
Sus raíces son venas,
nervios sus ramas,
sus confusos follajes pensamientos.
....
Allá adentro, en mi frente,
el árbol habla.
Acércate, ¿lo oyes?
Octavio Paz

The simplest example of a separable Banach space with non-separable dual is ℓ_1 and it was conjectured by Banach [1] that this was a sort of standard situation, namely that every separable Banach space with a non-separable dual had a subspace isomorphic to ℓ_1. This raises the question of when a given non-reflexive Banach space does not admit ℓ_1 or some other typical non-reflexive space as a subspace. In the previous chapter we saw that the space J has this kind of property, since neither c_0 nor ℓ_1 can be embedded in J; however, J is not a counterexample to Banach's conjecture, because J^* is separable. Nevertheless, using J as a building block, James [3] proved the conjecture to be false, by constructing the so-called James tree space JT, which is the subject matter of this chapter.

3.a. The space JT

Besides being the first example of a separable Banach space not containing ℓ_1 with a non-separable dual (Lemma 3.a.5), the space JT has many other remarkable features, some shared with J and some not, and this makes it another important test case for many conjectures in the geometry of Banach spaces. Among the main properties of JT discussed in this section we cite the following: first, it is a somewhat reflexive space and even more, every infinite dimensional subspace contains an infinite dimensional Hilbert space, disproving another conjecture stated by Davis and Singer [1], who believed that each separable somewhat reflexive space had a separable dual. This has several consequences: since neither c_0 nor ℓ_1 has an infinite dimensional subspace isomorphic to a Hilbert space, JT does not contain any subspace isomorphic to either of these spaces, and as a result no non-reflexive subspace of JT has an unconditional basis (Theorem 1.7). Finally, we will prove the result of Andrew [3] showing that JT, like J, is primary.

In this section we will consider the James space J with the summing basis $\{\xi_n\}_{n=1}^{\infty}$ and the norm given by

$$\|\sum_{i=1}^{\infty} a_i \xi_i\|_J = \sup \left(\sum_{n=1}^{k-1}\left(\sum_{i=p_n}^{p_{n+1}-1} a_i\right)^2\right)^{1/2},$$

where the sup is taken over all choices of k and all positive integers $p_1 <\ldots< p_k$.

Roughly speaking, the idea behind the definition of JT is to consider a binary tree and then attach to each branch of this tree a copy of J. Thus, to give the precise definition of JT, but also to fix some notation and terminology, we now recall the basic notions about trees.

Definition 3.a.1. The standard binary tree is

$$\mathcal{T} = \{(n,i) : 0 \le n < \infty,\ 0 \le i < 2^n\}.$$

The points (n,i) are called nodes. The node (0,0) is called the vertex of $\mathcal{T}$. We say that (n+1,2i) and (n+1,2i+1) are the offspring of (n,i).
A segment is a finite set

$$S = \{t_1, t_2,\ldots,t_n\}$$

of nodes, such that for each j, t_{j+1} is an offspring of t_j.

$\mathcal{T}$ is partially ordered by the relation <, with $t_1 < t_2$ if and only if $t_1 \ne t_2$ and there is a segment S with first element t_1 and last element t_2. We will denote by $[t_1,t_2]$, (t_1,t_2), $(t_1,t_2]$ and $[t_1,t_2)$ the segments in $\mathcal{T}$ depending on whether they include t_1 and t_2 or not.
If $t_1 < t_2$ we say t_2 is a descendant of t_1. Observe that an offspring is an immediate descendant.

The set

$$\{(n,i) : 0 \le i < 2^n\}$$

is called the n-th level of $\mathcal{T}$ and we say that the node (n, i) has level n; we denote the level of a node t by lev(t).

An n-branch is a totally ordered set $\{(m,i_m)\}_{m=n}^{\infty}$ and a branch is a set which is an n-branch for some n.

A tree is a partially ordered set $\mathcal{S}$ which is order isomorphic to $\mathcal{T}$. If $\mathcal{S}$ and $\mathcal{S}'$ are trees with $\mathcal{S}' \subset \mathcal{S}$, we say $\mathcal{S}'$ is a subtree of $\mathcal{S}$. If $\mathcal{S}$ is a tree and $\psi : \mathcal{S} \longrightarrow \mathcal{T}$ is an order isomorphism, we define

$$\mathrm{lev}_{\mathcal{S}}(s) = \mathrm{lev}(\psi(s)).$$

If ν is a node, $\mathcal{T}_\nu$ will denote the subtree of $\mathcal{T}$ given by

$$\mathcal{T}_\nu = \{s \in \mathcal{T} : s \ge \nu\}.$$

Definition 3.a.2. The James tree space JT is the completion of the space of functions $x : \mathcal{T} \longrightarrow \mathbb{R}$ so that

(1) $$\|x\| = \sup\left(\sum_{j=1}^{k}(\sum_{(n,i)\in S_j} x(n,i))^2\right)^{1/2} < \infty,$$

where the sup is taken over all finite collections of mutually disjoint segments $S_1,\dots,S_k$.

Clearly JT is separable, for if $\eta_t : \mathcal{T} \longrightarrow \mathbb{R}$ is given by $\eta_t(t) = 1$, $\eta_t(s) = 0$ if $s \ne t$, then $\{\eta_{t_n}\}_{n=1}^{\infty}$ where $t_{2^i+j} = (i,j)$ for $i = 0,1,\dots$, $0 \le j < 2^i$ is a boundedly complete basis for JT. We will call this enumeration of the nodes in $\mathcal{T}$ the "usual order".

The elements of this basis will be denoted either by η_{t_n}, $\eta_{i,j}$, η_{ij}, $\eta_{(i,j)}$ or simply by η_t, when t is the n-th node (i,j) and $\{\eta^*_{t_n}\}$, $\{\eta^*_{i,j}\}$, $\{\eta^*_{ij}\}$, $\{\eta^*_{(i,j)}\}$ or $\{\eta^*_t\}$ will denote the corresponding sequence of biorthogonal functionals.

If $x \in$ JT, the support of x, denoted by supp$x$, consists of the nodes $t \in \mathcal{T}$ for which $x(t) \ne 0$.

An immediate consequence from the definition of the norm in JT is the following:

Lemma 3.a.3. Let $x = \sum_{t\in\mathcal{T}} a_t \eta_t$, $y = \sum_{t\in\mathcal{T}} b_t \eta_t \in JT$ be such that if $t_1, t_2 \in \operatorname{supp} x$ and $t_1 \le t_2$ then $[t_1, t_2] \cap \operatorname{supp} y = \emptyset$ and also if $t_1, t_2 \in \operatorname{supp} y$ and $t_1 \le t_2$ then $[t_1, t_2] \cap \operatorname{supp} x = \emptyset$. Then

$$\|x\|^2 + \|y\|^2 \le \|x + y\|^2.$$

Next we introduce some necessary notation

Definition 3.a.4. Let S be a segment, B a branch, $t = (n,i)$ a node, N a non-negative integer and $x \in JT$.

(1) $$f_S(x) = \langle f_S, x\rangle = \sum_{t\in S} \langle \eta_t^*, x\rangle.$$

(2) $$f_B(x) = \langle f_B, x\rangle = \sum_{t\in B} \langle \eta_t^*, x\rangle.$$

(3) $$P_S x = \sum_{t\in S} \langle \eta_t^*, x\rangle \eta_t.$$

(4) $$P_B x = \sum_{t\in B} \langle \eta_t^*, x\rangle \eta_t.$$

(5) $$P_N x = \sum_{\{t:\mathrm{lev}(t)\le N\}} \langle \eta_t^*, x\rangle \eta_t, \quad Q_N = I - P_{N-1} \quad \text{for } N \ge 1.$$

(6) $$Q_{n,i} = Q_t x = \sum_{s\ge t} \langle \eta_s^*, x\rangle \eta_s.$$

All these functionals and projections have norm one. We will show here that the functional defined by (2) has norm one; the other proofs are similar.

If $B = \{t_1, t_2, \dots\}$ let S_i be the segment $\{t_1, t_2, \dots, t_i\}$ and

$$x = \sum_{n=1}^{i} \langle \eta_{t_n}^*, x\rangle \eta_{t_n}.$$

Then

$$\|x\| \ge \left|\sum_{n=1}^{i} \langle \eta_{t_n}^*, x\rangle\right|.$$

Therefore

$$|f_B(x)| = \left|\sum_{n=1}^{i} \langle \eta_{t_n}^*, x\rangle\right| \le \|x\| \text{ and } \|f_B\| \le 1.$$

On the other hand, if $x = \eta_{t_1}$ then $f_B(x) = 1$ and we conclude that $\|f_B\| = 1$.

Using the above, we can now prove one essential difference between the spaces J and JT.

Lemma 3.a.5. JT^* is not separable.

Proof: Let B_1 and B_2 be two different 1-branches, let $s \in B_1 \backslash B_2$, $t \in B_2 \backslash B_1$ and $x = \eta_s - \eta_t$. Then $\|x\| = \sqrt{2}$ and $(f_{B_1} - f_{B_2})(x) = 2$. Hence $\|f_{B_1} - f_{B_2}\| \geq \sqrt{2}$ and JT^* is not separable, since the set of 1-branches is uncountable.

As mentioned, we will see in Proposition 3.a.7 that if one restricts oneself to any branch of $\mathcal{T}$ then one obtains a space isometric to J. This was proved by Andrew [3], who also proved that if one restricts the index set to a subtree, a complemented isometric copy of JT is obtained; for this he used the following notion.

Definition 3.a.6. Let $\mathcal{S}$ be a subtree of $\mathcal{T}$ and let $\{\sigma_t\}_{t\in\mathcal{S}}$ be a collection of disjoint segments of $\mathcal{T}$. We will say that σ_t has no gaps if:
(i) $t \in \mathcal{S}$ implies $t \in \sigma_t$.
(ii) If $t_1, t_2 \in \mathcal{S}$, t_2 is an offspring of t_1 in $\mathcal{S}$ and $t \in \mathcal{T}$ satisfies $t_1 < t < t_2$, then either $t \in \sigma_{t_1}$ or $t \in \sigma_{t_2}$.

Proposition 3.a.7. (a) For any subtree $\mathcal{S}$ of $\mathcal{T}$, $[\eta_t]_{t\in\mathcal{S}}$ is isometric to JT and complemented in JT.
(b) For any branch $B \subset \mathcal{T}$, $P_B JT$ is isometric to $(J, \| \ \|_J)$.

Proof: (a) Let $X = [\eta_t]_{t\in\mathcal{S}}$. Since $\mathcal{S}$ is a tree, there exists an order isomorphism $\phi : \mathcal{S} \longrightarrow \mathcal{T}$. Let $x \in X$, $x = \sum_{s\in\mathcal{S}} a_s \eta_s$ and define $R : X \longrightarrow JT$ by

$$Rx = \sum_{s\in\mathcal{S}} a_s \eta_{\phi(s)}.$$

Then

$$\|Rx\| = \sup_{S_1,\dots,S_k \subset \mathcal{T}} \Big(\sum_{j=1}^{k} \big(\sum_{\phi(s)\in S_j} a_s\big)^2\Big)^{1/2} =$$

$$= \sup_{S'_1,\dots,S'_k \subset \mathcal{S}} \Big(\sum_{j=1}^{k} \big(\sum_{s\in S'_j} a_s\big)^2\Big)^{1/2} = \|x\|,$$

where the sup is taken over all finite collections $S_1,\dots,S_k$ of disjoint segments and $S'_j = \phi^{-1}(S_j)$.
Let $\{\sigma_t\}_{t\in\mathcal{S}}$ be a collection of disjoint segments of $\mathcal{T}$ such that σ_t has

no gaps. For $x = \sum_{t\in\mathcal{T}} a_t \eta_t$, let $P : JT \longrightarrow X$ be given by

$$Px = \sum_{t\in\mathcal{S}}(\sum_{s\in\sigma_t} a_s)\eta_t.$$

Then P is a projection onto X and if $S_1, S_2,\ldots,S_k$ are disjoint segments of $\mathcal{S}$, by (ii) of Definition 3.a.6, $T_j = \bigcup_{r\in S_j}\sigma_r$ is a segment of $\mathcal{T}$ and $T_i \cap T_j = \emptyset$ for $i \neq j$. Therefore

$$\sum_{j=1}^{k}(\sum_{r\in S_j}(\sum_{s\in\sigma_r} a_s))^2 = \sum_{j=1}^{k}(\sum_{t\in T_j} a_t)^2.$$

Hence $\|Px\| \le \|x\|$ and $\|P\| = 1$.

(b) Let B be any branch of $\mathcal{T}$, then $B = \{t_1, t_2,\ldots\}$ where t_{i+1} is an offspring of t_i. Let $x \in P_B JT$, $x = \sum_{n=1}^{\infty} a_n \eta_{t_n}$, let $\{\xi_n\}$ be the summing basis in J and define $R : P_B JT \longrightarrow J$ by

$$Rx = \sum_{n=1}^{\infty} a_n \xi_n.$$

Then obviously R is an isometry.

We now establish another deep property of JT, namely its somewhat reflexivity; this theorem was stated in James [3] but the proof given there had a minor mistake; this was afterwards corrected by James himself and it is this version we present here.

Theorem 3.a.8. Let $\theta > \sqrt{2}$ and $\varepsilon > 0$. Then each infinite dimensional subspace of JT contains an infinite dimensional subspace H for which there is a norm $|\|\ \ \||$ given by an inner product such that for $x \in H$

$$(1 - \varepsilon)|\|x\|| \le \|x\| \le \theta|\|x\||.$$

Proof: Let X be the subspace of JT consisting of those members with finite support. As we show below, it suffices to prove the theorem for X; in this case we will prove in fact that every infinite dimensional subspace Y contains an infinite dimensional subspace H with an inner product norm $|\|\ \ \||$ such that for $x \in H$

(1) $$|\|x\|| \le \|x\| \le \theta|\|x\||.$$

Suppose this is true, and let Z be an infinite dimensional closed subspace of JT. By a well known theorem proved by S. Mazur (see e.g. Lindenstrauss and Tzafriri [1]), there exists a normalized basic sequence $\{z_n\}$ belonging to Z. Let $\varepsilon > 0$, $\delta = \frac{\varepsilon}{2-\varepsilon}$ and $\{z'_n\} \subset X$ be such that $\sum_{n=1}^{\infty}\|z_n - z'_n\| < \delta/2K$, where K is the basis constant of $\{z_n\}$. Then by Proposition 1.8, $\{z'_n\}$ is a basic sequence equivalent to $\{z_n\}$ via an isomorphism T, and it can be shown that

(2) $$\|T\| \le 1 + \delta \quad \text{and} \quad \|T^{-1}\| \le 1/(1 - \delta).$$

Now let $H' \subset \text{span}\{z'_n\} \subset X$ and $|\|\ \||$ be an inner product norm in H' satisfying (1). Let H be the completion in $[z'_n]$ of H' with the norm $|\|\ \||$; then for every $x \in H$, (1) also holds. Consider the space $T^{-1}H$ equipped with the inner product norm $|\|T^{-1}x\||_1 = |\|x\||$ for every $x \in H$. Using (1) and (2) we get for every $x \in H$,

$$\frac{1}{1+\delta}|\|T^{-1}x\||_1 = \frac{1}{1+\delta}|\|x\|| \le \frac{1}{1+\delta}\|x\| \le \|T^{-1}x\| \le$$

$$\le \frac{1}{1-\delta}\|x\| \le \frac{\theta}{1-\delta}|\|x\|| = \frac{\theta}{1-\delta}|\|T^{-1}x\||_1,$$

and this proves the theorem.

Next we proceed to prove the theorem for X.

Let Y be an infinite dimensional subspace of X and let Y^k be the subspace of Y whose members are zero at the nodes with level less than k.

For each x in X, let

$$[x]_k = \sup(\sum_{j=0}^{2^k-1}(\sum_{t\in B_j} x(t))^2)^{1/2},$$

where the sup is taken over all collections $\{B_j\}_{j=0}^{2^k-1}$ of pairwise disjoint k-branches.

We will show first that

$$\lim_{k\to\infty} (\inf\{[x]_k : x \in Y^k \text{ and } \|x\| = 1\}) = 0.$$

Suppose this is false. Let

$$\omega = \lim\sup_{k\to\infty} (\inf\{[x]_k : x \in Y^k \text{ and } \|x\| = 1\}) \text{ and suppose } \omega > 0.$$

Let $q_1, q_2, \ldots$ be a subsequence of $\mathbb{N}$ such that

$$\lim_{n\to\infty} (\inf\{[x]_{q_n} : x \in Y^{q_n} \text{ and } \|x\| = 1\}) = \omega.$$

Let

(3) $$0 < \varepsilon < \frac{\omega^2}{737}$$

and choose $L = q_{n_0}$ so that for $q_n \ge L$

(4) $$\omega^2 - \varepsilon < \inf\{[x]^2_{q_n} : x \in Y^{q_n} \text{ and } \|x\| = 1\} < \omega^2 + \varepsilon.$$

Now construct a subsequence of $\{q_n\}$ called $\{m(k)\}_{k=1}^{\infty}$ and a sequence $\{y^k\}$ such that $m(1) = L$, $y^k \in Y^{m(k)}$, $\|y^k\| = 1$ and such that the support of y^k is contained in nodes with levels in the interval $[m(k), m(k+1))$ and

(5) $$[y^k]^2_{m(k)} < \omega^2 + \varepsilon.$$

This is done as follows:
There exists $y^1 \in Y^{m(1)}$ with $\|y^1\| = 1$ and $[y^1]^2_{m(1)} < \omega^2 + \varepsilon$. The support of y^1 consists of nodes t such that $m(1) \le \mathrm{lev}(t) < n(1)$. Let $m(2)$ be the first q_n such that $q_n \ge n(1)$. There exists $y^2 \in Y^{m(2)}$ with $\|y^2\| = 1$ and $[y^2]^2 < \omega^2 + \varepsilon$. The support of y^2 consists of nodes t such that $m(2) \le \mathrm{lev}(t) < n(2)$. Let $m(3)$ be the first q_n such that $q_n \ge n(2)$ and so forth.

(6) For every k there are 2^L disjoint L-branches $B^k_0, \ldots, B^k_{2^L-1}$ such that B^k_n starts at node (L,n), $n = 0, \ldots, 2^L - 1$, and

$$[y^k]^2_L = \sum_{n=0}^{2^L-1} \Big(\sum_{t\in B^k_n} y^k(t)\Big)^2.$$

This is true because $\mathrm{supp}\, y^k$ is finite. Now, since $y^k(t) = 0$ if $\mathrm{lev}(t) < m(k)$, there are 2^L $m(k)$-branches $T^k_0, \ldots, T^k_{2^L-1}$ with first node $(m(k), p^k_n)$, $n = 0, \ldots, 2^L - 1$, such that $T^k_n \subset B^k_n$ and

(7) $$\omega^2 - \varepsilon < [y^k]^2_L = \sum_{n=0}^{2^L-1} \Big(\sum_{t\in T^k_n} y^k(t)\Big)^2 < \omega^2 + \varepsilon.$$

If $\{\beta_n\}_{n=0}^{2^L-1}$ is a set of L-branches where β_n starts at node (L, n) and $\beta_n \cap T^k_n = \varnothing$ for $n = 0, \ldots, 2^L - 1$, then by (5) and (7)

$$\omega^2 - \varepsilon < [y^k]^2_L + \sum_{n=0}^{2^L-1} \Big(\sum_{t\in\beta_n} y^k(t)\Big)^2 \le [y^k]^2_{m(k)} < \omega^2 + \varepsilon$$

and thus, again by (7),

(8) $$\sum_{n=0}^{2^L-1} \Big(\sum_{t\in\beta_n} y^k(t)\Big)^2 < 2\varepsilon.$$

For each $n = 0, \ldots, 2^L - 1$ and every $k < \kappa$ define

$$\vartheta_n(k, \kappa) = \begin{cases} r & \text{if there exists } r < 2^L \text{ with } (m(k), p^k_n) < (m(\kappa), p^\kappa_r), \\ 2^L & \text{otherwise.} \end{cases}$$

This is well defined since for $r \ne s$, $(m(\kappa), \rho^\kappa_r)$ and $(m(\kappa), \rho^\kappa_s)$ are on the disjoint L-branches B_r^{κ} and B_s^{κ}.

We will choose an increasing sequence of positive integers $I_1 = \{k_1, k_2, \ldots\}$ so that for $n = 0,\ldots,2^L - 1$ and for i,j,u with $i < j$, $i < u$, $\vartheta_n(k_i, k_j) = \vartheta_n(k_i, k_u)$.

Let $k_1 = 1$; then there exists an increasing sequence $I_1^1 = \{q_1^1, q_2^1, \ldots\}$ such that for $i < 2^L$ and $j, s = 1, 2, \ldots,$ $\vartheta_i(k_1, q_j^1) = \vartheta_i(k_1, q_{j+s}^1)$, taking one of the values $0, 1, 2, \ldots, 2^L$.

Let $k_2 = q_1^1$. Let $I_1^2 = \{q_1^2, q_2^2, \ldots\}$ be a subsequence of I_1^1 such that $\vartheta_i(k_2, q_j^2) = \vartheta_i(k_2, q_{j+s}^2)$, $i = 0,\ldots,2^L - 1$ and $j, s = 1, 2, \ldots$.

Let $k_3 = q_2^2$ and so forth.

Next we choose a subsequence $I_2 = \{h_1, h_2, \ldots\}$ of I_1 so that if $h_i, h_j \in I_2$, then $\vartheta_n(h_i, h_{i+k}) = \vartheta_n(h_j, h_{j+s})$ for $n = 0,\ldots,2^L - 1$, and $i, j, k, s = 1, 2, \ldots$.

This is done as follows: for $n = 1$ there is a subsequence $R_1 = \{r_1^1, r_2^1, \ldots\}$ of I_1 such that for $k, v \in I_1$, $k > r_i^1$, $v > r_j^1$ we have $\vartheta_1(r_i^1, k) = \vartheta_1(r_j^1, v)$, $i, j = 1, 2, \ldots$. There is also a subsequence $R_2 = \{r_j^2\}$ of R_1 such that for $k, v \in R_1$, $k > r_i^2$, $v > r_j^2$ we have $\vartheta_2(r_i^2, k) = \vartheta_2(r_j^2, v)$, $i, j = 1, 2, \ldots$, and so forth; let $I_2 = R_{2^L-1}$.

Suppose now that n is such that $\vartheta_n(h_i, h_j) = 2^L$ for $h_i, h_j \in I_2$ and $i < j$. Then no L-branch that contains $(m(k), p_n^k)$ for some $k \in I_2$ can contain any other $(m(\kappa), p_j^{\kappa})$ for $\kappa \in I_2$.

On the other hand, if n is such that $\vartheta_n(h_i, h_j) = r < 2^L$ for $i < j$, then

$$(m(h_i), p_n^{h_i}) < (m(h_{i+s}), p_r^{h_{i+s}}) \text{ for } s > 0$$

and

$$(m(h_{i+1}), p_n^{h_{i+1}}) < (m(h_{i+s}), p_r^{h_{i+s}}) \text{ for } s > 1.$$

Thus $(m(h_i), p_n^{h_i}) < (m(h_{i+1}), p_n^{h_{i+1}})$ and $\vartheta_n(h_i, h_j) = n$. Hence all the nodes $(m(h_i), p_n^{h_i})$, $i = 1, 2, \ldots$, are contained in an L-branch B_n.

For $k \in I_2$ and $0 \le r \le 2^L - 1$, if B_r^k is as in (6) and if B is any L-branch starting at node (L,r) we have

(9) $$|\sum_{t\in B} y^k(t)| \le |\sum_{t\in B^k_r} y^k(t)| \le [y^k]_L < \sqrt{\omega^2 + \varepsilon}.$$

In particular $|\sum_{t\in B_n} y^k(t)| \le [y^k]_L < \sqrt{\omega^2 + \varepsilon}$ for every $k \in I_2$.

Hence we may choose a subsequence $I_3 = \{\mu_j\}$ of I_2 such that for every n with $\vartheta_n(h_i, h_j) = n$ and for every $k, \kappa \in I_3$ and for every $0 \le r \le 2^L - 1$, if B_n is the branch defined above,

(10) $$|\sum_{t\in B_n} y^k(t) - \sum_{t\in B_n} y^\kappa(t)| < 2^{-L/2}\varepsilon^{1/2}$$

and

(11) $$||\sum_{t\in B^k_r} y^k(t)|^2 - |\sum_{t\in B^\kappa_r} y^\kappa(t)|^2| < 2^{-L}\varepsilon.$$

Suppose B is any branch that starts at node (L,n). Then by (9) and (11), for every $k, \kappa \in I_3$

(12) $$|\sum_{t\in B} y^k(t)|^2 \le |\sum_{t\in B^k_n} y^k(t)|^2 \le |\sum_{t\in B^\kappa_n} y^\kappa(t)|^2 + 2^{-L}\varepsilon.$$

For $\mu_r \in I_3$, $r = 1,\dots,9$, and if B is as above, consider

$$\left[\sum_{t\in B}(\sum_{r=1}^{9}(-1)^r y^{\mu_r}(t))\right]^2 = \left[\sum_{r=1}^{9}(-1)^r(\sum_{t\in B} y^{\mu_r}(t))\right]^2 = *.$$

We want to estimate $*$. Denote the term $|\sum_{t\in B} y^{\mu_r}(t)|$ by $\rho_B^{\mu_r}$ or $\Delta_B^{\mu_r}$ according as B contains at least one of the branch points $(m(\mu_r), p_s^{\mu_r})$, $s = 0,\dots,2^L - 1$, or not. Since B starts at node (L,n), and for every k, B^k_m starts at node (L,m), and $T^k_m \subset B^k_m$, it follows that $T^k_m \cap B = \emptyset$ if $m \ne n$. Therefore we have the following three cases:

(a) $T^{\mu_r}_s \cap B = \emptyset$ for every $r \le 9$ and for every $s = 0,\dots,2^L - 1$.

(b) There exists $r \le 9$ such that $T^{\mu_r}_n \cap B \ne \emptyset$ and n is such that $\vartheta_n(\mu_i, \mu_j) = n$ for $\mu_i, \mu_j \in I_3$. Then there exists $1 \le \kappa \le 9$ such that $(m(\mu_j), p_n^{\mu_j}) \in B$ for every $1 \le j \le \kappa$ and $(m(\mu_k), p_s^{\mu_k}) \notin B$ for $\kappa < k \le 9$ and $s \le 2^L - 1$.

(c) There exists $r_0 \le 9$ such that $T^{\mu_{r_0}}_n \cap B \ne \emptyset$ and n is such that $\vartheta_n(\mu_i, \mu_j) = 2^L$, $\mu_i, \mu_j \in I_3$. Then only $(m(\mu_{r_0}), p_n^{\mu_{r_0}}) \in B$ and $(m(\mu_k), p_s^{\mu_k}) \notin B$ for any $s \ne n$ or for $s = n$ and $k \ne r_0$.

Case (a)

$$* \le (\sum_{r=1}^{9} \Delta_B^{\mu_r})^2.$$

Case (b)

Because B and B_n are both L-branches starting at node (L,n), they coincide at least up to the node $(m(\mu_\kappa), p_n^{\mu_\kappa})$ and, since the support of y^{μ_r} is contained in nodes whose level belongs to the interval $[m(\mu_r), m(\mu_r + 1))$, for $1 \le r \le \kappa - 1$ we have

$$\sum_{t\in B} y^{\mu_r}(t) = \sum_{t\in B_n} y^{\mu_r}(t).$$

If $\kappa = 1$ let $r_0 = 1$ and go to case (c); if $1 < \kappa \le 9$ is odd, using (10) and (12) we get

$$* \le 2(\sum_{t\in B} y^{\mu_\kappa}(t))^2 + 4(\sum_{r=1}^{\kappa-1}(-1)^r(\sum_{t\in B_n} y^{\mu_r}(t)))^2 +$$

$$+ 4(\sum_{r=\kappa+1}^{9}(-1)^r(\sum_{t\in B} y^{\mu_r}(t)))^2 \le 2(\rho_B^{\mu_\kappa})^2 + 4(\frac{\kappa - 1}{2} 2^{-L/2}\varepsilon^{1/2})^2 +$$

$$+ 4(\sum_{r=\kappa+1}^{9} \Delta_B^{\mu_r})^2 \le 2(\rho_{B_n^{\mu_1}}^{\mu_1})^2 + 2^{-L+1}\varepsilon + 4(4\cdot 2^{-L/2}\varepsilon^{1/2})^2 + 4(\sum_{r=1}^{9} \Delta_B^{\mu_r})^2.$$

If $\kappa \le 9$ is even

$$* = |\sum_{r=1}^{\kappa-2}(-1)^r \sum_{t\in B} y^{\mu_r}(t) - \sum_{t\in B} y^{\mu_{\kappa-1}}(t) + \sum_{t\in B} y^{\mu_\kappa}(t) +$$

$$+ \sum_{r=\kappa+1}^{9}(-1)^r \sum_{t\in B} y^{\mu_r}(t)|^2 \le$$

$$\le 2(\rho_B^{\mu_{\kappa-1}} + \rho_B^{\mu_\kappa})^2 + 4(\frac{\kappa - 2}{2} 2^{-L/2})^2\varepsilon + 4(\sum_{r=\kappa+1}^{9} \Delta_B^{\mu_r})^2 \le$$

$$\le 8(\rho_{B_n^{\mu_1}}^{\mu_1})^2 + 2^{-L+3}\varepsilon + 4(4\cdot 2^{-L/2})^2\varepsilon + 4(\sum_{r=1}^{9} \Delta_B^{\mu_r})^2.$$

Case (c)

$$* = |(-1)^{r_0} \sum_{t\in B} y^{\mu_{r_0}}(t) + \sum_{r\neq r_0, r\le 9}(-1)^r \sum_{t\in B} y^{\mu_r}(t)|^2 \le$$

$$\le 2\ (\rho_B^{\mu_{r_0}})^2 + 2(\sum_{r\neq r_0, r\le 9} \Delta_B^{\mu_r})^2 \le 2(\rho_{B_n^{\mu_1}}^{\mu_1})^2 + 2^{-L+1}\varepsilon + 2(\sum_{r=1}^{9} \Delta_B^{\mu_r})^2.$$

Putting all three cases together we obtain

(13) $$* \le 8(\rho_{B_n^{\mu_1}}^{\mu_1})^2 + 8\cdot 2^{-L}\varepsilon + 64\cdot 2^{-L}\varepsilon + 4(\sum_{r=1}^{9} \Delta_B^{\mu_r})^2.$$

Therefore, summing (13) over any 2^L disjoint L-branches $\{\beta_i\}_{i=0}^{2^L-1}$, using the triangle inequality in $\mathbb{R}^{2^L}$, (8), (7) and (3) we get that

(14) $$[\sum_{k=1}^{9}(-1)^k y^{\mu_k}]_L^2 \le 8[y^{\mu_1}]_L^2 + 72\varepsilon + 4\sum_{i=0}^{2^L-1}(\sum_{r=1}^{9}\Delta_{\beta_i}^{\mu_r})^2 \le$$

$$\le 8[y^{\mu_1}]_L^2 + 72\varepsilon + 4\left(\sum_{r=1}^{9}\left(\sum_{i=0}^{2^L-1}(\Delta_{\beta_i}^{\mu_r})^2\right)^{1/2}\right)^2 \le 8[y^{\mu_1}]_L^2 + 72\varepsilon + 648\varepsilon \le$$

$$\le 8\omega^2 + 728\varepsilon < 9(\omega^2 - \varepsilon).$$

If $S_1^r, S_2^r,\ldots,S_{m_r}^r$ are disjoint segments with $S_i^r \subset \operatorname{supp} y^{\mu_r}$, $i = 1,\ldots,m_r$, such that

$$1 = \|y^{\mu_r}\|^2 = \sum_{s=1}^{m_r}(\sum_{t\in S_s^r} y^{\mu_r}(t))^2,$$

then since for $j\ne k$, $\operatorname{supp} y^{\mu_j} \cap \operatorname{supp} y^{\mu_k} = \emptyset$, we have that $\operatorname{supp} y^{\mu_k} \cap S_s^r = \emptyset$ for $k\ne r$. Hence

$$\|\sum_{k=1}^{9}(-1)^k y^{\mu_k}\|^2 \ge \sum_{r=1}^{9}\sum_{s=1}^{m_r}(\sum_{t\in S_s^r}\sum_{k=1}^{9}(-1)^k y^{\mu_k}(t))^2 =$$

$$= \sum_{r=1}^{9}\sum_{s=1}^{m_r}(\sum_{t\in S_s^r}(-1)^r y^{\mu_r}(t))^2 = 9.$$

But this together with (14) contradicts (4). Hence $\omega = 0$.

Let $\varepsilon > 0$. Since $\omega = 0$, using the construction following (4), there is a sequence of integers $\{m(k)\}$ and a sequence $\{y^k\}$ in X such that for each k, $\|y^k\| = 1$, the support of y^k is the set of nodes t such that $m(k) \le \operatorname{lev}(t) < m(k+1)$ and $[y^k]_{m(k)}^2 < 2^{-k}\varepsilon^2$.

Let $a_1, a_2,\ldots,a_n$ be a finite sequence of real numbers. Since the support of y^k is finite, there exist disjoint segments $S_1^k,\ldots,S_{n_k}^k$ such that $S_j^k \subset \operatorname{supp} y^k$, $j = 1,\ldots,n_k$, and

$$1 = \|y^k\|^2 = \sum_{r=1}^{n_k}(\sum_{t\in S_r^k} y^k(t))^2.$$

Therefore

$$\|\sum_{k=1}^{n}a_k y^k\|^2 \ge \sum_{k=1}^{n}\sum_{r=1}^{n_k}(\sum_{t\in S_r^k}\sum_{s=1}^{n}a_s y^s(t))^2 =$$

$$= \sum_{k=1}^{n}\sum_{r=1}^{n_k}(\sum_{t\in S_r^k}a_k y^k(t))^2 = \sum_{k=1}^{n}a_k^2\sum_{r=1}^{n_k}(\sum_{t\in S_r^k}y^k(t))^2 = \sum_{k=1}^{n}a_k^2.$$

Now choose a finite set $A_1,\ldots,A_p$ of disjoint segments such that

$$\|\sum_{k=1}^{n} a_k y^k\|^2 = \sum_{r=1}^{p}(\sum_{t\in A_r}\sum_{s=1}^{n} a_s y^s(t))^2 = **.$$

For each r let j_r be the first index k such that $A_r \cap \operatorname{supp} y^k \neq \emptyset$ and let k_r be the last such index. Then A_r is the union of at most three disjoint segments C_r, D_r and E_r, where $C_r = A_r \cap \operatorname{supp} y^{j_r}$, $E_r = A_r \cap \operatorname{supp} y^{k_r}$ if $k_r > j_r$, $E_r = \emptyset$ if $k_r = j_r$, $D_r = \emptyset$ if $k_r - j_r \le 1$, and otherwise D_r is a segment whose first node is of level $m(j_r + 1)$ and whose last node is of level $m(k_r) - 1$. Hence $A_r \cap \operatorname{supp} y^s \subset D_r$ for $j_r + 1 \le s \le k_r - 1$ and $A_r \cap \operatorname{supp} y^s = \emptyset$ for $s < j_r$ and $s > k_r$.

Using the inequalities

$$(a + b + c)^2 \le (2 + \varepsilon)(a^2 + b^2) + (1 + 2/\varepsilon)c^2$$

and

$$(\sum_{i=1}^{n} a_i)^2 \le \sum_{i=1}^{n} 2^i a_i^2$$

we obtain

$$** \le (2 + \varepsilon)(\sum_{r=1}^{p}(\sum_{t\in C_r}\sum_{s=1}^{n} a_s y^s(t))^2 + \sum_{r=1}^{p}(\sum_{t\in E_r}\sum_{s=1}^{n} a_s y^s(t))^2) +$$

$$+ (1 + 2/\varepsilon)\sum_{r=1}^{p}(\sum_{t\in D_r}\sum_{s=1}^{n} a_s y^s(t))^2 =$$

$$= (2 + \varepsilon)(\sum_{r=1}^{p}((\sum_{t\in C_r} a_{j_r} y^{j_r}(t))^2 + (\sum_{t\in E_r} a_{k_r} y^{k_r}(t))^2)) +$$

$$+ (1 + 2/\varepsilon)\sum_{r=1}^{p}(\sum_{s=1}^{n}\sum_{t\in D_r} a_s y^s(t))^2 \le$$

$$\le (2 + \varepsilon)(\sum_{s=1}^{n}(\sum_{\{r:j_r=s\}}(\sum_{t\in C_r} a_s y^s(t))^2 + \sum_{\{r:k_r=s\}}(\sum_{t\in E_r} a_s y^s(t))^2)) +$$

$$+ (1 + 2/\varepsilon)\sum_{s=1}^{n} 2^s \sum_{\{r:j_r+1=s\}}(\sum_{t\in D_r} a_s y^s(t))^2 \le$$

$$\le (2 + \varepsilon)\sum_{s=1}^{n}\|a_s y^s\|^2 + (1 + 2/\varepsilon)\sum_{s=1}^{n} 2^s[a_s y^s(t)]_{m(s)}^2 \le$$

$$\le (2 + \varepsilon)\sum_{s=1}^{n} a_s^2 + (1 + 2/\varepsilon)\sum_{s=1}^{n} a_s^2\varepsilon^2 \le (2 + 3\varepsilon + \varepsilon^2)\sum_{s=1}^{n} a_s^2.$$

Since ε is arbitrary, for every $\theta > \sqrt{2}$ there exists an infinite sequence $\{y^k\} \subset Y$ such that for all sequences of real numbers $\{a_i\}$

$$(\sum_{i=1}^{\infty} a_i^2)^{1/2} \le \|\sum_{i=1}^{\infty} a_i y^i\| \le \theta(\sum_{i=1}^{\infty} a_i^2)^{1/2},$$

and this proves the theorem.

Next we will discuss the primarity of JT. The road towards this fact, Theorem 3.a.18, is a rather long one and will require several preliminary results, mainly of a technical nature. We start with some elementary facts about trees and bounded functions defined on trees. The remainder of this section is based on Andrew [3].

Proposition 3.a.9. Let $\mathcal{T}$ be a tree and A a subset of $\mathcal{T}$. Then either A or $\mathcal{T}\setminus A$ contains a subtree of $\mathcal{T}$.

Proof: Let $\mathcal{T}$ be a tree, $A \subset \mathcal{T}$. Suppose A does not contain any subtree of $\mathcal{T}$. For any node v let $\mathcal{T}_v$ be as in Definition 3.a.1. There is a node $u_{0,0} \in \mathcal{T}\setminus A$. If $v_{1,0}$ and $v_{1,1}$ are the offspring of $u_{0,0}$, then by our assumption there exist $u_{1,0} \in (\mathcal{T}\setminus A) \cap \mathcal{T}_{v_{1,0}}$ and $u_{1,1} \in (\mathcal{T}\setminus A) \cap \mathcal{T}_{v_{1,1}}$.

We proceed inductively: Suppose $\{u_{i,j}\}$ for $j = 0,\dots,2^i - 1$, $i = 0,\dots,n$ have been constructed. Let $v_{i+1,2j}$ and $v_{i+1,2j+1}$ be the offspring of $u_{i,j}$. There exist nodes

$$u_{i+1,2j} \in (\mathcal{T}\setminus A) \cap \mathcal{T}_{v_{i+1,2j}} \quad \text{and} \quad u_{i+1,2j+1} \in (\mathcal{T}\setminus A) \cap \mathcal{T}_{v_{i+1,2j+1}}.$$

By construction $\{u_{i,j}\}$, for $i = 0, 1,\dots,$ $j = 0,\dots,2^i - 1$, is a subtree of $\mathcal{T}\setminus A$.

Proposition 3.a.10. Let $\mathcal{T}$ be a tree and $f : \mathcal{T} \longrightarrow \mathbb{R}$ a bounded function. Then for every $\varepsilon > 0$ there exists a node $t \in \mathcal{T}$ such that

$$A_t = \{s \in \mathcal{T} : |f(s) - f(t)| < \varepsilon\}$$

contains a subtree of $\mathcal{T}$.

Proof: Suppose this is false and therefore there is $\varepsilon > 0$ such that for every $t \in \mathcal{T}$, A_t does not contain any subtree. Let $\mathcal{T}_0 = \mathcal{T}$ and $s_1 = (0,0)$. By Proposition 3.a.9 there is a subtree $\mathcal{T}_1$ of $\mathcal{T}_0$ such that

$$\mathcal{T}_1 \subset B_1 = \{s \in \mathcal{T}_0 : |f(s_1) - f(s)| \geq \varepsilon\}.$$

Let $s_2 \in \mathcal{T}_1$. Suppose that proceeding inductively we have constructed trees $\mathcal{T}_0 \supset \mathcal{T}_1 \supset \dots \supset \mathcal{T}_n$ and nodes $s_i \in \mathcal{T}_{i-1}$ for $i = 1,\dots,n$ such that

$$\mathcal{T}_i \subset B_i = \{s \in \mathcal{T}_{i-1} : |f(s_i) - f(s)| \geq \varepsilon\}.$$

Let $s_{n+1} \in \mathcal{T}_n$. By assumption and by Proposition 3.a.9 there exists a subtree $\mathcal{T}_{n+1}$ of $\mathcal{T}_n$ such that $\mathcal{T}_{n+1} \subset B_{n+1}$. But then $\{f(s_i)\}_{i=1}^{\infty}$ is a sequence such that $|f(s_i) - f(s_j)| \geq \varepsilon$ for every $i \neq j$, which contradicts the fact that $\{f(s_i)\}_{i=1}^{\infty}$ must contain a convergent subsequence.

Corollary 3.a.11. Let $\mathcal{T}$ be a tree and $f : \mathcal{T} \longrightarrow \mathbb{R}$ a bounded function. Then for every $\varepsilon > 0$ there exists a subtree $\mathcal{S} \subset \mathcal{T}$ such that for every $s, t \in \mathcal{S}$, $|f(s) - f(t)| < \varepsilon$.

Proof: Let $\varepsilon > 0$. Then by Proposition 3.a.10 there exist $t_0 \in \mathcal{T}$ and a subtree $\mathcal{S} \subset \mathcal{T}$ such that $|f(s) - f(t_0)| < \varepsilon/2$ for every $s \in \mathcal{S}$.

Proposition 3.a.12. Let f be a bounded real valued function defined on a tree $\mathcal{T}$. Then for every $\varepsilon > 0$, there exists a subtree $\mathcal{S}$ of $\mathcal{T}$ such that for any branch B of $\mathcal{S}$

(a) $\lim_{t\to\infty; t\in B} f(t) = L_B$ exists,

(b) $\sum_{t\in B} |f(t) - L_B| < \varepsilon$.

Proof: We construct $\mathcal{S}$ applying the previous corollary several times.
Let $\mathcal{U}_{0,0} = \mathcal{T}$. There is a subtree $\mathcal{S}_{0,0}$ of $\mathcal{T}$ such that $|f(s) - f(t)| < \varepsilon/2$ for every $s, t \in \mathcal{S}_{0,0}$. Let $s_{0,0}$ be the vertex of $\mathcal{S}_{0,0}$. Suppose we have constructed trees $\mathcal{S}_{i,j}$ and a sequence $\{s_{i,j}\}$ of nodes such that for $j = 0,\dots,2^i - 1$, $i = 0,\dots,n$,

(i) $s_{i,j}$ is the vertex of $\mathcal{S}_{i,j}$,

(ii) $\mathcal{S}_{i,j} \subset \mathcal{S}_{i-1,j/2}$ if j is even,

(iii) $\mathcal{S}_{i,j} \subset \mathcal{S}_{i-1,(j-1)/2}$ if j is odd,

(iv) $\mathcal{S}_{i,j} \cap \mathcal{S}_{i,k} = \emptyset$ if $j \neq k$,

(v) $|f(s) - f(t)| < \varepsilon/2^{i+1}$ if $s, t \in \mathcal{S}_{i,j}$.

For $r = 0,\dots,2^{n+1} - 1$ we construct $\mathcal{S}_{n+1,r}$ as follows:
For $j = 0,\dots,2^n - 1$ let $u_{n+1,2j}$ and $u_{n+1,2j+1}$ be the offspring of $s_{n,j}$ in $\mathcal{S}_{n,j}$, and let $\mathcal{U}_{n+1,2j}$ and $\mathcal{U}_{n+1,2j+1}$ be the subtrees of $\mathcal{S}_{n,j}$ formed respectively by $u_{n+1,2j}$ and its descendants and by $u_{n+1,2j+1}$ and its descendants. Then by Corollary 3.a.11 there are subtrees

$$\mathcal{S}_{n+1,2j} \subset \mathcal{U}_{n+1,2j} \text{ and } \mathcal{S}_{n+1,2j+1} \subset \mathcal{U}_{n+1,2j+1}$$

such that $|f(s) - f(t)| < \varepsilon/2^{n+2}$ if $s,t \in \mathcal{S}_{n+1,2j}$ or if $s,t \in \mathcal{S}_{n+1,2j+1}$.
Let $s_{n+1,2j}$ and $s_{n+1,2j+1}$ be the vertices of $\mathcal{S}_{n+1,2j}$ and $\mathcal{S}_{n+1,2j+1}$ respectively.
Let $\mathcal{S}$ be the tree $\{s_{i,j}\}$.
If $B = \{t_k, t_{k+1}, \dots\}$ is a k-branch in $\mathcal{S}$ and if $j > i$,

$$|f(t_{k+i}) - f(t_{k+j})| < \varepsilon/2^{k+i+1}.$$

Hence $\{f(t_i)\}_{i=k}^{\infty}$ is a Cauchy sequence and $\lim_{t\to\infty;t\in B} f(t) = L_B$ exists.
Furthermore

$$\sum_{i=0}^{\infty} |f(t_{k+i}) - L_B| \le \sum_{i=0}^{\infty} (\varepsilon/2^{k+i+1}) = \varepsilon/2^k \le \varepsilon$$

and this finishes the proof.

To prove that JT is primary, Andrew [3] showed that for each bounded linear operator U on JT there exists a subspace X such that U (or I - U) acts as an isomorphism on X, X is isometric to JT and UX (or (I - U)X) is complemented in JT. To this end we need first some results about bounded linear operators on JT.

Proposition 3.a.13. Let $U : JT \to JT$ be a bounded linear operator, $\varepsilon > 0$, N an integer, $\mathcal{S}$ a subtree of $\mathcal{T}$ and $t_0 \in \mathcal{S}$. Then there is $t_1 \in \mathcal{S}$, $t_1 > t_0$, such that $\|P_N U\eta_{t_1}\| < \varepsilon$.

Proof: If this is false, there exist N, ε and t_0 such that for every $t_1 \in \mathcal{S}$ with $t_1 > t_0$, $\|P_N U\eta_{t_1}\| \ge \varepsilon$. Then, since

$$\|P_N U\eta_t\| = \|\sum_{(s:\mathrm{lev}(s)\le N)} \langle \eta_s^*, P_N U\eta_t\rangle \eta_s\|,$$

for any descendant $t \in \mathcal{S}$ of t_0, there exists t' with lev(t') ≤ N and

$$(1) \qquad |\langle \eta_{t'}^*, P_N U\eta_t\rangle| \ge \varepsilon/K,$$

where $K = 2^{N+1} - 1$ is the number of nodes of level ≤ N.
Thus for any L and any collection $\{t_r\}_{r=1}^{L}$ of descendants in $\mathcal{S}$ of t_0, if [ ] denotes the greatest integer function, there exists a node t' such that at least [L/K] of the t_r satisfy (1) for this node t'. Hence there is a choice of signs $\{\theta_r = \pm 1\}$ such that

(2) $$\|\sum_{r=1}^{L} P_N U(\theta_r \eta_{t_r})\| \geq \langle \eta^*_{t'}, \sum_{r=1}^{L} P_N U(\theta_r \eta_{t_r})\rangle \geq \frac{\varepsilon}{K}\left[\frac{L}{K}\right].$$

We may choose $\{t_r\}_{r=1}^{L}$ to be mutually incomparable with respect to the order on $\mathcal{T}$, in which case it follows from the definition of the norm in JT that

(3) $$\|\sum_{r=1}^{L} P_N U(\theta_r \eta_{t_r})\| \leq \|U\| \|\sum_{r=1}^{L} \theta_r \eta_{t_r}\| = \|U\| L^{1/2}.$$

Since (2) and (3) are contradictory for sufficiently large L, the proposition is proved.

Proposition 3.a.14. Let $U : JT \longrightarrow JT$ be a bounded linear operator, let $\varepsilon > 0$, $\mathcal{S}$ be a subtree of $\mathcal{T}$ and $t_0, t_1, \ldots, t_r$ mutually incomparable nodes of $\mathcal{S}$; let L and N be positive integers with $L \geq \max_i \mathrm{lev}_{\mathcal{T}}(t_i)$. Then there exists $t > t_0$, $t \in \mathcal{S}$ with $\mathrm{lev}_{\mathcal{T}}(t) > L$, such that

(a) $\|P_N U \eta_t\| < \varepsilon/2$.

If $M > L$ is an integer, then there exist segments $S_0, S_1, \ldots, S_r$ of $\mathcal{T}$ such that:

(b) For each i, S_i starts at node t_i, ends at level M and there is $t'_i \in \mathcal{S}$ with $t'_i > s$ for all $s \in S_i$.

(c) For each i, $\|P_{S_i} U \eta_t\| < \varepsilon$.

Proof: Let K satisfy $2^{-K/2}\|U\| < \varepsilon/2$. For each $i \leq r$ there are 2^K segments in $\mathcal{S}$ starting at t_i and ending at each of the 2^K descendants of t_i whose level in $\mathcal{S}$ is $\mathrm{lev}_{\mathcal{S}}(t_i) + K$. Consider any 2^K branches of $\mathcal{S}$ starting at t_i containing these segments. These branches of $\mathcal{S}$ in turn are contained in 2^K branches of $\mathcal{T}$ also starting at t_i: $B_i^1, B_i^2, \ldots, B_i^{2^K}$.

Let

$$A = \{u \in \mathcal{S} : \mathrm{lev}_{\mathcal{S}}(u) = \mathrm{lev}_{\mathcal{S}}(t_i) + K \text{ for some } 0 \leq i \leq r\}$$

$$\text{and } N_1 = \max(N, \max_{u \in A} \mathrm{lev}_{\mathcal{T}}(u)).$$

Let $s_0 \in \mathcal{S}$, $s_0 > t_0$, be such that $\mathrm{lev}_{\mathcal{T}}(s_0) \geq L$. By Proposition 3.a.13 there exists $t \in \mathcal{S}$, $t > s_0$, such that

(1) $$\|P_{N_1} U \eta_t\| < \varepsilon/2,$$

and hence (a) holds.

Define for $j = 1, \ldots, 2^K$

$$S_i^j = B_i^j \cap \{u \in \mathcal{T} : \mathrm{lev}_{\mathcal{T}}(u) \le M\}.$$

Then $t_i \in S_i^j$ and for each fixed $i \le r$, there exists $j(i) \le 2^K$ such that

(2) $$\|P_{S_i^{j(i)}} (I - P_{N_1})U\eta_t\| < \varepsilon/2.$$

For, if this were not the case, since

$$S_i^j \cap S_i^r \cap \{t \in \mathcal{T} : \mathrm{lev}_{\mathcal{T}}(t) > N_1\} = \emptyset$$

for $j \ne r$, then

$$\frac{\varepsilon^2}{4}2^K \le \sum_{j=1}^{2^K} \|P_{S_i^j} (I - P_{N_1})U\eta_t\|^2 \le \|(I - P_{N_1})U\eta_t\|^2 \le \|U\|^2 < \frac{\varepsilon^2}{4}2^K,$$

and this is a contradiction. Denoting $S_i^{j(i)}$ by S_i, we obtain by (1) and (2)

$$\|P_{S_i} U\eta_t\| \le \|P_{S_i} (I - P_{N_1})U\eta_t\| + \|P_{S_i} P_{N_1} U\eta_t\| \le \frac{\varepsilon}{2} + \|P_{N_1} U\eta_t\| < \varepsilon.$$

Thus (b) and (c) are satisfied.

We now come to the main step in our way towards proving that JT is primary. The proof of the next theorem is rather long since it is based on an inductive application of the previous proposition, and in order to make the proof understandable, we detail several steps of the construction.

Theorem 3.a.15. Let U be a bounded linear operator on JT. Then there exists a subspace X of JT such that

(I) X is isometric to JT,

(II) $U|_X$ (or $(I - U)|_X$) is an isomorphism,

(III) UX (respectively (I - U)X) is complemented in JT.

Proof: We will construct a subtree $\mathcal{S} \subset \mathcal{T}$ such that either $\{U\eta_t\}_{t\in\mathcal{S}}$ or $\{(I - U)\eta_t\}_{t\in\mathcal{S}}$ is equivalent to $\{\eta_t\}_{t\in\mathcal{T}}$ and has complemented span. The desired subspace is then $X = [\eta_t]_{t\in\mathcal{S}}$.

Let $V = I - U$. For each $t \in \mathcal{T}$, $t = (n,i)$, let B_t be the 0-branch containing t which consists of the interval $\{s \in \mathcal{T} : (0,0) \le s \le t\}$ and all nodes of the form $(n + k, 2^k i)$, $k \ge 1$. Choose $N_t > \mathrm{lev}_{\mathcal{T}}(t)$ such that

(1) $$\sum_{t\in\mathcal{T}} \|P_{N_t} U\eta_t - U\eta_t\| < \delta_1 \quad\text{and}\quad \sum_{t\in\mathcal{T}} \|P_{N_t} V\eta_t - V\eta_t\| < \delta_1,$$

where δ_1 will be determined later.

Let S_t be the segment

$$S_t = B_t \cap \{s \in \mathcal{T} : \operatorname{lev}_{\mathcal{T}}(s) \le N_t\},$$

and denote its last element by $\ell(t)$. We construct a subtree $\mathcal{T}_1$ of $\mathcal{T}$ inductively. Place (0,0) in $\mathcal{T}_1$ and assume the n-th level of $\mathcal{T}_1$ is already constructed. The (n + 1)st level of $\mathcal{T}_1$ consists of all nodes in $\mathcal{T}$ which are offspring in $\mathcal{T}$ of nodes $\ell(t)$, where t belongs to the n-th level of $\mathcal{T}_1$.

Let f_{B_t} and f_{S_t} be as in Definition 3.a.4 and $0 < \gamma < 1/2$.

For each $t \in \mathcal{T}$, $1 = \langle f_{B_t}, U\eta_t\rangle + \langle f_{B_t}, V\eta_t\rangle$; so

$$\langle f_{B_t}, U\eta_t\rangle \ge 1/2 \quad \text{or} \quad \langle f_{B_t}, V\eta_t\rangle \ge 1/2.$$

We may assume, increasing N_t if necessary, that

(2) $$\langle f_{S_t}, U\eta_t\rangle > \gamma \quad \text{or} \quad \langle f_{S_t}, V\eta_t\rangle > \gamma.$$

Define $\gamma_t = \langle f_{S_t}, U\eta_t\rangle$ and $\beta_t = \langle f_{S_t}, V\eta_t\rangle$. Let $A = \{t \in \mathcal{S}_1 : \gamma_t > \gamma\}$.

By Proposition 3.a.9 there exists a subtree $\mathcal{S}_2$ of $\mathcal{S}_1$ such that either $\mathcal{S}_2 \subset A$ or $\mathcal{S}_2 \subset \mathcal{S}_1 \setminus A$. We shall assume $\mathcal{S}_2 \subset A$, and hence discuss the operator U, rather than V.

By Proposition 3.a.12 there exists a subtree $\mathcal{S}_1$ of $\mathcal{T}_1$ such that for each branch B of $\mathcal{S}_1$

(3) $$\lim_{t\to\infty; t\in B} \gamma_t = \gamma_B \ge \gamma,$$

(4) $$\sum_{t\in B} |\gamma_t - \gamma_B| < \gamma/3.$$

For $t \in \mathcal{S}_2$ define $y_t = P_{N_t} U\eta_t$. Then

(5) $$\|y_t\| \le \|U\|.$$

The desired subtree $\mathcal{S} \subset \mathcal{S}_2$ will be constructed inductively by repeatedly using Proposition 3.a.14. For that we need the following concepts: we will say that a node t follows a segment S if $t > s$ for all $s \in S$, that a segment S_1 follows a segment S_2 ($S_1 \succ S_2$), if every node in S_1 follows S_2, and that a node t is between the segments S_1 and S_2, if t follows S_1 and $t < s$ for every $s \in S_2$.

Let $\{\varepsilon_t\}_{t\in\mathcal{S}_2}$ be a sequence such that $\varepsilon_t > 0$,

(6) $$\langle f_{S_t}, U\eta_t\rangle - \varepsilon_t = \langle f_{S_t}, y_t\rangle - \varepsilon_t > \gamma,$$

(7) $$\sum_{t\in B}|\gamma_t - \gamma_B| + \sum_{t\in\mathcal{S}_2}\varepsilon_t < \gamma/2 \text{ for every branch } B \text{ in } \mathcal{S}_1,$$

and

(8) $$\sum_{t\in\mathcal{S}_2}\varepsilon_t < \delta_2,$$

where δ_2 will be determined later.

Place the vertex of $\mathcal{S}_2$ in $\mathcal{S}$ and call it $t(0,0)$. Let $u_0(0,0)$ and $u_1(0,0)$ be two incomparable nodes in $\mathcal{S}_2$ following $\ell(t(0,0))$.

Let $L(0,0) = \max_i \text{lev}_{\mathcal{T}}(u_i(0,0))$. By Proposition 3.a.14, there exist $t(1,0) \in \mathcal{S}_2$, $t(1,0) > u_0(0,0)$ with $\text{lev}_{\mathcal{T}}(t(1,0)) > L(0,0)$, and a segment $S_1(1,0)$ in $\mathcal{T}$ starting at $u_1(0,0)$ such that $S_1(1,0)$ ends at level $N_{t(1,0)}$ and such that

(a) $\|P_{L(0,0)}y_{t(1,0)}\| < \varepsilon_{t(1,0)}/8$,

(b) $\|P_{S_1(1,0)}y_{t(1,0)}\| < \varepsilon_{t(1,0)}/8$.

There also exists a node $u_2(1,0) \in \mathcal{S}_2$ following $S_1(1,0)$. Let $S_0(1,0)$ be the interval starting at $u_0(0,0)$ and ending at $\ell(t(1,0))$ and let $u_0(1,0)$ and $u_1(1,0)$ be two incomparable nodes in $\mathcal{S}_2$ following $\ell(t(1,0))$. Put $t(1,0)$ in $\mathcal{S}$.

Let $L(1,0) = \max_i \text{lev}_{\mathcal{T}}\, u_i(1,0)$. By Proposition 3.a.14, there exists $t(1,1) \in \mathcal{S}_2$, $t(1,1) > u_2(1,0)$ with $\text{lev}_{\mathcal{T}}(t(1,1)) > L(1,0)$, and for $i=0,1$ there are segments $S_i(1,1)$ in $\mathcal{T}$ starting at $u_i(1,0)$ and ending at level $N_{t(1,1)}$ such that

(a) $\|P_{L(1,0)}y_{t(1,1)}\| < \varepsilon_{t(1,1)}/8$,

(b) $\|P_{S_i(1,1)}y_{t(1,1)}\| < \varepsilon_{t(1,1)}/8$.

For $i = 0, 1$, there exist nodes $u_i(1,1) \in \mathcal{S}_2$ following $S_i(1,1)$. Let $S_2(1,1)$ be the interval starting at $u_2(1,0)$ and ending at $\ell(t(1,1))$, and let $u_2(1,1)$ and $u_3(1,1)$ be two incomparable nodes in $\mathcal{S}_2$ following $\ell(t(1,1))$. Complete the first level by placing $t(1,1)$ in $\mathcal{S}$.

We will proceed one level further to illustrate the construction.

Let $L(1,1)=\max_i \text{lev}_{\mathcal{T}}(u_i(1,1))$. By Proposition 3.a.14 there is $t(2,0) \in\mathcal{S}_2$, $t(2,0) > u_0(1,1)$ with $\text{lev}_{\mathcal{T}} t(2,0) \geq L(1,1)$, and for $i = 1, 2, 3$ there are segments $S_i(2,0)$ in $\mathcal{T}$ starting at $u_i(1,1)$, ending at level $N_{t(2,0)}$ such that

(a) $\|P_{L(1,1)}y_{t(2,0)}\| < \varepsilon_{t(2,0)}/16$,

(b) $\|P_{S_i(2,0)}y_{t(2,0)}\| < \varepsilon_{t(2,0)}/16$.

For $i = 1, 2, 3$, there also exist nodes $u_{i+1}(2,0) \in \mathcal{S}_2$ following

$S_i(2,0)$. Let $S_0(2,0)$ be the interval starting at $u_0(1,1)$ and ending at $\ell(t(2,0))$ and let $u_0(2,0)$ and $u_1(2,0)$ be two incomparable nodes in $\mathcal{S}_2$ following $\ell(t(2,0))$. Put $t(2,0)$ in $\mathcal{S}$.

Let $L(2,0) = \max_i \mathrm{lev}_{\mathcal{T}}(u_i(2,0))$. By Proposition 3.a.14 there is $t(2,1)$ in $\mathcal{S}_2$, $t(2,1) > u_2(2,0)$ with $\mathrm{lev}_{\mathcal{T}} t(2,1) \geq L(2,0)$, and for $i = 0, 1, 3, 4$ there are segments $S_i(2,1)$ in $\mathcal{T}$ starting at $u_i(2,0)$ and ending at level $N_{t(2,1)}$ such that

(a) $\|P_{L(2,0)} y_{t(2,1)}\| < \varepsilon_{t(2,1)}/16$,

(b) $\|P_{S_i(2,1)} y_{t(2,1)}\| < \varepsilon_{t(2,1)}/16$.

For $i = 0, 1$ there exist nodes $u_i(2,1) \in \mathcal{S}_2$ following $S_i(2,1)$ and for $i = 3, 4$ nodes $u_{i+1}(2,1)$ following $S_i(2,1)$. Let $S_2(2,1)$ be the interval starting at $u_2(2,0)$ and ending at $\ell(t(2,1))$ and let $u_2(2,1)$ and $u_3(2,1)$ be two incomparable nodes in $\mathcal{S}_2$ following $\ell(t(2,1))$. Place $t(2,1)$ in $\mathcal{S}$.

Let $L(2,1) = \max_i \mathrm{lev}_{\mathcal{T}}(u_i(2,1))$. By Proposition 3.a.14 there exists a node $t(2,2) \in \mathcal{S}_2$, $t(2,2) > u_4(2,1)$ with $\mathrm{lev}_{\mathcal{T}} t(2,2) \geq L(2,1)$, and for $i = 0, 1, 2, 3, 5$, there are segments $S_i(2,2)$ in $\mathcal{T}$ starting at $u_i(2,1)$ and ending at level $N_{t(2,2)}$ such that

(a) $\|P_{L(2,1)} y_{t(2,2)}\| < \varepsilon_{t(2,2)}/16$,

(b) $\|P_{S_i(2,2)} y_{t(2,2)}\| < \varepsilon_{t(2,2)}/16$.

For $i = 0, 1, 2, 3$, there also exist nodes $u_i(2,2) \in \mathcal{S}_2$ following $S_i(2,2)$ and there is $u_6(2,2)$ following $S_5(2,2)$. Let $S_4(2,2)$ be the interval starting at $u_4(2,1)$ and ending at $\ell(t(2,2))$ and let $u_4(2,2)$ and $u_5(2,2)$ be two incomparable nodes in $\mathcal{S}_2$ following $\ell(t(2,2))$. Place $t(2,2)$ in $\mathcal{S}$.

Let $L(2,2) = \max_i \mathrm{lev}_{\mathcal{T}}(u_i(2,2))$. By Proposition 3.a.14 there exists a node $t(2,3) \in \mathcal{S}_2$, $t(2,3) > u_6(2,2)$ with $\mathrm{lev}_{\mathcal{T}} t(2,3) \geq L(2,2)$, and for $i = 0,\dots,5$ there are segments $S_i(2,3)$ in $\mathcal{T}$ starting at $u_i(2,2)$ and ending at level $N_{t(2,3)}$ such that

(a) $\|P_{L(2,2)} y_{t(2,3)}\| < \varepsilon_{t(2,3)}/16$,

(b) $\|P_{S_i(2,3)} y_{t(2,3)}\| < \varepsilon_{t(2,3)}/16$.

For $i = 0,\dots,5$, there also exist nodes $u_i(2,3) \in \mathcal{S}_2$ following $S_i(2,3)$. Let $S_6(2,3)$ be the interval starting at $u_6(2,2)$ and ending at $\ell(t(2,3))$. Place $t(2,2)$ in $\mathcal{S}$. This finishes the second level.

Continuing in this manner we construct the tree $\mathcal{S} = \{t(n,i)\}$, a sequence of integers $\{L(n,i)\}$ and a sequence of segments $\{S_j(n,i)\}$ in $\mathcal{T}$ for

$i = 0,\ldots,2^n-1$, $n = 0, 1, 2,\ldots$ and $j = 0,\ldots,2^n - 1 + i$, so that:

(i) $N_{t(n,i-1)} < L(n,i - 1) < N_{t(n,i)}$ for $i = 0,\ldots 2^n - 1$, $n = 1, 2,\ldots$.

Here $t(n,-1)$ means $t(n - 1,2^{n-1} - 1)$.

(ii) $S_j(n,i)$ starts at a level less than or equal to $L(n, i - 1)$ and ends at level $N_{t(n,i)}$.

(iii) $\ell(t(n,i))$ is the last node of $S_{2i}(n,i)$.

(iv) $S_j(n,i + 1)$ follows $S_j(n,i)$ if $j \le 2i$ and $S_j(n,0)$ follows $S_j(n - 1,2^{n-1}- 1)$ if $j \le 2^n - 2$.

(v) $S_{j+1}(n,i + 1)$ follows $S_j(n,i)$ if $j \ge 2i$ and $S_{2^n-1}(n,0)$ follows $S_{2^n-2}(n - 1,2^{n-1} - 1)$.

(vi) $\|P_{L(n,i-1)}y_{t(n,i)}\| < \varepsilon_{t(n,i)}/2^{n+2}$.

(vii) $\|P_{S_j(n,i)}y_{t(n,i)}\| < \varepsilon_{t(n,i)}/2^{n+2}$ if $j \ne 2i$.

Define the following interval:

$$S_0(0,0) = S_{t(0,0)}.$$

For $n \ge 1$, $i \ge 1$ and $j \le 2i - 2$ let

$$T_j(n,i) = S_j(n,i) \cup \{t \in \mathcal{T} : t \text{ is between } S_j(n,i - 1) \text{ and } S_j(n,i)\}.$$

For $n \ge 1$, $i \ge 1$ and $2i - 1 \le j \le 2^n + i - 1$ let

$$T_j(n,i) = S_j(n,i) \cup \{t \in \mathcal{T} : t \text{ is between } S_{j-1}(n,i - 1) \text{ and } S_j(n,i)\}.$$

For $n \ge 1$, $j \le 2^n - 2$ let

$$T_j(n,0) = S_j(n,0) \cup \{t \in \mathcal{T} : t \text{ is between } S_j(n-1,2^{n-1}- 1) \text{ and } S_j(n,0)\}.$$

For $n \ge 1$ let

$$T_{2^n-1}(n,0) =$$

$$= S_{2^n-1}(n,0) \cup \{t \in \mathcal{T} : t \text{ is between } S_{2^n-2}(n-1,2^{n-1}-1) \text{ and } S_{2^n-1}(n,0)\}.$$

Then $T_j(n,i)$ begins at level $N_{t(n,i-1)} + 1$ and ends at level $N_{t(n,i)}$, $\{T_j(n,i)\}$ has properties (iii) to (v) and by (vi) and (vii)

(viii) $\|P_{T_j(n,i)}y_{t(n,i)}\| < \varepsilon_{t(n,i)}/2^{n+1}$ for $j = 0,\ldots,2^n - 1 + i$, $j \ne 2i$,

$i = 0,\ldots,2^n - 1$, $n = 0, 1\ldots$.

From (iv) and (v) we obtain for $i < 2^{n+1} - 1$

$$T_{2i}(n+1,i) \succ T_{2i-1}(n+1,i-1) \succ\ldots\succ T_i(n+1,0) \succ T_i(n,2^{n-1}-1) \succ$$

$$\succ T_i(n,2^{n-1}-2) \succ\ldots\succ T_i(n,[i/2]+1) \succ T_{2[i/2]}(n,[i/2])$$

and if $i = 2^{n+1} - 1$

$$T_{2i}(n+1,i) = T_{2^{n+2}-2}(n+1,2^{n+1}-1) \succ T_{2^{n+2}-3}(n+1,2^{n+1}-2) \succ\ldots\succ$$

$$\succ T_{2^{n+1}-1}(n+1,0) \succ T_{2^{n+1}-2}(n,2^n-1) = T_{2[i/2]}(n,[i/2]).$$

Since $S_{t(n+1,i)} - S_{t(n,[i/2])}$ is the interval

$$(\ell(t(n,[i/2])), \ell(t(n+1,i))],$$

we obtain

$$\text{(ix)} \qquad S_{t(n+1,i)} - S_{t(n,[i/2])} = \bigcup_{r=1}^{2^n-1-[i/2]} T_i(n,[i/2]+r) \cup \bigcup_{k=0}^{i} T_{i+k}(n+1,k),$$

where the first union is empty if $2^n - 1 - [i/2] = 0$.

Observe that the only interval of the form $T_{2k}(m,k)$ appearing in (ix) is $T_{2i}(n+1,i)$.

Let

$$\mathfrak{A}^i_{n+1} = \{S : S \text{ is a segment in } \mathcal{T} \text{ starting at level } N_{t(n+1,i-1)} + 1,$$
$$\text{ending at level } N_{t(n+1,i)}, \text{ and } S \neq T_j(n+1,i),\ j = 0,\ldots,2^n - 1 + i\},$$

let $z_{t(0,0)} = y_{t(0,0)}$ and for $n = 0, 1,\ldots,\ i = 0,\ldots,2^n - 1$,

$$z_{t(n+1,i)} = P_{T_{2i}(n+1,i)} y_{t(n+1,i)} + \textstyle\sum_{S\in\mathfrak{A}^i_{n+1}} P_S y_{t(n+1,i)}.$$

Then

$$y_{t(n+1,i)} - z_{t(n+1,i)} = P_{N_{t(n+1,i-1)}} y_{t(n+1,i)} + \textstyle\sum_{j=0,\, j\neq 2i}^{2^{n+1}-1+i} P_{T_j(n+1,i)} y_{t(n+1,i)}$$

and by (i), (vi), (viii) and (8)

$$\text{(9)} \qquad \textstyle\sum_{t(n,i)\in\mathcal{S}} \|y_{t(n,i)} - z_{t(n,i)}\| < \sum_{t(n,i)\in\mathcal{S}} \varepsilon_{t(n,i)} < \delta_2.$$

Also, if for $t \in \mathcal{S}$ we define γ'_t by $\gamma'_t = \langle f_{S_t}, z_t\rangle$, we get by (5) and (6)

$$\text{(10)} \quad \begin{cases} \gamma'_{t(n,i)} = \langle f_{S_{t(n,i)}}, z_{t(n,i)}\rangle \geq \langle f_{S_{t(n,i)}}, y_{t(n,i)}\rangle - \varepsilon_{t(n,i)} > \gamma \\ \text{and } \gamma'_{t(n,i)} \leq \|f_{S_{t(n,i)}}\| \|z_{t(n,i)}\| \leq \|y_{t(n,i)}\| < \|U\|. \end{cases}$$

For $t(n,i) \in \mathcal{S}$, $(n,i) \neq (0,0)$, let $S'_{t(n,i)} = S_{t(n,i)} - S_{t(n-1,[i/2])}$. Then by (ix), $\langle f_{S'_{t(n,i)}}, z_{t(m,k)}\rangle = \langle f_{S'_{t(n,i)}}, P_{T_{2k}(m,k)} z_{t(m,k)}\rangle$

and therefore

$$(11) \qquad \langle f_{S'_{t(n,i)}}, z_{t(m,k)}\rangle = \begin{cases} \gamma'_{t(n,i)} & \text{if } (m,k) = (n,i), \\ 0 \text{ otherwise.} \end{cases}$$

To show that $\{z_t\}_{t\in\mathcal{S}}$ is equivalent to $\{\eta_t\}_{t\in\mathcal{S}}$ we need the following: For $t \in \mathcal{T}\setminus\mathcal{S}$ define γ'_t as $\gamma'_t = \gamma_t$.

Since $\|y_{t(n,i)} - z_{t(n,i)}\| < \varepsilon_{t(n,i)}$ we have $|\gamma_t - \gamma'_t| < \varepsilon_t$ for every $t \in \mathcal{T}$, and by (3), (4) and (7) it follows that for every branch B in $\mathcal{S}_1$, $\lim_{t\to\infty;t\in B} \gamma'_t = \gamma_B$ exists and $\sum_{t\in B}|\gamma'_t - \gamma_B| < \gamma/2$ if δ_2 is small enough. By the last condition the operator T_B on $[\eta_t]_{t\in B}$ defined for $t \in B$ by $T_B\eta_t = (\gamma'_t/\gamma_B)\eta_t$ satisfies $\|I - T_B\| < 1/2$. In fact if $B = \{t_1, t_2,\dots\}$, writing a_i and γ'_i instead of a_{t_i} and γ'_{t_i} we get

$$\|(I - T_B)\textstyle\sum_{t\in B}a_t\eta_t\| = \|\sum_{t\in B}(1 - \gamma'_t/\gamma_B)a_t\eta_t\| = \|\sum_{t\in B}a_t(\gamma_B - \gamma'_t)\gamma_B^{-1}\eta_t\| =$$

$$= \sup\gamma_B^{-1}\Big(\textstyle\sum_{n=1}^{k}\big(\sum_{i=p_n+1}^{p_{n+1}}a_i(\gamma_B - \gamma'_i)\big)^2\Big)^{1/2} \le$$

$$\le \gamma_B^{-1}\sup\Big(\textstyle\sum_{n=1}^{k}\big(\max_{p_n+1\le i\le p_{n+1}} a_i^2\big)\big(\sum_{i=p_n+1}^{p_{n+1}}|\gamma_B - \gamma'_i|\big)^2\Big)^{1/2} <$$

$$< \gamma(2\gamma_B)^{-1}\big(\textstyle\sum_{t\in B}a_t^2\big)^{1/2} \le (1/2)\|\sum_{t\in B}a_t\eta_t\|,$$

where the sup is taken over all $p_1 <\dots< p_{k+1}$ and the last inequality follows from the definition of the norm $\|\ \|_J$ and by Proposition 3.a.7(b). Hence T_B is invertible and

$$\|T_B^{-1}\| - 1 \le \|T_B^{-1} - I\| \le \|T_B^{-1}\|\|I - T_B\| < (1/2)\|T_B^{-1}\|.$$

Therefore

$$(12) \qquad \|T_B^{-1}\| < 2.$$

Finally we can prove the equivalence of $\{z_t\}$ and $\{\eta_t\}$ for $t \in \mathcal{S}$. Let $\{a_t\}_{t\in C}$ be a finite set of scalars, where $C \subset \mathcal{S}$. There exist disjoint segments $S_1,\dots,S_k$ in $\mathcal{S}_1$ such that

$$\|\textstyle\sum_{t\in C}a_t\eta_t\| = \big(\sum_{j=1}^{k}(\sum_{s\in S_j\cap C} a_s)^2\big)^{1/2}.$$

Furthermore, there exist disjoint branches $B_1,\dots,B_r$ in $\mathcal{T}$ such that each S_j is a subset of some B_i. For $t \in \mathcal{S}$, if $z_t = \sum_{s\in\mathcal{S}}b^t_s\eta_s$ then by (11) $\gamma'_t = \sum_{s\in S'_t}b^t_s$. Using (12) and the fact that $\gamma \le \gamma_B$ for every branch B in

$\mathcal{T}$,

$$\|\sum_{t\in C} a_t\eta_t\| = (\sum_{j=1}^{r}(\sum_{\{i:S_i \subset B_j\}} (\sum_{t\in S_i \cap C} a_t)^2))^{1/2} =$$

$$= (\sum_{j=1}^{r}\|\sum_{t\in B_j \cap C} a_t\eta_t\|^2)^{1/2} \le \frac{1}{\gamma} (\sum_{j=1}^{r}\|\sum_{t\in B_j \cap C} \gamma_{B_j} a_t\eta_t\|^2)^{1/2} =$$

$$= \frac{1}{\gamma} (\sum_{j=1}^{r}\|\sum_{t\in B_j \cap C} a_t\gamma'_t T_{B_j}^{-1}\eta_t\|^2)^{1/2} \le \frac{2}{\gamma}(\sum_{j=1}^{r}\|\sum_{t\in B_j \cap C} a_t\gamma'_t\eta_t\|^2)^{1/2} =$$

$$= \frac{2}{\gamma}(\sum_{j=1}^{r}\sum_{\{u:S''_u \subset B_j\}}(\sum_{t\in S''_u \cap C} a_t\gamma'_t)^2)^{1/2} =$$

$$= \frac{2}{\gamma}(\sum_{j=1}^{r}\sum_{\{u:S''_u \subset B_j\}}(\sum_{t\in S''_u \cap C} a_t \sum_{s\in S'_t} b_s^t)^2)^{1/2} \le \frac{2}{\gamma}\|\sum_{t\in C} a_t \sum_{s\in S'_t} b_s^t\eta_s\| =$$

$$= \frac{2}{\gamma}\|P_{S'_t}\sum_{t\in C} a_t z_t\| \le \frac{2}{\gamma}\|\sum_{t\in C} a_t z_t\|,$$

for some choice of disjoint segments S''_u.

Since $\{z_t\}$ is a block basic sequence, using (1) and (9), if δ_1 and δ_2 are small enough, applying Proposition 1.8 we get that for $t \in \mathcal{S}$, $\{z_t\}$ is equivalent to $\{y_t\}$ and $\{y_t\}$ is equivalent to $\{U\eta_t\}$ and that

$$\|\sum_{t\in\mathcal{S}} a_t z_t\| \le 4\|\sum_{t\in\mathcal{S}} a_t U\eta_t\|.$$

Thus for any scalar sequence $\{a_t\}$, we have

$$\|\sum_{t\in\mathcal{S}} a_t\eta_t\| \le \frac{2}{\gamma} \|\sum_{t\in\mathcal{S}} a_t z_t\| \le \frac{8\|U\|}{\gamma} \|\sum_{t\in\mathcal{S}} a_t\eta_t\|, \tag{13}$$

so that $\{\eta_t\}_{t\in\mathcal{S}}$ and $\{z_t\}_{t\in\mathcal{S}}$ are equivalent. Hence we have proved that $\{\eta_t\}_{t\in\mathcal{S}}$ and $\{U\eta_t\}_{t\in\mathcal{S}}$ are equivalent. Thus U acts as an isomorphism on $X = [\eta_t]_{t\in\mathcal{S}}$. But by Proposition 3.a.7 X is isometric to JT and complemented in JT, hence (I) and (II) hold.

To see that $[U\eta_t]_{t\in\mathcal{S}}$ is complemented, we note that there are no gaps in $\mathcal{T}$ between the S'_t. In fact, let $t \in \mathcal{T}$; then using (ix):

if $t(n,[i/2]) < t < \ell(t(n,[i/2])) < t(n+1,i)$, then $t \in S'_{t(n,[i/2])}$;

if $t(n,[i/2]) < \ell(t(n,[i/2])) < t < t(n+1,i)$, then $t \in S'_{t(n+1,i)}$.

Let P be the projection of JT onto $[\eta_t]_{t\in\mathcal{S}}$ defined in the proof of Proposition 3.a.7, using the segments S'_t. Then, if $x \in [z_t]$ with $x = \sum_{t\in\mathcal{S}} a_t z_t$, we have

$$Px = \sum_{t\in\mathcal{S}} a_t(\sum_{s\in S'_t} b_s^t)\eta_t = \sum_{t\in\mathcal{S}} a_t\gamma'_t\eta_t.$$

Let $D \subset \mathcal{S}$ be a finite set. Then there exists a finite number of branches

$B_1,\dots,B_r$ in $\mathcal{T}$ such that

$$\|\sum_{t\in D}(a_t/\gamma'_t)\eta_t\| = (\sum_{j=1}^r\|\sum_{t\in D\cap B_j}(a_t/\gamma'_t)\eta_t\|^2)^{1/2}.$$

Hence using (13) and recalling that $\gamma \le \gamma_B$ and $\|T_B^{-1}\| \le 2$ for every branch $B \subset \mathcal{S}_1$, we get

$$\|\sum_{t\in D}(a_t/\gamma'_t)z_t\| \le 4\|U\|\|\sum_{t\in D}(a_t/\gamma'_t)\eta_t\| = 4\|U\|(\sum_{j=1}^r \|\sum_{t\in D\cap B_j}(a_t/\gamma'_t)\eta_t\|^2)^{1/2}$$

$$= 4\|U\|(\sum_{j=1}^r\|\sum_{t\in D\cap B_j}(\gamma_{B_j})^{-1}T_{B_j}^{-1}(a_t\eta_t)\|^2)^{1/2} \le \frac{8}{\gamma}\|U\|(\sum_{j=1}^r\|\sum_{t\in D\cap B_j}a_t\eta_t\|^2)^{1/2}$$

$$\le \frac{8}{\gamma}\|U\|\|\sum_{t\in D}a_t\eta_t\|.$$

Thus if we define $R(\sum_{t\in\mathcal{S}}a_t\eta_t) = \sum_{t\in\mathcal{S}}(a_t/\gamma'_t)z_t$, then $[z_t]_{t\in\mathcal{S}}$ is complemented in JT by RP. Hence, taking δ_1 and δ_2 even smaller than before if necessary, applying Proposition 1.8 again, we get that $[U\eta_t]_{t\in\mathcal{S}}$ is complemented in JT.

Taking U as a projection in Theorem 3.a.15, we obtain the following corollary.

Corollary 3.a.16. If JT = Y ⊕ Z then Y (or Z) is isomorphic to JT ⊕ W for some subspace W of Y (or Z).

Proof: Let P : JT ⟶ Y be an onto projection. Applying Theorem 3.a.15 we may assume that there exists a subspace X of JT with PX isomorphic to JT and such that PX is complemented in JT via a projection Q. Then, since PX ⊂ Y, $Q|_Y$ is a projection from Y onto PX. Hence Y = PX ⊕ W for some subspace W of Y.

In the following lemma we give another representation of JT. This enables us to prove that JT is isomorphic to its square, which is also an essential difference between JT and J.

Lemma 3.a.17. Let $B \subset \mathcal{T}$ be the branch $\{(n,0) : n = 0, 1, 2,\dots\}$. Then JT is isomorphic to $\mathcal{JT} = P_B JT \oplus (\sum_{n=1}^\infty Q_{n,1}JT)_{\ell_2}$, where P_B and $Q_{n,1}$ are as in Definition 3.a.4.

Proof: For $(n,i) \in \mathcal{T}$ let

$$\mathcal{T}_{n,i} = \{t \in \mathcal{T} : t \geq (n,i)\}$$

as in Definition 3.a.1 and let $S_{n,i}$ be the interval starting at (0,0) and ending at (n,i). Let $(m_{n,i},0)$ be the last element of the segment $S_{n,i} \cap B$. If $(m_{n,i},0) \neq (n,i)$ then $(m_{n,i}+1,1) \in S_{n,i}$. Hence either $(n,i) \in B$ or $(n,i) \in \mathcal{T}_{m_{n,i}+1,1}$.

Let $C \subset \mathcal{T}$ be a finite set. Since for n, m = 1, 2,..., $B \cap \mathcal{T}_{n,1} = \emptyset$, and $\mathcal{T}_{n,1} \cap \mathcal{T}_{m,1} = \emptyset$ if $n \neq m$,

$$\|\textstyle\sum_{t\in C} a_t\eta_t\|^2_{\mathcal{JT}} = \|\sum_{t\in C\cap B} a_t\eta_t\|^2 + \sum_{n=1}^{\infty}\|\sum_{t\in\mathcal{T}_{n,1}\cap C} a_t\eta_t\|^2 \leq \|\sum_{t\in C} a_t\eta_t\|^2.$$

On the other hand there is a finite number of disjoint intervals $S_1,\dots,S_r$ such that

$$\|\textstyle\sum_{t\in C} a_t\eta_t\| = (\sum_{j=1}^{r}\|\sum_{t\in S_j\cap C} a_t\eta_t\|^2)^{1/2}.$$

Let $B \cap S_j = T_j$ and $S_j\setminus T_j = T'_j$. Suppose $T'_j \neq \emptyset$ and let s(j) be its first element. Then, by the above arguments, there exists an integer r_j such that $s(j) \in \mathcal{T}_{r_j,1}$ and therefore $T'_j \subset \mathcal{T}_{r_j,1}$. Let $\{\nu_1,\dots,\nu_s\}$ be the set of different r_j. Then for every $j = 1,\dots,r$ such that $T'_j \neq \emptyset$, there is some $k \leq s$ with $T'_j \subset \mathcal{T}_{\nu_k,1}$. Thus

$$\|\textstyle\sum_{t\in C} a_t\eta_t\| = (\sum_{j=1}^{r}\|\sum_{t\in S_j\cap C} a_t\eta_t\|^2)^{1/2} \leq$$

$$\leq (\textstyle\sum_{j=1}^{r}\|\sum_{t\in T_j\cap C} a_t\eta_t\|^2)^{1/2} + (\sum_{j=1}^{r}\|\sum_{t\in T'_j\cap C} a_t\eta_t\|^2)^{1/2} \leq$$

$$\leq \|\textstyle\sum_{t\in C\cap B} a_t\eta_t\| + (\sum_{k=1}^{s}\|Q_{\nu_k,1}\sum_{t\in C} a_t\eta_t\|^2)^{1/2} \leq \sqrt{2}\|\sum_{t\in C} a_t\eta_t\|_{\mathcal{JT}}.$$

This finishes the proof.

At last we have reached our goal:

Theorem 3.a.18. The James tree space JT is primary.

Proof: By Lemma 3.a.17 and Proposition 3.a.7

$$JT \approx P_B JT \oplus (\textstyle\sum_{n=1}^{\infty} Q_{n,1} JT)_{\ell_2} \approx J \oplus (\sum_{n=1}^{\infty} JT)_{\ell_2}.$$

From this it follows that JT is isomorphic to its square, since

$$JT \approx J \oplus (\sum_{n=1}^{\infty} JT)_{\ell_2} \approx JT \oplus J \oplus (\sum_{n=1}^{\infty} JT)_{\ell_2} \approx JT \oplus JT.$$

Now, if $JT = Y \oplus Z$, by Corollary 3.a.16 we may assume that $Y \approx W \oplus JT$. Then using the Pelczynski decomposition method

$$Y \approx W \oplus JT \approx W \oplus JT \oplus JT \approx Y \oplus JT \approx Y \oplus (\sum_{n=1}^{\infty} JT)_{\ell_2} \oplus J \approx$$

$$\approx Y \oplus (\sum_{n=1}^{\infty} Y \oplus Z)_{\ell_2} \oplus J \approx (\sum_{n=1}^{\infty} Y \oplus Z)_{\ell_2} \oplus J \approx JT.$$

As we have seen, two of the main properties of JT proved in this section, namely the somewhat reflexivity and the primarity, are shared by J; in fact the somewhat reflexivity of J is inherited from JT, although it was proved independently in Section 2.d.

3.b. The fixed point property

In this section we will show that JT has the fixed point property, which is a relevant subject for the theory of operator equations. In order to do this, we shall in fact show that JT has weak normal structure, which by a theorem by Kirk [1] implies the above assertion. The concept of normal structure was introduced by Brodskii and Milman [1] and turned out to be an important tool in the fixed point theory of non-expansive mappings in Banach spaces. The proof of JT having weak normal structure is based on Khamsi [2].

Definition 3.b.1. Let K be a non-empty weakly compact convex subset of a Banach space X. We say that K has the fixed point property, if every $T : K \longrightarrow K$ satisfying $\|Tx - Ty\| \leq \|x - y\|$ for all x, $y \in K$ has a fixed point.

We say that X has the fixed point property, if every weakly compact convex subset of X has the fixed point property.

The notion of weak normal structure is based on those of diametral point and diametral sequence, which we define next.

Definition 3.b.2. A point x of a bounded convex subset K of a Banach space X is called diametral if

$$\sup\{\|y - x\| : y \in K\} = \text{diam } K,$$

where diam K denotes the diameter of K.

A non-constant bounded sequence $\{x_n\}$ in X is called diametral if

$$\lim_{n\to\infty} d(x_n, \text{co}\{x_i : i < n\}) = \text{diam }\{x_i\}_{i=1}^{\infty},$$

where coA denotes the convex hull of A and $d(x, A)$ denotes the distance from x to A.

Lemma 3.b.3. Any subsequence of a diametral sequence $\{x_i\}_{i=1}^{\infty}$ is also diametral.

Proof: Let $c = \text{diam }\{x_i\}_{i=1}^{\infty}$ and $\{x_{n_k}\}_{k=1}^{\infty}$ be a subsequence of $\{x_i\}_{i=1}^{\infty}$. Since

$$c \geq d(x_{n_k}, \text{co}\{x_{n_i} : i < k\}) \geq d(x_{n_k}, \text{co}\{x_i : i < n_k\})$$

the result follows.

Proposition 3.b.4. If $\{x_i\}$ is a diametral sequence with $c = \text{diam}\{x_i\}_{i=1}^{\infty}$, then every x in the closure of $\text{co}\{x_i\}_{i=1}^{\infty}$ is diametral and

$$\lim_{n\to\infty} \|x_n - x\| = c.$$

Proof: Let $x \in \text{co}\{x_i\}_{i=1}^{\infty}$, then there exists N_0 such that for every $n > N_0$ $x \in \text{co}\{x_i\}_{i=1}^{n}$. Let $n > N_0$ and suppose $x = \sum_{i=1}^{n}\lambda_i x_i$, $\sum_{i=1}^{n}\lambda_i = 1$. Then

(1)
$$c \geq \sum_{i=1}^{n}\lambda_i\|x_{n+1} - x_i\| \geq \|x_{n+1} - \sum_{i=1}^{n}\lambda_i x_i\| =$$

$$= \|x_{n+1} - x\| \geq d(x_{n+1}, \text{co}\{x_i\}_{i=1}^{n}) \geq c,$$

and passing to the limit as n tends to ∞ we get that x is diametral, since the diameter of a set is equal to the diameter of its convex hull.

If x is in the closure of $\text{co}\{x_i\}_{i=1}^{\infty}$, then it follows easily that

$$\lim_{n\to\infty} \|x_n - x\| = c.$$

If in (1) we take $x = x_n$ and pass to the limit, we obtain

(2)
$$\lim_{n\to\infty} \|x_{n+1} - x_n\| = \text{diam }\{x_i\}_{i=1}^{\infty}.$$

We now give the definitions of normal and weak normal structure:

Definition 3.b.5. We say that a bounded convex subset K of a Banach space X has normal structure, if every convex subset C of K containing more than one point contains a non-diametral point.

The space X has normal structure if every non-trivial bounded convex subset of X has normal structure.

If every weakly compact convex subset K of X which contains more than one point has normal structure, we say that X possesses weak normal structure.

It is easy to see that normal structure implies weak normal structure; however, the converse is not true; in fact we will show that JT and $(J, \|\ \|)$ are counterexamples to this.

The following equivalence yields a very convenient tool, which in many cases provides the easiest way to see if a bounded convex subset of a Banach space has normal structure; it is also due to Brodskii and Milman [1].

Theorem 3.b.6. A bounded convex subset K of X has normal structure if and only if K does not contain a diametral sequence.

Proof: If K contains a diametral sequence $\{x_n\}$, by Proposition 3.b.4 every $x \in \mathrm{co}\{x_i\}_{i=1}^{\infty}$ is a diametral point and therefore K doesn't have normal structure.

Conversely suppose that K doesn't have normal structure. Then there exists a convex subset C of K containing at least two points, such that every element of C is a diametral point. Let c = diam C; start with any $x_1 \in C$ and construct the sequence $\{x_n\}_{n=1}^{\infty}$ inductively as follows: let $x_{n+1} \in C$ be so that

$$\left\| \frac{x_1 + x_2 + \ldots + x_n}{n} - x_{n+1} \right\| > c - 1/n^2.$$

This is possible since $\frac{x_1 + x_2 + \ldots + x_n}{n} \in C$ is a diametral point.

We will show that $\{x_n\}_{n=1}^{\infty}$ is a diametral sequence. Let $x = \sum_{i=1}^{n} \lambda_i x_i$ with $0 \le \lambda_i \le 1$, $\sum_{i=1}^{n} \lambda_i = 1$. Then we may write

$$\frac{x_1 + x_2 + \ldots + x_n}{n} = \frac{1}{n} x + \sum_{i=1}^{n} \left(\frac{1}{n} - \frac{\lambda_i}{n} \right) x_i,$$

and by the induction hypothesis

$$c - 1/n^2 \le \left\| \frac{x_1 + x_2 + \ldots + x_n}{n} - x_{n+1} \right\| \le$$

$$\le \left\| \frac{1}{n} x - \frac{1}{n} x_{n+1} \right\| + \sum_{i=1}^{n} \left(\frac{1}{n} - \frac{\lambda_i}{n} \right) \| x_i - x_{n+1} \| \le$$

$$\le \frac{1}{n} \| x - x_{n+1} \| + \left(1 - \frac{1}{n} \right) c.$$

Thus

$$c - \frac{1}{n} \le \| x - x_{n+1} \| \le c.$$

Hence $\lim_{n \to \infty} \| x - x_{n+1} \| = c$, c is the diameter of $\{x_n\}_{n=1}^{\infty}$, and this sequence is diametral.

We shall now prove Kirk's theorem which loosely speaking says that having weak normal structure is a stronger property than the fixed point property. The proof we present here is taken from Diestel [1].

Theorem 3.b.7. If K is a weakly compact convex subset of a Banach space X and K possesses normal structure, then K has the fixed point property. In particular, if X has weak normal structure, then X has the fixed point property.

Proof: Let K be a non-trivial weakly compact convex subset of X which possesses normal structure. Let $T : K \longrightarrow K$ satisfy $\|Tx - Ty\| \le \|x - y\|$ for all x, $y \in K$. We will see that T has a fixed point.
First we need to introduce some notation:

$$\rho(x, K) = \sup \{ \|x - y\| : y \in K \},$$

$$\rho(K) = \inf \{ \rho(x, K) : x \in K \},$$

$$\hat{K} = \{ x \in K : \rho(K) = \rho(x, K) \}.$$

Observe that $\mathfrak{K}$ is a non-empty closed convex subset of K, since if we consider $K_n(x) = \{y \in K : \|x - y\| \le \rho(K) + \frac{1}{n}\}$, this is a non-empty convex closed subset of K. Let $x_0 \in K$ be such that $\rho(K) + \frac{1}{n} > \rho(x_0, K)$. Since $\rho(x_0,K) \ge \|x_0 - x\|$ for every $x \in K$, we have that $x_0 \in \bigcap_{x \in K} K_n(x) = K_n$. Hence for every n, K_n is a non-empty closed convex subset of K, which is thus weakly closed and therefore weakly compact. Clearly $K_{n+1} \subset K_n$, and $\mathfrak{K} = \bigcap_n K_n$ is also a non-empty closed convex subset of K. Next we will show that diam $\mathfrak{K}$ < diam K. Since K has normal structure, there exists $x \in K$ with $\rho(x, K) <$ diam K. Let $y, z \in \mathfrak{K}$; then

$$\|y - z\| \le \rho(z, K) = \rho(K).$$

Hence

(1) $$\operatorname{diam} \mathfrak{K} = \sup\{\|y - z\| : y, z \in \mathfrak{K}\} \le \rho(K) \le \rho(x,K) < \operatorname{diam} K.$$

Now let $\mathfrak{F}$ denote the family of non-empty closed convex subsets of K that are invariant under T, ordered by inclusion. Since K is weakly compact, clearly every descending chain of elements of $\mathfrak{F}$ has a least element which also belongs to $\mathfrak{F}$. Therefore, applying Zorn's lemma, there exists a minimal element G of $\mathfrak{F}$. We will show that G consists of one point which is a fixed point for T. Let $\mathfrak{G} = \{x \in G : \rho(G) = \rho(x, G)\}$, which is, as we showed before, a non-empty closed convex subset of G. Let $x \in \mathfrak{G}$, then for all $y \in G$

$$\|Tx - Ty\| \le \|x - y\| \le \rho(G).$$

Thus T(G) is contained in the ball with center Tx and radius $\rho(G)$, which we will denote by B; hence $T(G \cap B) \subset G \cap B$, and because G is minimal, we must have $G \subset B$. Therefore for all $y \in G$, $\|Tx - y\| \le \rho(G)$. Since $Tx \in G$ it follows that $Tx \in \mathfrak{G}$, that is $T(\mathfrak{G}) \subset \mathfrak{G}$ and $\mathfrak{G}$ belongs to $\mathfrak{F}$. If diam G > 0, by (1) we have diam $\mathfrak{G}$ < diam G, and this contradicts the minimality of G. It follows that diam G = 0 and hence $G = \{x\}$ and $Tx = x$.

It was shown by Karlovitz [1] that the converse to the above theorem is not true, and Alspach [1] proved that weak compactness is not sufficient to have the fixed point property.

Now we come back to JT and study its normal structures. We will see that although JT does not possess normal structure, it possesses weak normal structure. In order to do this, we prove first that $(J, \| \|_J)$, where $\| \|_J$ is the norm in J induced by the norm in JT (see Proposition 3.a.7), does not possess normal structure.

Theorem 3.b.8. $(J, \| \|_J)$ and JT do not possess normal structure.

Proof: We will show that the summing basis $\{\xi_n\}$ is a diametral sequence in J, and this by Theorem 3.b.6 proves the result.

In fact, $\|\xi_n - \xi_m\|_J = \sqrt{2}$ if $n \neq m$. Hence diam $\{\xi_n\}_{n=1}^{\infty} = \sqrt{2}$.

Now let $0 \leq \lambda_i \leq 1$ with $\sum_{i=1}^{n-1}\lambda_i = 1$. Then a typical sum for calculating $\|\xi_n - \sum_{i=1}^{n-1}\lambda_i\xi_i\|_J$ is the following:

$$\sum_{j=1}^{k-1}\left(\sum_{i=p_j}^{p_{j+1}-1}\lambda_i\right)^2 + (1 - \sum_{i=p_k}^{n-1}\lambda_i)^2 \leq 2,$$

where $1 = p_1 < p_2 <\ldots< p_k \leq n$ and $\sum_{i=n}^{n-1}\lambda_i = 0$. Here we used the inequality

$$\sum_i a_i^2 \leq (\sum_i a_i)^2 \text{ if } a_i \geq 0.$$

But taking $k = 2$, $p_1 = 1$ and $p_2 = n$, we get $(\sum_{i=1}^{n-1}\lambda_i)^2 + 1^2 = 2$.

Therefore $\|\xi_n - \sum_{i=1}^{n-1}\lambda_i\xi_i\|_J = \sqrt{2}$ and hence

$$d(\xi_n, \operatorname{co}\{\xi_i\}_{i=1}^{n-1}) = \sqrt{2};$$

this proves that $\{\xi_n\}_{n=1}^{\infty}$ is a diametral sequence in $(J, \| \|_J)$ and thus, by Proposition 3.a.7(b), $\{\eta_{(n,0)}\}_{n=0}^{\infty}$ is a diametral sequence in JT.

Theorem 3.b.9. JT has weak normal structure and in particular possesses the fixed point property. From this it follows that $(J, \| \|_J)$ has the same properties.

Proof: Suppose this were false. Then by Theorem 3.b.6 there would exist a weakly compact convex subset K of JT containing a diametral sequence $\{x_n\}$. We may assume, by passing to a subsequence, that $\{x_n\}$ is weakly convergent, and this sequence by Lemma 3.b.3 is also diametral. Since the diametral property is invariant under translations, we may suppose that $\{x_n\}$ converges weakly to 0. Then by Proposition 3.b.4

(1) $$\lim_{n\to\infty} \|x_n\| = c.$$

Thus, by passing again to a subsequence if necessary, we can assume that

$\{x_n\}$ is seminormalized and as in the proof of Corollary 1.9 we can find a block basic sequence $\{u_n\}$ of the basis $\{\eta_{t_n}\}$ of JT and a subsequence $\{x_{m_n}\}$ of $\{x_n\}$ so that

$$\lim_{n\to\infty} \|x_{m_n} - u_n\| = 0.$$

Then by (1)

$$\lim_{n\to\infty} \|u_n\| = \lim_{n\to\infty} \|x_{m_n}\| = c,$$

and by (2) of the proof of Proposition 3.b.4,

$$\lim_{n\to\infty} \|u_{n+1} - u_n\| = \lim_{n\to\infty} \|x_{m_{n+1}} - x_{m_n}\| = c.$$

By Lemma 3.a.3,

$$\|u_{n+1}\|^2 + \|u_n\|^2 \leq \|u_{n+1} - u_n\|^2,$$

and therefore

$$\sqrt{2}c = \lim_{n\to\infty} \left(\|u_{n+1}\|^2 + \|u_n\|^2\right)^{1/2} \leq \lim_{n\to\infty} \|u_{n+1} - u_n\| \leq c,$$

which is a contradiction.

It is easy to see that J equipped with the norm $\| \ \|$ also has weak normal structure and thus has the fixed point property; but it is interesting to note that for other equivalent norms there remain open problems in this direction; apparently it is not known whether (J, $|||\ |||$) has weak normal structure, although Khamsi in [1], using ultraproduct techniques, proved that it also has the fixed point property. This shows the strong dependence of weak normality on the specific norm of the space, in fact, van Dulst in [1] proved that every Banach space may be equivalently renormed so as to fail normal structure and on the other hand Day, James and Swaminathan in [1] showed that every separable Banach space can be renormed to have normal structure.

3.c. The conjugates of JT

In this section we give an explicit description of the conjugates of JT and some of their properties, and we show in addition that (1) jJT is w^*-sequentially dense in JT^{**}, where j is the canonical embedding from JT into JT^{**}, and (2) every bounded sequence in JT has a weakly Cauchy subsequence. These results can be obtained from the theorems of Odell and Rosenthal and of Rosenthal; however, we give a direct proof. These properties of JT answer negatively another question that goes back to Banach, of whether a separable space which has the property (1) or (2) must also have a separable dual. All results in this section are due to Lindenstrauss and Stegall [1].

Our first objective is the explicit description of JT^* and to start, we describe a special instance in which the calculation of the norm in JT^* is very easy.

Lemma 3.c.1. Let $T_k : JT \longrightarrow JT$, $k = 1,\dots,j$, be continuous linear operators such that there exists $\sigma_k \subset \mathcal{T}$ such that

(i) for every $x \in JT$, $\langle \eta_t^*, T_k x\rangle = 0$ if $t \notin \sigma_k$,

(ii) any segment in $\mathcal{T}$ intersects at most one σ_k,

(iii) the intersection of any segment in $\mathcal{T}$ with σ_k is again a segment.

Then for every $x^* \in JT^*$

$$\sum_{k=1}^{j} \|T_k^* x^*\|^2 = \|\sum_{k=1}^{j} T_k^* x^*\|^2.$$

Proof: If $x \in JT$ and $x = \sum_{k=1}^{j} a_k x_k$ where x_k has its support contained in σ_k, then by (ii), (iii) and by the definition of the norm in JT,

$$\|x\|^2 = \sum_{k=1}^{j} a_k^2 \|x_k\|^2.$$

Let $\varepsilon > 0$, let $x_k \in JT$ with $\|x_k\| = 1$, with support contained in σ_k and such that for $k = 1, 2,\dots,j$,

$$\|T_k^* x^*\|^2 \le \langle T_k^* x^*, x_k\rangle^2 + \varepsilon/j;$$

this is possible because of (i).

Let $x = \sum_{k=1}^{j} \langle T_k^* x^*, x_k\rangle x_k$. Then $\|x\|^2 = \sum_{k=1}^{j} \langle T_k^* x^*, x_k\rangle^2$ and therefore

$$\|\sum_{k=1}^{j} T_k^* x^*\|^2 \|x\|^2 \ge |\langle \sum_{k=1}^{j} T_k^* x^*, x\rangle|^2 = \left(\sum_{k=1}^{j} \langle T_k^* x^*, x_k\rangle^2\right)^2 =$$

$$= \sum_{k=1}^{j} \langle T_k^* x^*, x_k\rangle^2 \|x\|^2 \ge (\sum_{k=1}^{j} \|T_k^* x^*\|^2 - \varepsilon)\|x\|^2.$$

Thus

$$\sum_{k=1}^{j} \|T_k^* x^*\|^2 \le \|\sum_{k=1}^{j} T_k^* x^*\|^2.$$

Now let $x \in JT$ and $x_i = \sum_{t \in \sigma_i} \langle \eta_t^*, x\rangle \eta_t$. Then

$$\langle \sum_{k=1}^{j} T_k^* x^*, x\rangle = \sum_{k=1}^{j}\langle T_k^* x^*, x\rangle = \sum_{k=1}^{j}\langle T_k^* x^*, x_k\rangle \le \sum_{k=1}^{j}\|T_k^* x^*\|\|x_k\| \le$$

$$\le (\sum_{k=1}^{j}\|T_k^* x^*\|^2)^{1/2}(\sum_{k=1}^{j}\|x_k\|^2)^{1/2} \le (\sum_{k=1}^{j}\|T_k^* x^*\|^2)^{1/2}\|x\|.$$

Hence

$$\|\sum_{k=1}^{j} T_k^* x^*\|^2 \le \sum_{k=1}^{j}\|T_k^* x^*\|^2.$$

The next lemma is needed for the description of the dual of JT.

Lemma 3.c.2. Let $Y = \{x^* \in JT^* : \lim_{t\in B} x^*(\eta_t) = 0 \text{ for all branches } B\}$. Then for $x^* \in Y$, if Q_{ni} is as in Definition 3.a.4,

$$\lim_{n\to\infty} (\max_{0\le i<2^n} \|Q_{ni}^* x^*\|) = 0.$$

Proof: Suppose there are $x^* \in Y$, an $\alpha > 0$ and a sequence (n_k, i_k) such that for $k = 1, 2, \ldots,$

$$\|Q_{n_k i_k}^* x^*\| > \alpha. \tag{1}$$

We show first that among the (n_k, i_k) there exists only a finite number of mutually incomparable nodes. Indeed, assume that for $k = 1, 2, \ldots, j$, the (n_k, i_k) are mutually incomparable. By (1) and since $\|Q_{n_k i_k}\| = 1$, for every k there is an $x_k \in Q_{n_k i_k} JT$ with $\|x_k\| = 1$ and $x^*(x_k) \ge \alpha$. Then $\|\sum_{k=1}^{j} x_k\| = j^{1/2}$ and hence

$$j\alpha \le x^*(\sum_{k=1}^{j} x_k) \le \|x^*\| j^{1/2},$$

or equivalently,

$$j < (\|x^*\|/\alpha)^2.$$

Thus there is no loss of generality in assuming that the sequence (n_k, i_k) satisfying (1) is totally ordered and determines a unique branch $B = \{t_1, t_2, t_3, \ldots\}$ of $\mathcal{T}$. We can further assume that $n_{k+1} > n_k$ and

$$\|Q_{n_k i_k}^* x^* - Q_{n_{k+1} i_{k+1}}^* x^*\| \ge \tfrac{3}{4}\alpha \quad \text{for all } k. \tag{2}$$

This can be done since for every $x \in JT$ and any choice of $0 \le i_n < 2^n$, $\lim_{n\to\infty} Q_{n,i_n} x = 0$. Therefore if x_0 is such that $\|x_0\| = 1$ and $\langle Q_{n_k i_k}^* x^*, x_0\rangle > \alpha$, we may suppose that (n_{k+1}, i_{k+1}) satisfies

$$\|Q_{n_{k+1} i_{k+1}} x_0\| < \alpha/(4\|x^*\|).$$

Then

$$\langle Q^*_{n_k i_k} x^* - Q^*_{n_{k+1} i_{k+1}} x^*, x_0\rangle = \langle Q^*_{n_k i_k} x^*, x_0\rangle - \langle x^*, Q_{n_{k+1} i_{k+1}} x_0\rangle > \alpha - \frac{\alpha}{4} = \frac{3}{4}\alpha.$$

Hence we get (2).

On the other hand, if P_B and Q_n are as in Definition 3.a.4, the operator $T : P_B JT \longrightarrow J$ given by $T\eta_{t_n} = \xi_n$ is an isometry by Proposition 3.a.7 and since $x^* \in Y$, we have that

$$(3) \qquad 0 = \lim_{n\to\infty} x^*(\eta_{t_n}) = \lim_{n\to\infty} x^* T^{-1}\xi_n .$$

Hence by Corollary 2.f.11 we get that $x^* T^{-1} = \sum_{i=1}^{\infty} b_{t_i} \xi^*_i \in \tilde{J}$ and

$$P^*_B x^* = \sum_{i=1}^{\infty} b_{t_i} \eta^*_{t_i} = \sum_{k=1}^{\infty} \sum_{\{t_i : n_k \le \mathrm{lev}(t_i) < n_{k+1}\}} b_{t_i} \eta^*_{t_i} = \sum_{k=1}^{\infty} (Q^*_{n_k} - Q^*_{n_{k+1}}) P^*_B x^* .$$

Thus for sufficiently large k (and therefore without loss of generality for every k),

$$(4) \qquad \|(Q^*_{n_k} - Q^*_{n_{k+1}}) P^*_B x^*\| < \frac{1}{2}\alpha.$$

Set now for k = 1, 2,...

$$U_k = Q_{n_k i_k} - Q_{n_{k+1} i_{k+1}} - P_B(Q_{n_k} - Q_{n_{k+1}})$$

which is a projection. If $x \in U_k JT$, its support is contained in σ_k, where

$$\sigma_k = \{(n,i) \in \mathcal{T} : (n,i) \ge (n_k, i_k),\ (n,i) \text{ is not} \ge (n_{k+1}, i_{k+1}),\ (n,i) \notin B\}.$$

That is, σ_k is the finite union of disjoint trees whose vertex is the offspring not belonging to B of the nodes in the interval $[(n_k,i_k), (n_{k+1},i_{k+1}))$. It is easy to check that $\{U_k\}_{k=1}^{\infty}$ and $\{\sigma_k\}_{k=1}^{\infty}$ satisfy the conditions of Lemma 3.c.1. Thus

$$(5) \qquad \sum_{k=1}^{j} \|U^*_k x^*\|^2 = \|\sum_{k=1}^{j} U^*_k x^*\|^2.$$

However, by (2) and (4), $\|U^*_k x^*\| > \frac{1}{4}\alpha$ for every k, while for all j

$$4 \ge \|Q^*_{n_1 i_1} - Q^*_{n_{j+1} i_{j+1}} - (Q^*_{n_1} - Q^*_{n_{j+1}}) P^*_B\| = \|\sum_{k=1}^{j} U^*_k\|.$$

This contradicts (5) for $j > 256\|x^*\|^2/\alpha^2$, and concludes the proof.

The characterization of the dual JT^* of JT depends heavily on the set of all 0-branches of $\mathcal{T}$ which will be denoted by Γ; clearly Γ has the cardinality of the continuum, a fact that will be important in what follows.

Define the map $S : JT^* \longrightarrow \ell_2(\Gamma)$ by

$$Sx^* = \{\lim_{t\in B} x^*(\eta_t)\}_{B\in\Gamma} .$$

By Corollary 2.f.11 all the above limits exist, so S is well defined, and in the proof of the next theorem we will see that S has norm ≤ 1. Also observe that the space Y in the above lemma is the kernel of S, and this is crucial for the proof of Theorem 3.c.3.

Theorem 3.c.3. Let $\mathfrak{B} = [\eta_t^*]_{t\in\mathcal{T}}$. The operator $S : JT^* \longrightarrow \ell_2(\Gamma)$ defined above is an onto map whose kernel Y is equal to $\mathfrak{B}$. Also $JT^*/\mathfrak{B}$ is isometric to $\ell_2(\Gamma)$ and JT^* is the closed linear span of $\mathfrak{B}$ and the elements f_B where B is a branch.

Proof: We check first that S is an operator of norm ≤ 1, that is

$$\|Sx^*\|^2 = \sum_{B\in\Gamma}(\lim_{t\in B} x^*(\eta_t))^2 \le \|x^*\|^2.$$

Let $\{B_p\}_{p=1}^q$ be distinct branches of $\mathcal{T}$. There is an integer m such that for $p = 1,\ldots,q$ the sets

$$B_p \cap \{t \in \mathcal{T} : \text{lev}(t) \ge m\}$$

are pairwise disjoint. It follows that if $t_p \in B_p$ with $\text{lev}(t_p) \ge m$ and if $x^* \in JT^*$, taking $x = \sum_{p=1}^q \langle x^*, \eta_{t_p}\rangle \eta_{t_p}$, we get

$$\|x\|^2 = \sum_{p=1}^q (\langle x^*, \eta_{t_p}\rangle)^2.$$

Hence for every $x^* \in JT^*$ and x as above,

$$\sum_{p=1}^q (\langle x^*, \eta_{t_p}\rangle)^2 = x^*(x) \le \|x^*\|\|x\| = \|x^*\|\left(\sum_{p=1}^q (\langle x^*, \eta_{t_p}\rangle)^2\right)^{1/2}.$$

Therefore

(1) $$\sum_{p=1}^q (\langle x^*, \eta_{t_p}\rangle)^2 \le \|x^*\|^2.$$

From this, if for $p = 1,\ldots,q$, t_p tends to ∞ along B_p, we obtain that S is bounded and of norm ≤ 1. On the other hand Sf_B is the unit vector e_B in $\ell_2(\Gamma)$ given by $e_B(B) = 1$ and $e_B(B') = 0$ if $B' \in \Gamma$ and $B' \ne B$. Since $\|f_B\| = \|e_B\| = 1$, it follows that

(2) $$\|S\| = 1.$$

Next we will prove that S is an onto map. Indeed, let $\{B_p\}_{p=1}^q$ and m be as above. Let $\{a_p\}_{p=1}^q$ also be given. Define $x^* \in JT^*$ by

$$x^*(\sum_{t\in\mathcal{T}} b_t \eta_t) = \sum_{p=1}^{q} a_p \sum_{\{t\in B_p : \mathrm{lev}(t)>m\}} b_t .$$

Then $\|x^*\| = (\sum_{p=1}^{q} a_p^2)^{1/2}$. In fact, using the definition of norm in JT,

$$\|x^*(\sum_{t\in\mathcal{T}} b_t \eta_t)\| = |\sum_{p=1}^{q} a_p \sum_{\{t\in B_p : \mathrm{lev}(t)>m\}} b_t| \le$$

$$\le (\sum_{p=1}^{q} a_p^2)^{1/2} (\sum_{p=1}^{q} (\sum_{\{t\in B_p : \mathrm{lev}(t)>m\}} b_t)^2)^{1/2} \le (\sum_{p=1}^{q} a_p^2)^{1/2} \|\sum_{t\in\mathcal{T}} b_t \eta_t\|.$$

On the other hand, if $t_p \in B_p$ with $\mathrm{lev}(t_p) > m$, then $\langle x^*, \eta_{t_p}\rangle = a_p$, and using (1) we get the reverse inequality. Also, by the definition of S, $Sx^*(B_p) = \lim_{t\in B_p} x^*(\eta_t) = a_p$ and $Sx^*(B) = 0$ if B is not any of the B_p. Thus S is an onto map.

It is evident that $\mathfrak{B}$ is contained in Y, the kernel of S. The main point in the proof of this theorem is to verify the reverse inclusion.

Assume that $\mathfrak{B}$ is a proper subspace of Y. Let $\delta > 0$ be such that

(3) $$\frac{7}{2} < 4(1 - \delta)^2.$$

Let $y^* \in Y$ with $\|y^*\| = 1$ and $d(y^*, \mathfrak{B}) = a > 0$, where d denotes the distance function, and let $b_0 \in \mathfrak{B}$ be such that $\|b_0 + y^*\| < a/(1 - \delta)$. Then if $x^* = (b_0 + y^*)/\|b_0 + y^*\|$, we get that

(4) $$d(x^*,\mathfrak{B}) = \frac{1}{\|b_0+y^*\|}\inf_{b\in\mathfrak{B}}\| y^* + b_0 - b\|b_0+y^*\| \| = \frac{a}{\|b_0+y^*\|} > 1 - \delta.$$

Let $x \in$ JT with $\|x\| = 1$ such that $\langle x^*, x\rangle > 1 - \delta$ and let r be an integer such that $\langle P_r^* x^*, x\rangle = \langle x^*, P_r x\rangle > 1 - \delta$. Then

(5) $$\|P_r^* x^*\| > 1 - \delta.$$

Now let $\varepsilon > 0$ be such that

(6) $$2^{r+2}\varepsilon^2 < (1 - \delta)^2.$$

By Lemma 3.c.2 there is a $q > r$ such that for $0 \le j < 2^q$

(7) $$\|Q_{qj}^* x^*\| \le \varepsilon.$$

Since

$$(I - Q_q^*)x^* = \sum_{\{t:\mathrm{lev}(t)<q\}} \langle x^*, \eta_t\rangle \eta_t^* \in \mathfrak{B},$$

by (4), $\|Q_q^* x^*\| > 1 - \delta$, and hence by Lemma 3.c.1 we obtain

$$\sum_{j=0}^{2^q-1} \|Q_{qj}^* x^*\|^2 = \|Q_q^* x^*\|^2 > (1 - \delta)^2.$$

It follows that for $0 \le j < 2^q$, there exists $x_j \in JT$ with $\|x_j\| = 1$, $Q_{qj}x_j = x_j$ and

(8) $$C^2 = \sum_{j=0}^{2^q-1} |Q^*_{qj}x^*(x_j)|^2 = \sum_{j=0}^{2^q-1} |x^*(x_j)|^2 > (1 - \delta)^2.$$

Define next

(9) $$x = C^{-1}\sum_{j=0}^{2^q-1} x^*(x_j)x_j.$$

Then the support of x is contained in the set of nodes with level greater than or equal to q and by (7), (8) and (9) we have for $0 \le j < 2^q$

(10) $$\begin{cases} \|x\| = 1, \ C = x^*(x) > 1 - \delta, \\ \|Q_{qj}x\| = C^{-1}|x^*(x_j)| = C^{-1}|Q^*_{qj}x^*(x)| \le C^{-1}\|Q^*_{qj}x^*\| \le \varepsilon(1 - \delta)^{-1}. \end{cases}$$

Since $\lim_{n\to\infty} Q_n x = 0$ there is no loss of generality if we assume that $Q_p(x) = 0$ for some $p > q$. By (5) there is a $y \in JT$ with

(11) $$\|y\| = 1, \quad Q_{r+1}y = 0, \quad x^*(y) > 1 - \delta,$$

and thus, in particular,

(12) $$x^*(x + y) > 2(1 - \delta).$$

Our next aim is to obtain an estimate for $\|x + y\|$ and use this to obtain a contradiction to (3). Let

$$x + y = \sum_{n=0}^{p}\sum_{i=0}^{2^n-1} a_{ni}\eta_{ni}.$$

Then by (11) and since $Q_{qj}x_j = x_j$, $a_{ni} = 0$ for $r < n < q$.

By the definition of the norm in JT there are pairwise disjoint segments S_k, $R_{k'}$, $U_{k''}$ with $1 \le k \le L$, $1 \le k' \le M$, $1 \le k'' \le N$, such that each $R_{k'}$ contains no element (n,i) with $n \ge q$, each $U_{k''}$ does not contain an element (n,i) with $n \le r$, and each S_k contains nodes of the form (r,k) and (q,j) such that

(13) $$\|x + y\|^2 = \sum_{k=1}^{L}(\sum_{t\in S_k} a_t)^2 + \sum_{k'=1}^{M}(\sum_{t\in R_{k'}} a_t)^2 + \sum_{k''=1}^{N}(\sum_{t\in U_{k''}} a_t)^2 =$$
$$= s + w + u.$$

First note that by (11)

(14) $$u \le \|x\|^2 = 1.$$

From $(a + b)^2 \le 2(a^2 + b^2)$ for all a and b, we get that

(15) $$s \le 2\left(\sum_{k=1}^{L}(\sum_{\{t\in S_k:\mathrm{lev}(t)<q\}} a_t)^2 + \sum_{k=1}^{L}(\sum_{\{t\in S_k:\mathrm{lev}(t)\ge q\}} a_t)^2\right) = 2(s' + s'').$$

Note that since by the definition of x, $\langle x, \eta_t^* \rangle = 0$ if lev(t) < q,

(16) $$2s' + w \leq 2(s' + w) \leq 2\|y\|^2 = 2.$$

Observe also that L, the number of the S_k's, is less than or equal to 2^r. For $1 \leq k \leq L \leq 2^r$, let $(q, j_k) \in S_k$. By (11), (10) and (6)

(17) $$s'' = \sum_{k=1}^{L} \Big(\sum_{\{t \in S_k : t \geq (q, j_k)\}} a_t \Big)^2 \leq \sum_{k=1}^{L} \|Q_{qj_k} x\|^2 \leq 2^r \varepsilon^2 (1-\delta)^{-2} < \frac{1}{4}.$$

Combining (13)-(17), we get that

(18) $$\|x + y\|^2 \leq u + 2(s' + w) + 2s'' \leq 1 + 2 + \frac{1}{2} = \frac{7}{2}.$$

This contradicts (3), since by (18) and (12) we have

$$1 = \|x^*\| \geq |x^*(x + y)|(\|x + y\|)^{-1} > 2(1 - \delta)(\tfrac{7}{2})^{-1/2}.$$

Hence the kernel of S is $\mathcal{B}$.

Let $\tilde{S}$ be the map from $JT^* / \mathcal{B}$ onto $\ell_2(\Gamma)$ induced by S and for $f \in JT^*$ let $[\![f]\!]$ denote its equivalence class in $JT^* / \mathcal{B}$. Then by (2),

$$\|\tilde{S}[\![f]\!]\| = \|Sf\| \leq \|f\|$$

for every $f \in [\![f]\!]$. Hence

$$\|\tilde{S}[\![f]\!]\| \leq \|[\![f]\!]\|.$$

On the other hand, if $\sum_{i=1}^{j} c_i e_{B_i} \in \ell_2(\Gamma)$, let N be such that the B_i are pairwise disjoint above level N and let $S_i = \{t \in B_i : \text{lev}(t) \leq N\}$. Then

$$S\Big(\sum_{i=1}^{j} c_i f_{B_i} - \sum_{i=1}^{j} c_i \sum_{t \in S_i} \eta_t^* \Big) = \sum_{i=1}^{j} c_i e_{B_i},$$

and for $x = \sum_{t \in \mathcal{J}} a_t \eta_t \in JT$

$$\Big| \langle \sum_{i=1}^{j} c_i f_{B_i} - \sum_{i=1}^{j} c_i \sum_{t \in S_i} \eta_t^*, x \rangle \Big| = \Big| \sum_{i=1}^{j} c_i \sum_{t \in B_i \setminus S_i} a_t \Big| \leq \Big(\sum_{i=1}^{j} c_i^2 \Big)^{1/2} \|x\|.$$

Hence

$$\Big\|\Big[\!\!\Big[\sum_{i=1}^{j} c_i f_{B_i} \Big]\!\!\Big]\Big\| \leq \Big(\sum_{i=1}^{j} c_i^2 \Big)^{1/2};$$

therefore

$$\Big\| \tilde{S}^{-1}\Big(\sum_{i=1}^{j} c_i e_{B_i} \Big) \Big\| \leq \Big\| \sum_{i=1}^{j} c_i e_{B_i} \Big\|,$$

and $\tilde{S}$ is an isometry. Since $\tilde{S}([\![f_B]\!]) = e_B$ for every branch B, we conclude that JT^* is the closed linear span of $\mathcal{B}$ and $\{f_B\}_{B \in \Gamma}$.

Observe that from the above, since $\mathfrak{B} = [\eta^*_t]_{t\in\mathcal{T}}$ and the basis $\{\eta_t\}_{t\in\mathcal{T}}$ of JT is monotone and boundedly complete, it follows by Theorem 1.6 that JT is canonically isometric to $\mathfrak{B}^*$, via the isomorphism $j_{\mathfrak{B}} : \mathrm{JT} \to \mathfrak{B}^*$ given by $\langle j_{\mathfrak{B}}x, b\rangle = b(x)$.

From Theorem 3.a.8 we now that JT does not contain ℓ_1 and thus Rosenthal's dichotomy theorem implies that every bounded sequence in JT has a weak Cauchy sequence. However, here we obtain this result directly as a consequence of the previous theorem.

Corollary 3.c.4. Every bounded sequence in JT has a weak Cauchy subsequence.

Proof: Let $\{y_m\}_{m=1}^{\infty}$ be contained in $\mathrm{JT} = \mathfrak{B}^*$ with $\|y_m\| = 1$ for all m. There is a subsequence of $\{y_m\}$ (which we may assume to be the sequence itself), which w^*-converges to some $y \in \mathrm{JT}$ with $\|y\| \le 1$.
Put $x_m = y_m - y$; then $\|x_m\| \le 2$. Let $\{B_j\}_{j=1}^{k}$ be distinct branches of $\mathcal{T}$, and L be such that the branches B_j do not intersect above the L-th level; let $S_j = B_j \cap \{t \in \mathcal{T} : \mathrm{lev}(t) \le L\}$ and let $B'_j = B_j \setminus S_j$ for $j = 1,\dots,k$. Since for every $t \in \mathcal{T}$, $\langle \eta^*_t, x_m\rangle$ tends to zero as m tends to ∞, $\lim_{m\to\infty} f_{S_j}(x_m) = 0$ for $j = 1,\dots,k$. Also

$$\sum_{j=1}^{k} |f_{B'_j}(x_m)|^2 \le \|x_m\|^2 \le 4.$$

Therefore

(1) $$\limsup_{m\to\infty} \sum_{j=1}^{k} |f_{B_j}(x_m)|^2 = \limsup_{m\to\infty} \sum_{j=1}^{k} |f_{S_j}(x_m) + f_{B'_j}(x_m)|^2 \le 4.$$

If there is a branch B_{j_1} such that $\limsup_{m\to\infty} |f_{B_{j_1}}(x_m)| \ge \frac{1}{2}$, take a subsequence $\{x_{1,r}\}_{r=1}^{\infty}$ of $\{x_m\}$ so that $\lim_{r\to\infty} |f_{B_{j_1}}(x_{1,r})| \ge \frac{1}{2}$ exists. If there is a branch $B_{j_2} \ne B_{j_1}$ such that $\limsup_{r\to\infty} |f_{B_{j_2}}(x_{1,r})| \ge \frac{1}{2}$, take a subsequence $\{x_{2,r}\}_{r=1}^{\infty}$ of $\{x_{1,r}\}_{r=1}^{\infty}$ so that $\lim_{r\to\infty} |f_{B_{j_2}}(x_{2,r})| \ge \frac{1}{2}$ exists and so forth.
By (1) we can continue this procedure at most 16 times, that is, there

are a subsequence $\{m_{1,i}\}_{i=1}^{\infty}$ of the integers and an integer $0 \le k_1 \le 16$ such that, for k_1 distinct branches B_j, $\lim_{i\to\infty} f_{B_j}(x_{m_{1,i}})$ exists and is of absolute value bigger than or equal to $\frac{1}{2}$, while for all other branches $\lim\sup_i |f_B(x_{m_{1,i}})| < \frac{1}{2}$.

By using (1) again we can find a subsequence $\{m_{2,i}\}_{i=1}^{\infty}$ of $\{m_{1,i}\}_{i=1}^{\infty}$ and a k_2 with $k_1 \le k_2 \le 2^6$ such that for k_2 different branches B_j, $\lim_{i\to\infty} |f_{B_j}(x_{m_{2,i}})| \ge \frac{1}{4}$ exists, while for all other branches the absolute value of the corresponding sequence has lim sup at most $\frac{1}{4}$. Continuing in the same manner and passing to the diagonal sequence, we get finally a subsequence $\{m_i\}$ of the integers such that $\lim_{i\to\infty} f_B(x_{m_i})$ exists for all branches B of $\mathcal{T}$. Since by Theorem 3.c.3 the elements of the form f_B together with $\mathfrak{B}$ span JT^*, it follows that $\lim_{i\to\infty} x^*(y_{m_i})$ exists for every $x^* \in JT^*$ and this proves the corollary.

McWilliams [1] proved that every quasi-reflexive Banach space X is w^*-sequentially dense in X^{**}, so that J and J^* have this property. Since neither J nor J^* nor JT contains ℓ_1, the w^*-sequential density of X in X^{**} in these cases also follows from Odell and Rosenthal's theorem (see e.g. Lindenstrauss and Tzafriri [1]). Here we give a direct proof of this for JT.

Corollary 3.c.5. JT is w^*-sequentially dense in JT^{**}.

Proof: Let j, j_0 and j_1 be the canonical embeddings $j : JT \longrightarrow JT^{**}$, $j_0 : \mathfrak{B} \longrightarrow \mathfrak{B}^{**}$, $j_1 : \mathfrak{B}^* \longrightarrow \mathfrak{B}^{***}$ and let $j_{\mathfrak{B}} : JT \longrightarrow \mathfrak{B}^*$ be the canonical isometry mentioned before.
Define $T : (\mathfrak{B}^{**}/j_0\mathfrak{B})^* \longrightarrow (j_0\mathfrak{B})^{\perp}$ by

$$(Tf)b^{**} = f(\dashv b^{**} \vdash),$$

where $b^{**} \in \mathfrak{B}^{**}$, $f \in (\mathfrak{B}^{**}/j_0\mathfrak{B})^*$ and $\dashv b^{**} \vdash$ denotes the equivalence class of b^{**} in $\mathfrak{B}^{**}/j_0\mathfrak{B}$. Then T is an isometry; this can be seen for example in Beauzamy [1].

Let $U : (\mathfrak{B}^{**}/j_0\mathfrak{B}) \longrightarrow (JT^*/\mathfrak{B})$ be the isometry given by

$$U(\dashv b^{**} \vdash) = [\![j_{\mathcal{B}}^{*}(b^{**})]\!],$$

where $[\![\ ]\!]$ denotes the equivalence class in $JT^{*}/\mathcal{B}$, and let $\tilde{S} : JT^{*}/\mathcal{B} \longrightarrow \ell_2(\Gamma)$ be the map induced by S, as in the proof of Theorem 3.c.3. Since T, U and $\tilde{S}$ are onto isometries, for every $h \in (j_0\mathcal{B})^{\perp}$ there are a sequence of branches $\{B_j\}_{j=1}^{\infty}$ and a sequence of reals $\{s_j\}_{j=1}^{\infty}$ with

$$h = TU^{*}\tilde{S}^{*}(\textstyle\sum_{j=1}^{\infty} s_j e_{B_j}^{*}) \quad \text{and} \quad \sum_{j=1}^{\infty}(s_j)^2 < \infty,$$

where $\{e_B^{*}\}_{B\in\Gamma}$ are the biorthogonal functionals associated to $\{e_B\}_{B\in\Gamma}$ which were defined in the proof of Theorem 3.c.3. Since JT^{*} is the span of $\mathcal{B}$ and $\{f_B\}_{B\in\Gamma}$, it is not difficult to see that

$$(j_{\mathcal{B}}^{**})^{-1}(h) = \textstyle\sum_{j=1}^{\infty} s_j F_{B_j},$$

where for every branch B, $F_B \in JT^{**}$ is defined by

$$F_B(\eta_t^{*}) = 0 \text{ for all } t \in \mathcal{T},\ F_B(f_B) = 1 \text{ and } F_B(f_{B'}) = 0 \text{ if } B \neq B'.$$

In fact, if $B_1 \in \Gamma$, $b^{**} \in \mathcal{B}^{**}$ and $j_{\mathcal{B}}^{*} b^{**} = \sum_{B\in\Gamma} \lambda_B f_B + \sum_{t\in\mathcal{T}} c_t \eta_t^{*}$, then

$$\langle j_{\mathcal{B}}^{**} F_{B_1},\ b^{**}\rangle = \langle F_{B_1},\ j_{\mathcal{B}}^{*} b^{**}\rangle = \lambda_{B_1}$$

and

$$\langle TU^{*}\tilde{S}^{*} e_{B_1}^{*},\ b^{**}\rangle = \langle U^{*}\tilde{S}^{*} e_{B_1}^{*},\ \dashv b^{**}\vdash\rangle = \langle \tilde{S}^{*} e_{B_1}^{*},\ [\![j_{\mathcal{B}}^{*} b^{**}]\!]\rangle =$$

$$= \langle e_{B_1}^{*},\ \textstyle\sum_{B\in\Gamma} \lambda_B e_B\rangle = \lambda_{B_1}.$$

Also for every $b^{*} \in \mathcal{B}^{*}$ we have that $b^{*} = j_{\mathcal{B}}(x)$ for some $x \in JT$ and if $b^{**} \in \mathcal{B}^{**}$

$$\langle j_1 b^{*},\ b^{**}\rangle = \langle b^{**}, b^{*}\rangle = \langle b^{**},\ j_{\mathcal{B}} x\rangle = \langle j_{\mathcal{B}}^{*} b^{**},\ x\rangle =$$

$$= \langle jx,\ j_{\mathcal{B}}^{*} b^{**}\rangle = \langle j_{\mathcal{B}}^{**} jx, b^{**}\rangle.$$

Therefore $(j_{\mathcal{B}}^{**})^{-1}(j_1 b^{*}) = jx$ and since by Lemma 1.12, $\mathcal{B}^{***} = (j_0\mathcal{B})^{\perp} \oplus j_1\mathcal{B}^{*}$ and since $JT^{**} = (j_{\mathcal{B}}^{**})^{-1}\mathcal{B}^{***}$, we get that every $x^{**} \in JT^{**}$ has a unique representation of the form

(1) $$x^{**} = \textstyle\sum_{j=1}^{\infty} s_j F_{B_j} + jx_0,$$

where $x_0 \in JT$, the B_j are distinct branches of $\mathcal{T}$, $\sum_{j=1}^{\infty} s_j^2 < \infty$, and $s_j = 0$ implies $s_{j+1} = 0$ for all j.

Let $x^{**} \in JT^{**}$ and consider its representation as in (1). Choose a sequence of integers $n_1 < n_2 <\ldots$ so that for $j = 1,\ldots,k$ the branches B_j do not intersect on or above the n_k-th level. Pick $i_{k(j)}$ so that $(n_k, i_{k(j)}) \in B_j$, $j = 1,\ldots,k$, and put $x_k = \sum_{j=1}^{k} s_j \eta_{n_k i_{k(j)}}$.

We will show that $j(x_0 + x_k)$ w^*-converges to x^{**} or equivalently we will see that for $y^* \in JT^*$ $y^*(x_0 + x_k) \longrightarrow x^{**}(y^*)$ as $k \longrightarrow \infty$.

Indeed, if $n_k > n$ then

(2) $$\eta^*_{ni}(x_0 + x_k) = \eta^*_{ni}(x_0) = x^{**}(\eta^*_{ni}).$$

Now let $B \in \Gamma$. If B intersects only a finite number of B_j's or if $B = B_j$ for some j, then

$$f_B(x_0 + x_k) = x^{**}(f_B)$$

if k is large enough. If B intersects an infinite number of B_j's but $B \neq B_j$ for every j, then

$$f_B(x_0 + x_k) = f_B(x_0) + f_B(x_k) = x^{**}(f_B) + f_B(x_k),$$

and $f_B(x_k)$ tends to zero as k tends to infinity. From (2), the above and Theorem 3.c.3 we get the desired result.

In the remainder of this section we study some additional properties of the conjugates of JT.

Definition 3.c.6. A Banach space X is weakly compactly generated (WCG for short) if it is the closed linear span of a weakly compact set.

The importance of WCG spaces lies in the fact that they have a structure that is similar to that of a finite dimensional decomposition (see for example Day [1]). As we have seen, the conjugates of JT are not separable, so they cannot have a Schauder basis; however, in the next corollary it is shown that the odd conjugates of $\mathfrak{B}$ are weakly compactly generated.

To prove this result we need the following well known properties about WCG spaces which can also be found in the books of Diestel [1] and Day [1]:

(i) Separable and reflexive spaces are WCG.

(ii) Complemented subspaces of WCG are WCG.

(iii) Non-separable conjugates of separable spaces are not WCG.

Corollary 3.c.7. For every integer k > 1,

$$\mathfrak{B}^{(2k)} \approx \mathfrak{B}^{**} \oplus \ell_2(\Gamma), \quad \mathfrak{B}^{(2k-1)} \approx \mathfrak{B}^{*} \oplus \ell_2(\Gamma),$$

and none of the conjugates of $\mathfrak{B}$ contains a subspace isomorphic to c_0 or ℓ_1. The conjugates of odd order of $\mathfrak{B}$ are all WCG while those of even order (except $\mathfrak{B}$ itself) are not WCG.

Proof: Let j_0 and j_1 be the canonical embeddings of $\mathfrak{B}$ in $\mathfrak{B}^{**}$ and $\mathfrak{B}^{*}$ in $\mathfrak{B}^{***}$ respectively. Since JT is canonically isometric to $\mathfrak{B}^{*}$, $\mathfrak{B}^{**}$ is isometric to JT^{*}, and hence $\mathfrak{B}^{**}/j_0\mathfrak{B}$ is isometric to $JT^{*}/\mathfrak{B}$ and by Theorem 3.c.3, to $\ell_2(\Gamma)$. Hence, if $\cong$ denotes isometric equivalence, using Lemma 1.12,

$$\mathfrak{B}^{***} = (j_0\mathfrak{B})^{\perp} \oplus j_1\mathfrak{B}^{*} \cong (\mathfrak{B}^{**}/j_0\mathfrak{B})^{*} \oplus \mathfrak{B}^{*} \cong \ell_2(\Gamma)^{*} \oplus \mathfrak{B}^{*} \cong \ell_2(\Gamma) \oplus \mathfrak{B}^{*}.$$

Then, by induction on k, using the fact that $\ell_2(\Gamma) \oplus \ell_2(\Gamma) \approx \ell_2(\Gamma)$, we obtain $\mathfrak{B}^{(2k)} \approx \mathfrak{B}^{**} \oplus \ell_2(\Gamma)$ and $\mathfrak{B}^{(2k-1)} \approx \mathfrak{B}^{*} \oplus \ell_2(\Gamma)$.

Since by Theorem 3.a.8 $\mathfrak{B}^{*}$ does not contain a subspace isomorphic to ℓ_1, we may apply the theorem stating that if X is a Banach space such that X^{*} contains c_0, then X has a complemented subspace isomorphic to ℓ_1 (see e.g. Lindenstrauss and Tzafriri [1]), and from this we conclude that $\mathfrak{B}^{**}$ does not contain a subspace isomorphic to c_0. Also, since in addition $\mathfrak{B}^{*}$ is separable, we now apply Odell and Rosenthal's theorem saying that if X is separable then X does not contain ℓ_1 if and only if the cardinality of X^{**} is equal to that of X (see e.g. Lindenstrauss and Tzafriri [1]). Thus $\mathfrak{B}^{***}$ has the cardinality of the continuum which is strictly less than the cardinality of ℓ_∞^{*}. Hence

$$\text{card } \mathfrak{B}^{(k)} < \text{card } \ell_\infty^{*} = \text{card } \ell_1^{**}$$

for every k, and therefore it is not possible that $\ell_1 \subset \mathfrak{B}^{(k)}$, and it follows that $\mathfrak{B}^{(k+1)}$ cannot contain a subspace isomorphic to c_0. By the properties of WCG spaces preceding Corollary 3.c.7, we have the last assertion.

Let $\Delta = \{0,1\}^{\aleph_0}$ be the usual Cantor set and $C(\Delta)$ the space of continuous functions on Δ with the sup norm. Since Γ may be viewed as a Cantor set, we have another useful representation for JT, which we now describe:
For $\theta \in \Delta$ denote by $\theta(n)$ its n-th coordinate. Let $n \in \mathbb{N}$ be fixed and $0 \le i \le 2^n - 1$. If

$$i = \theta_{n-1} + 2\theta_{n-2} + \ldots + 2^{n-1}\theta_0$$

with $\theta_j \in \{0,1\}$ for $0 \le j \le n-1$, define the function $h_{n,i}$ as the characteristic function of the closed and open subsets $\Delta_{n,i}$ of Δ given by

$$\Delta_{0,0} = \Delta,$$

$$\Delta_{n,i} = \{\theta \in \Delta : \theta(j) = \theta_j,\ j = 0,1,\ldots,n-1\}.$$

Then

(1) $$\Delta_{n,i} = \Delta_{n+1,2i} \cup \Delta_{n+1,2i+1},$$

(2) $$h_{n,i} = h_{n+1,2i} + h_{n+1,2i+1}, \quad n = 0, 1, \ldots,\ i = 0,\ldots,2^n - 1.$$

Clearly there is a bijection between Δ and Γ: for $\theta \in \Delta$ let $\phi(\theta) = B$, where $B = \{(n,i) \in \mathcal{T} : \theta \in \Delta_{n,i}\} \cup \{(0,0)\}$. From (1) it follows that B is a 0-branch. Then if $R : JT \longrightarrow C(\Delta)$ is defined by

(3) $$R\eta_{ni} = h_{n,i}$$

for $n = 0, 1, \ldots,$ and $i = 0,\ldots,2^n - 1$, R is an operator of norm one from JT into $C(\Delta)$: Let $x = \sum_{(n,i)\in\mathcal{T}} a_{n,i}\eta_{ni}$, then

$$\|Rx\| = \sup_{\theta\in\Delta}\|(Rx)(\theta)\| = \sup_{\theta\in\Delta}\left|\sum_{(n,i)\in\mathcal{T}} a_{n,i}h_{n,i}(\theta)\right| =$$

$$= \sup_{\theta\in\Delta}\left|\sum_{\{(n,i):\theta\in\Delta_{n,i}\}} a_{n,i}\right| = \sup_{B\in\Gamma}|f_B(x)| \le \|x\|.$$

Hence $\|R\| \le 1$, and since $\|R\eta_{n,i}\| = 1$, we have $\|R\| = 1$.

As an illustration of the usefulness of this representation, we will show that all the even conjugates of JT, including JT, have the Radon-Nikodym property while the odd ones don't. First recall the following concepts.

Definition 3.c.8. A Banach space X has the Radon-Nikodym property (RNP for short), if for any finite measure space $(\mathfrak{S},\Sigma,\mu)$, any μ-continuous X valued measure m on Σ of finite total variation is the indefinite integral with respect to μ of an X valued Bochner measurable function on $\mathfrak{S}$.

Definition 3.c.9. A subset S of a Banach space will be called dentable, if for every $\varepsilon > 0$ there is an $x \in S$ not contained in the closed convex hull of the set $S\setminus B(x,\varepsilon)$, where $B(x,\varepsilon)$ is the closed ball about x of radius ε.

For a detailed study of spaces having the RNP we refer the reader to the book of Diestel [1]. For our purposes the following two results are required:

(i) A Banach space X has the RNP if and only if every bounded subset of X is dentable.

(ii) Separable conjugate spaces and reflexive spaces have the RNP.

Proposition 3.c.10. For k = 0, 1,..., all the spaces $\mathfrak{B}^{(2k+1)}$ have the RNP, while the spaces $\mathfrak{B}^{(2k)}$ fail to have the RNP. In particular, $\mathfrak{B}$ is not a subspace of a separable conjugate space.

Proof: By (ii) above, the first assertion follows immediately from Corollary 3.c.7. In order to prove the second (and third) assertion it is enough to show that $\mathfrak{B}$ does not have the RNP. Let R as before be given by $R\eta_{ni} = h_{n,i}$ and let μ be the Haar measure on Δ defined by

$$\int h_{n,i}\, d\mu = 2^{-n} \text{ for all n and i.}$$

For $f \in C(\Delta)$ define

$$H_{n,i}(f) = \int_{\Delta_{n,i}} f d\mu.$$

Then $H_{n,i} \in C(\Delta)^*$; in fact $|H_{n,i}(f)| \le 2^{-n}\|f\|$ and $H_{n,i}(h_{n,i}) = 2^{-n}$. Hence $\|H_{n,i}\| = 2^{-n}$. Now let $y^*_{n,i} = 2^n R^*(H_{n,i}) \in JT^*$. Then

$$y^*_{n,i}(\eta_{n,i}) = 2^n H_{n,i}(h_{n,i}) = 1$$

and for $0 \le j(m) \le 2^m - 1$,

$$|y^*_{n,i}(\eta_{m,j(m)})| = 2^n H_{n,i}(h_{m,j(m)}) = 2^n \mu(\Delta_{n,i} \cap \Delta_{m,j(m)}) \le 2^{n-m}.$$

Hence, as $m \longrightarrow \infty$ $y^*_{n,i}(\eta_{m,j(m)}) \longrightarrow 0$, and by Theorem 3.c.3, $y^*_{n,i} \in \mathfrak{B}$ for all n and i. Also

(1) $$\|y^*_{n,i}\| \le 2^n \|R^*\| \|H_{n,i}\| = 1.$$

By (2) after Corollary 3.c.7 we have that

(2) $$y^*_{n,i} = (y^*_{n+1,2i} + y^*_{n+1,2i+1})/2.$$

Finally

(3) $$\|y^*_{n,i} - y^*_{n+1,j}\| \ge |y^*_{n+1,j}(\eta_{n+1,j})| - |y^*_{n,i}(\eta_{n+1,j})| \ge 1 - 2^{n-(n+1)} = \frac{1}{2}.$$

Therefore $S = \{y^*_{n,i}\}$ is a bounded non-dentable set, since for every $\varepsilon < \frac{1}{2}$, $y^*_{n,i}$ is contained in the closed convex hull of $S \setminus B(y^*_{n,i},\varepsilon)$ for every n and i. By (i) above we have the desired result.

This last result gives us another essential difference between J and JT, since the predual I of J is isomorphic to J^* which is a separable conjugate space.

By a theorem due to Lindenstrauss (see e.g. Diestel [1]), it is known that if a Banach space X has the RNP, then every closed bounded convex subset of X is the norm closed convex hull of its extreme points. This is known as the Krein-Milman property (KMP) and thus, as a corollary to the above proposition we obtain that JT has KMP.

3.d. The norms of JT and JT* have the Kadec-Klee property

A classical result by M. Kadec, V. Klee and E. Asplund (see e.g. Diestel [1]) states that every Banach space X with a separable dual X^* has an equivalent norm, such that the dual norm on X^* is locally uniformly convex. This in turn implies that every point of the unit sphere is strongly exposed (see e.g. Day [1]), and from this it follows that on the unit sphere of X^* the weak and strong topologies coincide. In this section we will show that although JT^* is not separable, both the norm of JT and the norm of JT^* still have the Kadec-Klee property. The proof for JT is very simple, while the one for JT^* requires several technical intricacies. Most of the results here can be found in Schachermayer [1].

We start with the formal definition of the Kadec-Klee property:

Definition 3.d.1. The norm in a Banach space X has the Kadec-Klee property, if the weak topology and the norm topology coincide on the unit sphere of X.

As we mentioned at the beginning, if a space X is locally uniformly convex, then its norm has the Kadec-Klee property. We will prove that JT and JT^* with their usual norms are not locally uniformly convex, but as was shown in Corollary 3.c.7, JT is weakly compactly generated and this, by Troyanski's theorem (see e.g. Diestel [1]), implies that JT admits an equivalent locally uniformly convex norm which consequently also has the Kadec-Klee property.

Definition 3.d.2. A Banach space X is strictly convex, if for every $x, y \in X$ with $x \neq y$ and $\|x\| = \|y\| = 1$ we have $\|(x + y)/2\| < 1$, that is, every x on the unit sphere S_X of X is extreme.

A Banach space X is locally uniformly convex, if for every $x \in S_X$ and every sequence $\{x_n\}_{n=1}^{\infty} \subset S_X$ such that $\lim_{n\to\infty} \|(x + x_n)/2\| = 1$, we have that $\lim_{n\to\infty} \|x - x_n\| = 0$.

Clearly every locally uniformly convex space is strictly convex.

Lemma 3.d.3. JT and JT^* are not strictly convex and therefore not locally uniformly convex.

Proof: Let $x_1 = \frac{1}{4}\eta_{(0,0)} + \frac{3}{4}\eta_{(1,0)}$ and $x_2 = \frac{3}{4}\eta_{(0,0)} + \frac{1}{4}\eta_{(1,0)}$ be elements in JT; then clearly

$$\|x_1\| = \|x_2\| = 1 \text{ and } \|(x_1 + x_2)/2\| = 1.$$

In JT^* let $x_1^* = \eta^*_{(0,0)} + \eta^*_{(1,0)}$ and $x_2^* = \eta^*_{(0,0)} + \eta^*_{(1,1)}$; it is not difficult to show that

$$\|x_1^*\| = \|x_2^*\| = 1,$$

and if $x^* = (x_1^* + x_2^*)/2 = \eta^*_{(0,0)} + (\eta^*_{(1,0)} + \eta^*_{(1,1)})/2$, then $\|x^*\| = 1$.

To prove that JT has the Kadec-Klee property we apply the definition of this property directly.

Theorem 3.d.4. The norm in JT has the Kadec-Klee property.

Proof: We want to show that every ball in S_{JT} contains a weak neighborhood. So let $x \in JT$ with $\|x\| = 1$ and $1 > \varepsilon > 0$. Let $\delta < \varepsilon^2/8$ and $n \in \mathbb{N}$ be such that

(1) $$n > 2/\delta,$$

(2) $$\|P_n x\| > 1 - \delta/2,$$

(3) $$\|Q_{n+1} x\| < \delta,$$

where P_n and Q_n are as in Definition 3.a.4. Consider the weak neighborhood of x in B_{JT} given by

$$\mathcal{V} = \{z \in JT : \|z\| \le 1, \|P_n(x - z)\| < 1/n\}.$$

First we show that $\mathcal{V}$ is indeed a relative weak neighborhood of x: Since the unit ball B of $P_n^* JT^*$ is compact, there exists a sequence $\{y_i^*\}_{i=1}^k$ in $P_n^* JT^*$ so that B is contained in the union of the balls with center y_i^* and radius $\frac{1}{4n}$. Then, since $\|P_n(x - z)\| = \sup_{y^* \in B} |y^*(x - z)|$, it is easy to see that the weak neighborhood of x

$$\{z \in JT : \|z\| \le 1 \text{ and } |y_i^*(x - z)| < \frac{1}{3n} \text{ for } i = 1,\dots,k\}$$

is contained in $\mathcal{V}$. Now, if $z \in \mathcal{V}$,

$$\|P_n z\| = \|P_n x - P_n(z - x)\| \ge \|P_n x\| - \|P_n(z - x)\| > 1 - \delta.$$

Then, using Lemma 3.a.3 we obtain

$$1 \ge \|z\|^2 \ge \|P_n z\|^2 + \|Q_{n+1} z\|^2 > (1 - \delta)^2 + \|Q_{n+1} z\|^2.$$

Thus $\|Q_{n+1} z\|^2 < 2\delta$. Hence

$$\|x - z\| \le \|P_n(x - z)\| + \|Q_{n+1} x\| + \|Q_{n+1} z\| < \varepsilon.$$

Therefore V is contained in the ball with center x and radius ε, and this finishes the proof.

Observe that the above theorem in particular yields that the norm $\| \|_J$ of J defined in Section 3.a also has the Kadec-Klee property.

To prove the Kadec-Klee property for the norm of JT^*, we will start by introducing some notations:

For N = 0, 1,... and $t \in \mathcal{T}$ let P_N, Q_t, Q_N be as in Definition 3.a.4, let Γ be the set of 0-branches of $\mathcal{T}$ and $S : JT^* \longrightarrow \ell_2(\Gamma)$ be the surjective, norm-one operator defined in Theorem 3.c.3 as $Sx^* = \{\lim_{t\in B} x^*(\eta_t)\}_{B\in\Gamma}$. Let

$$\Pi_N = P_N^* - P_{N-1}^*,\ \Pi_{[M,N]} = P_N^* - P_{M-1}^*,\ \Pi_t = Q_t^*,\ \Pi_{[N,\infty)} = Q_N^*,$$

where we understand that $P_{-1}^* \equiv 0$; clearly all of these operators are projections of norm one on JT^* and

$$\Pi_N x^* = \sum_{\mathrm{lev}(t)=N} x^*(\eta_t)\eta_t^*,\quad \Pi_{[M,N]} x^* = \sum_{M\le \mathrm{lev}(t)\le N} x^*(\eta_t)\eta_t^*,$$

$$\Pi_t x^* = \sum_{(s:s\ge t)} x^*(\eta_s)\eta_s^* \quad\text{and}\quad \Pi_{[N,\infty)} x^* = \sum_{\mathrm{lev}(t)\ge N} x^*(\eta_t)\eta_t^*.$$

Finally let $\Pi_\infty = S$.

In this section, when talking about segments in $\mathcal{T}$ we will mean either a finite segment or any n-branch as in Definition 3.a.1; such a branch we will call an infinite segment.

Next we will define the concept of a simple function on JT, and we will show that the set of these functions is dense in JT^*. Some of the theorems we are going to state for JT^* are relatively easy to prove for simple functions, and then it is only a matter of density arguments, to conclude the proof for the general case.

Definition 3.d.5. A simple function is a function $m : JT \longrightarrow \mathbb{R}$ of the form

$$m = \sum_{j=1}^{n} \lambda_j f_{S_j},$$

where $\{S_j\}_{j=1}^n$ is a set of pairwise disjoint finite or infinite segments in $\mathcal{T}$, $\{\lambda_j\}_{j=1}^n \subset \mathbb{R}$ and f_S is the function defined in 3.a.4, that is for $x \in$ JT,

$$m(x) = \sum_{j=1}^n \lambda_j \sum_{t\in S_j} \langle \eta_t^*, x\rangle.$$

Observe that the representation of a simple function is not unique and that the sum of simple functions can be expressed as a simple function. Also

$$|m(x)| \le (\sum_{j=1}^n \lambda_j^2)^{1/2} (\sum_{j=1}^n (\sum_{t\in S_j} \langle \eta_t^*, x\rangle)^2)^{1/2} \le (\sum_{j=1}^n \lambda_j^2)^{1/2} \|x\|,$$

and thus

(*) $$\|m\| \le (\sum_{j=1}^n \lambda_j^2)^{1/2}.$$

Let $\mathcal{M}$ be the set

$$\{\sum_{j=1}^n \lambda_j f_{S_j} : \sum_{j=1}^n \lambda_j^2 \le 1, \text{ where the } S_j \text{ are pairwise disjoint segments of } \mathcal{T}\}$$

and let $\mathcal{MI}$ be the set

$$\{\sum_{j=1}^\infty \lambda_j f_{S_j} : \sum_{j=1}^\infty \lambda_j^2 \le 1 \text{ where the } S_j \text{ are pairwise disjoint segments of } \mathcal{T}\}.$$

It is clear that $\mathcal{M} \subset \mathcal{MI} \subset B_{JT^*}$, where B_{JT^*} denotes the closed unit ball in JT^*.

From here on, $\overline{A}$ will denote the closure of A, $\overline{A}^{w^*}$ the w^*-closure of A, coA the convex hull of A, $\overline{co}A$ the norm closure of coA and $\overline{co}^{w^*}A$ the w^*-closure of coA.

Lemma 3.d.6. $\mathcal{MI}$ is the w^*-closure of $\mathcal{M}$ and this implies that $\mathcal{M}$ is $\|\ \|$-dense in $\overline{\mathcal{M}}^{w^*}$.

Proof: First we will see that $\mathcal{MI}$ is w^*-closed in JT^*. Let $\{m_a\}_{a\in A} \subset \mathcal{MI}$, given by

$$m_a = \sum_{j=1}^\infty \lambda_j^a f_{S_j^a},$$

be a w^*-convergent net.

Consider the basis $\{\eta_{t_n}\}$ of JT in the usual order. We may suppose by re-arranging terms that if $t_{r_j^a}$ is the first element of S_j^a, then $r_j^a < r_{j+1}^a$.

For every $t\in\mathcal{T}$ let $S_{j(a,t)}^a$ be the segment such that $t\in S_k^a$ if $m_a(\eta_t) \neq 0$

and let $S^a_{j(a,t)} = \emptyset$ and $\lambda^a_{j(a,t)} = 0$ if $m_a(\eta_t) = 0$.

If $\{m_a(\eta_t)\}_{a\in A}$ converges to $\lambda_t \neq 0$, it follows that $\{\lambda^a_{j(a,t)}\}_{a\in A}$ converges to λ_t and $\{f_{S^a_{j(a,t)}}(\eta_t)\}_{a\in A}$ converges to 1.

Let $\{t_{n_i}\}$ be the set of nodes such that $\{m_a(\eta_{t_{n_i}})\}_{a\in A}$ converges to $\lambda_{t_{n_i}} \neq 0$ taken in the usual order (see Definition 3.a.2). By our assumption, for every i and for every subnet of $\{m_a\}_{a\in A}$ there is a subnet such that $j(a,t_{n_i}) \leq n_i$.

Hence for t_{n_1} there exists a subnet $\{m_a\}_{a\in A_1}$ of $\{m_a\}_{a\in A}$ and there exists $j_1 \leq n_1$ such that $j(a,t_{n_1}) = j_1$ for $a \in A_1$. This implies that $\{\lambda^a_{j_1}\}_{a\in A_1}$ converges to $\lambda_{t_{n_1}}$.

Proceeding inductively, for t_{n_i} there exists a subnet $\{m_a\}_{a\in A_i}$ of $\{m_a\}_{a\in A_{i-1}}$ and there exists $j_i \leq n_i$ such that $j(a,t_{n_i}) = j_i$ for $a \in A_i$, which implies that $\{\lambda^a_{j_i}\}_{a\in A_i}$ converges to $\lambda_{t_{n_i}}$.

Now let $S_k = \{t_{n_i} : j(a,t_{n_i}) = k$ for $a \in A_i\}$. If $t_{n_i}, t_{n_j} \in S_k$ and $n_j > n_i$, then $t_{n_i}, t_{n_j} \in S^a_k$ for $a \in A_j$ and thus S_k is a segment. It also follows from this that $\{\lambda^a_k\}_{a\in A_j}$ converges to $\lambda_{t_{n_i}} = \lambda_{t_{n_j}}$ and we will call this common limit λ_k. If $S_k = \emptyset$, let $\lambda_k = 0$.

Consequently if $m = \sum_{j=1}^{\infty} \lambda_j f_{S_j}$, m is the w^*-limit of the net $\{m_a\}_{a\in A}$.

Let N be fixed, $R = \{k \leq N : S_k \neq \emptyset\}$ and for $k \in R$ let t_{r_k} be the first element in S_k. Let $M = \max\{r_i : i \in R\}$. Then we have that for $k \in R$ the net $\{\lambda^a_k\}_{a\in A_M}$ converges to λ_k and since $\sum_{k=1}^{N}(\lambda^a_k)^2 \leq 1$, this implies that $\sum_{k=1}^{N}(\lambda_k)^2 \leq 1$. Thus $m \in \mathcal{MS}$, $\mathcal{MS}$ is w^*-closed,

$$\mathcal{MS} \subset \overline{\mathcal{M}} \subset \overline{\mathcal{M}}^{w^*} \subset \overline{\mathcal{MS}}^{w^*} = \mathcal{MS}$$

and therefore $\overline{\mathcal{M}}^{w^*} = \mathcal{MS}$.

Proposition 3.d.7. The unit ball of JT^* is the norm closure of the convex hull of $\mathcal{M}$.

Proof: Let $x \in \mathrm{JT}$, $x = \sum_{t\in\mathcal{T}} a_t \eta_t \neq 0$. Let $\varepsilon > 0$ and $S_1,\dots,S_k$ be pairwise disjoint segments such that

$$\|x\|^2 - \varepsilon < \sum_{j=1}^{k} (\sum_{t\in S_j} a_t)^2.$$

Define

$$m = \|x\|^{-1}\left(\sum_{j=1}^{k} (\sum_{t\in S_j} a_t) f_{S_j}\right).$$

Then by (*) after Definition 3.d.5, $\|m\| \leq 1$ and thus

$$m(x) = \|x\|^{-1}\left(\sum_{j=1}^{k} (\sum_{t\in S_j} a_t)^2\right) > \|x\|^{-1}(\|x\|^2 - \varepsilon) = \|x\| - \varepsilon\|x\|^{-1}.$$

Therefore, since for every $m \in \mathcal{M}$, $|m(x)| \leq \|x\|$, we get

$$\|x\| = \sup_{m\in\mathcal{M}} |m(x)|. \tag{1}$$

Now we will see that the w^*-closure of the convex hull of $\mathcal{M}$ is equal to B_{JT^*}.

Let $x^* \in \mathrm{JT}^*$ with $x^* \notin \overline{\mathrm{co}}^{w^*} \mathcal{M}$; by the Hahn-Banach theorem there exists x in the w^*-dual of JT^* with $\|x\| = 1$ such that

$$\sup\left\{\langle x, y^*\rangle : y^* \in \overline{\mathrm{co}}^{w^*} \mathcal{M}\right\} < \langle x, x^*\rangle \leq \|x^*\|.$$

By (1), the left side of this inequality is greater than or equal to $\|x\| = 1$ and hence $\|x^*\| > 1$, $x^* \notin B_{\mathrm{JT}^*}$, and thus $B_{\mathrm{JT}^*} \subset \overline{\mathrm{co}}^{w^*} \mathcal{M}$. On the other hand, as $\mathcal{M} \subset B_{\mathrm{JT}^*}$ and B_{JT^*} is convex and w^*-closed, we get that $\overline{\mathrm{co}}\ \overline{\mathcal{M}}^{w^*} \subset B_{\mathrm{JT}^*}$ and $\overline{\mathrm{co}}^{w^*} \mathcal{M} \subset B_{\mathrm{JT}^*}$. Thus

$$\overline{\mathrm{co}}^{w^*} \mathcal{M} = B_{\mathrm{JT}^*}. \tag{2}$$

Let x^* be an extreme point of B_{JT^*}; by (2) we can find a directed set A and for $\alpha \in A$, $y^*_\alpha, z^*_\alpha \in \mathcal{M}$ and $0 \leq \lambda_\alpha \leq 1$, so that the net $\lambda_\alpha y^*_\alpha + (1-\lambda_\alpha) z^*_\alpha$ w^*-converges to x^*. Since B_{JT^*} is w^*-compact, by passing to a subnet we can find $y^*, z^* \in \overline{\mathcal{M}}^{w^*}$ and $0 \leq \lambda \leq 1$ such that $x^* = \lambda y^* + (1-\lambda) z^*$. Since x^* is extreme, either $x^* = y^*$ or $x^* = z^*$, and hence $x^* \in \overline{\mathcal{M}}^{w^*}$. But by Corollary 3.c.7, JT does not contain ℓ_1, and thus we may apply the Theorem of Odell, Rosenthal and Haydon (see e.g. Diestel [2]), stating that a Banach space X does not contain ℓ_1 if and only if every w^*-compact convex subset of X^* is the norm-closure of the convex hull of its extreme points. From this and (2) it follows that the norm closure of $\mathrm{co}\overline{\mathcal{M}}^{w^*}$ is equal to B_{JT^*}. As by Lemma 3.d.6 $\mathcal{M}$ is norm dense in $\overline{\mathcal{M}}^{w^*}$, we get $B_{\mathrm{JT}^*} = \overline{\mathrm{co}}\ \overline{\mathcal{M}}^{w^*} = \overline{\mathrm{co}}\ \overline{\mathcal{M}} = \overline{\mathrm{co}}\ \mathcal{M}$ and this proves the theorem.

Next we show some properties of the norm in JT^*.

Lemma 3.d.8. Let $x^* \in JT^*$, then

(a) $\|x^*\| = \lim_{N\to\infty} \|\Pi_{[0,N]}x^*\|$,

(b) $\|\Pi_\infty x^*\| = \lim_{N\to\infty} \|\Pi_{[N,\infty)}x^*\| = \lim_{N\to\infty} \|\Pi_N x^*\|$.

Proof: (a) Let $\delta > 0$ and $x \in JT$ with $\|x\| = 1$ be such that

$$\langle x^*, x\rangle > \|x^*\|(1 - \delta/2)$$

and M such that for every $N \geq M$, $\|P_N x - x\| < \delta/2$. Then

$$\|x^*\| \geq \|\Pi_{[0,N]}x^*\| \geq |\langle \Pi_{[0,N]}x^*, x\rangle| = |\langle P_N^* x^*, x\rangle| = |\langle x^*, P_N x\rangle| =$$

$$= |\langle x^*, x\rangle + \langle x^*, P_N x - x\rangle| \geq |\langle x^*, x\rangle| - |\langle x^*, P_N x - x\rangle| >$$

$$> \|x^*\|(1 - \delta/2) - \|x^*\|\delta/2 = (1 - \delta)\|x^*\|.$$

Hence $\lim_{N\to\infty} \|\Pi_{[0,N]}x^*\| = \|x^*\|$.

(b) Let x^* be a simple function given by $x^* = \sum_{j=1}^k a_j f_{S_j}$ and

$$B = \{1 \leq j \leq k : S_j \text{ is a finite segment}\}.$$

Let M be large enough that $\Pi_{[M,\infty)} f_{S_j} \equiv 0$ for $j \in B$. Then, if $N \geq M$, since for $j \notin B$, S_j is an infinite segment, we can apply Lemma 3.c.1 taking $T_j x = Q_N P_{S_j} x$ and $\sigma_j = S_j$, to obtain

$$\|\Pi_{[N,\infty)}x^*\|^2 = \sum_{j\notin B} a_j^2 = \|\Pi_\infty x^*\|^2,$$

where the last equality follows by Theorem 3.c.3. On the other hand it is trivial that for every $N \geq M$,

$$\|\Pi_N x^*\|^2 = \sum_{j\notin B} a_j^2.$$

Hence the result is true for simple functions and by Proposition 3.d.7 also for every $x^* \in JT^*$.

The next three technical results give conditions for the norm of an element x^* in $\Pi_{[N,\infty)}JT^*$ to depend only on $\Pi_N x^*$.

Lemma 3.d.9. Let $N \in \mathbb{N}$ and $x^* \in JT^*$. For every node $(n,i) \in \mathcal{T}$ let $a(n,i) = \langle x^*, \eta_{(n,i)} \rangle$. Suppose that for every $(n,i) \in \mathcal{T}$ with $n \geq N$

$$a(n,i) \geq 0 \quad \text{and} \quad a(n,i) \geq a(n + 1,2i) + a(n + 1,2i + 1).$$

Then

$$\|\Pi_{[N,\infty)} x^*\| = \|\Pi_N x^*\|.$$

Proof: Let $x^* \in JT^*$. We will show that for every $M \geq N$

$$\|\Pi_{[N,M]} x^*\| = \|\Pi_N x^*\|,$$

and this, by Lemma 3.d.8(a) applied to $\Pi_{[N,\infty)} x^*$, proves the result. For every $j = 0,\dots,2^N - 1$ let

$$y_j^* = \sum_{n=0}^{M-N} \sum_{i=0}^{2^n-1} a(N + n,\ 2^n j + i)\eta^*_{(N+n,2^n j+i)} = \Pi_{(N,j)} \Pi_{[N,M]} x^*.$$

Then, if $S_{(N+n,2^n j+i)}$ denotes the segment starting at the node (N,j) and ending at the node $(N + n, 2^n j + i)$, we see that

$$y_j^* = \sum_{n=0}^{M-N} \sum_{i=0}^{2^n-1} \lambda_j(n,i) a(N,j) f_{S_{(N+n,2^n j+i)}},$$

where, if $a(N,j) \neq 0$,

$$\lambda_j(n,i) = \frac{a(N+n,2^n j+i) - a(N+n+1,2^{n+1} j+2i) - a(N+n+1,2^{n+1} j+2i+1)}{a(N,j)} \quad \text{if } n < M-N,$$

$$\lambda_j(M - N,\ i) = \frac{a(M, 2^{M-N} j+i)}{a(N,j)},$$

$$\lambda_j(n,i) = 0 \text{ otherwise.}$$

By hypothesis $\lambda_j(n,i) \geq 0$ and if $a(N,j) \neq 0$, then $\sum_{n=0}^{M-N} \sum_{i=0}^{2^n-1} \lambda_j(n,i) = 1$. Therefore $\|y_j^*\| \leq a(N,j)$ and since $y_j^*(\eta_{(N,j)}) = a(N,j)$, we have that $\|y_j^*\| = a(N,j)$. Applying Lemma 3.c.1 for $T_j = \Pi_{(N,j)} \Pi_{[N,M]} x^*$ and $\sigma_j = \{t \geq (N,j) : \text{lev}(t) \leq M\}$ we obtain

$$\|\Pi_{[N,M]} x^*\|^2 = \Big\|\sum_{j=0}^{2^N-1} \Pi_{(N,j)} \Pi_{[N,M]} x^*\Big\|^2 = \sum_{j=0}^{2^N-1} \|y_j^*\|^2 =$$
$$= \sum_{j=0}^{2^N-1} a(N,j)^2 = \|\Pi_N x^*\|^2,$$

and this finishes the proof of the lemma.

Corollary 3.d.10. Let $x^* = \sum_{i=1}^{n} \mu_i m_i$, where $\mu_i \geq 0$, m_i is a simple function of the form $m_i = \sum_{j=1}^{s(i)} \lambda_{i,j} f_{S_{i,j}}$ with

$$S_{i,j} \cap \{t \in \mathcal{T} : \text{lev}(t) \leq N\} \neq \emptyset$$

for $j = 1,\ldots,s(i)$, $i = 1,\ldots,n$ and such that

(1) $$\lambda_{i,j} \geq 0 \text{ if } S_{i,j} \cap \{t \in \mathcal{T} : \text{lev}(t) = N\} \neq \emptyset.$$

Then

$$\|\Pi_{[N,\infty)} x^*\| = \|\Pi_N x^*\|.$$

Proof: For every (i,j), if $m \geq N$ and $(m,k) \in S_{i,j}$, then at most one of $(m + 1,2k)$ and $(m + 1,2k + 1)$ can belong to $S_{i,j}$; thus by (1) we may apply Lemma 3.d.9 and the result follows.

Lemma 3.d.11. Let $N \in \mathbb{N}$, $x^* \in JT^*$ with $x^* = \sum_{i=1}^{n} \mu_i m_i$, where $\mu_i \geq 0$, and the m_i are simple functions of the form $m_i = \sum_{j=1}^{s(i)} \lambda_{i,j} f_{S_{i,j}}$ such that for every $j = 1,\ldots,s(i)$, $i = 1,\ldots,n$ and every node t with $\text{lev}(t) = N$,

(1) $S_{i,j} \cap \{s \in \mathcal{T} : \text{lev}(s) \leq N\} \neq \emptyset$,

(2) if $x^*(\eta_t) \geq 0$, then $\lambda_{i,j} \geq 0$ for every i,j such that $t \in S_{i,j}$,

(3) if $x^*(\eta_t) \leq 0$, then $\lambda_{i,j} \leq 0$ for every i,j such that $t \in S_{i,j}$.

Then

$$\|\Pi_{[N,\infty)} x^*\| = \|\Pi_N x^*\|.$$

Proof: Let $t \in \mathcal{T}$ with $\text{lev}(t) = N$ and define the operator $L_t : JT^* \to \Pi_{[N,\infty)} JT^*$ by

$$(L_t x^*)(\eta_s) = \begin{cases} 0 & \text{if } \text{lev}(s) < N, \\ x^*(\eta_s) & \text{if } s \in \sigma_1 = \{s : \text{lev}(s) \geq N \text{ and } s \text{ is not a descendant of } t\}, \\ -x^*(\eta_s) & \text{if } s \in \sigma_2 = \{s : \text{lev}(s) \geq N \text{ and } s \text{ is a descendant of } t\}, \end{cases}$$

where in this case we regard t as a descendant of itself.

By Lemma 3.c.1 applied to $T_1 = (I - Q_t)Q_N$, $T_2 = Q_t Q_N$, σ_1 and σ_2,

$$\|L_t x^*\|^2 = \|(T_1^* - T_2^*)x^*\|^2 = \|T_1^* x^*\|^2 + \|T_2^* x^*\|^2 = \|(T_1^* + T_2^*)x^*\|^2 =$$

$$= \|Q_N^* x^*\|^2 = \|\Pi_{[N,\infty)} x^*\|^2$$

and thus L_t is an isometric isomorphism on $\Pi_{[N,\infty)}JT^*$. Let

$$\{t : \text{lev}(t) = N,\ x^*(\eta_t) < 0\} = \{t_1, t_2, \dots, t_k\}.$$

Then $y^* = L_{t_1}L_{t_2}\dots L_{t_k}x^*$ satisfies the conditions of Corollary 3.d.10 and therefore

$$\|\Pi_{[N,\infty)}x^*\| = \|\Pi_{[N,\infty)}y^*\| = \|\Pi_N y^*\| = \|\Pi_N x^*\|.$$

In the general case it is not always true that $\|\Pi_{[N,\infty)}x^*\| = \|\Pi_N x^*\|$, however, Lemma 3.d.8 shows that $\|\Pi_{[N,\infty)}x^*\|$ and $\|\Pi_N x^*\|$, as well as $\|\Pi_{[0,N]}x^*\|$ and $\|x^*\|$, get closer as N increases. In Proposition 3.d.15 we will show how the N's in the two cases are related, and these results will be very useful in the task we are pursuing. In order to do this, we need the following Jensen type inequality and a combinatorial lemma which are both necessary to prove a result about the "improvement" of representations of simple functions in the following sense: if

$$x^* = \sum_{i=1}^n \mu_i m_i = \sum_{i=1}^k \rho_i m'_i$$

with $m_i, m'_i \in \mathcal{M}$, and $\sum_{i=1}^n \mu_i, \sum_{i=1}^k \rho_i \le 1$, $\mu_i, \rho_i > 0$, then the first representation is "better" than the second, if

$$\sum_{i=1}^n \mu_i < \sum_{i=1}^k \rho_i.$$

Observe that in this case $\|x^*\| \le \sum_{i=1}^n \mu_i < \sum_{i=1}^k \rho_i$.

Lemma 3.d.12. Let $\mu_i, r_i \in [0,1]$ for $i = 1,\dots,n$, with $0 < \sum_{i=1}^n \mu_i = \mu \le 1$. Then

$$\sum_{i=1}^n \mu_i(1 - r_i^2/2) \le \mu - ((\sum_{i=1}^n \mu_i r_i)^2/2).$$

Proof: Let $f(r) = 1 - (r^2/2)$ for $0 \le r \le 1$. Since $\sum_{i=1}^n \mu_i/\mu = 1$, we may apply Jensen's inequality to the concave function f (see e.g. Rudin [2]) to obtain

$$\sum_{i=1}^n (\mu_i/\mu)(1 - r_i^2/2) \le 1 - (\sum_{i=1}^n \mu_i r_i/\mu)^2/2.$$

Hence

$$\sum_{i=1}^n \mu_i(1 - r_i^2/2) \le \mu - \mu^{-1}(\sum_{i=1}^n \mu_i r_i)^2/2 \le \mu - (\sum_{i=1}^n \mu_i r_i)^2/2.$$

Lemma 3.d.13. Suppose that $a_i>0$ for $i = 1,\dots,n$, $b_r>0$ for $r = 1,\dots,m$ and

$$\sum_{i=1}^{n} a_i = \sum_{r=1}^{m} b_r.$$

Then for $i = 1,\dots,n$, $r = 1,\dots,m$ there exist $p(i), q(r) \in \mathbb{N}$, and for $j = 1,\dots,p(i)$, $k = 1,\dots,q(r)$ there are $\lambda_{ij} > 0$ and $\gamma_{rk} > 0$, such that

$$\sum_{j=1}^{p(i)} \lambda_{ij} = 1, \quad \sum_{k=1}^{q(r)} \gamma_{rk} = 1 \quad \text{and} \quad \sum_{i=1}^{n} p(i) = \sum_{r=1}^{m} q(r).$$

There also exists a bijective function

$f : \{(i,j) : 1 \le j \le p(i),\ 1 \le i \le n\} \longrightarrow \{(r,k) : 1 \le k \le q(r),\ 1 \le r \le m\}$

so that if $f(i,j) = (r,k)$, then

$$\lambda_{ij} a_i = \gamma_{rk} b_r,$$

and if we order the sets $\{(i,j): j \le p(i),\ i \le n\}$ and $\{(r,k): 1 \le k \le q(r),\ 1 \le r \le m\}$ with the order $(i,j) < (l,m)$ if $i < l$ or if $i = l$ and $j < m$, then

(1) $$(i,j) \le (i',j') \text{ implies } f(i,j) \le f(i',j').$$

Proof: Let

$$\{a_1, a_1+a_2, \dots, a_1+\dots+a_n\} \cup \{b_1, b_1+b_2, \dots, b_1+\dots+b_m\} = \{d_j : j = 1,\dots,s\}$$

where $d_1 < d_2 < \dots < d_s = \sum_{i=1}^{n} a_i = \sum_{r=1}^{m} b_r$.

Let $c_1 = d_1$, $c_2 = d_2 - d_1, \dots,$ $c_s = d_s - d_{s-1}$. Then obviously $d_j = \sum_{k=1}^{j} c_k$ for every $j = 1,\dots,s$. By the definition of $\{d_j\}_{j=1}^{s}$, for every i, $1 \le i \le n$, there exists $u(i)$ such that $\sum_{j=1}^{i} a_j = d_{u(i)}$. Define $u(0) = 0$ and $d_0 = 0$. Then

$$a_i = d_{u(i)} - d_{u(i-1)} = \sum_{k=u(i-1)+1}^{u(i)} c_k,$$

Let $p(i) = u(i) - u(i-1)$ and for $j = 1,\dots,p(i)$ let λ_{ij} satisfy $c_{u(i-1)+j} = \lambda_{ij} a_i$. Clearly $0 < \lambda_{ij}$ and $\sum_{j=1}^{p(i)} \lambda_{ij} = 1$. Observe that every $k = 1,\dots,s$ is of the form $u(i-1) + j$ with $1 \le j \le p(i)$ for some $1 \le i \le n$.

Applying the same reasoning to $\{b_r\}$, there exist r, k, m and γ_{rk} such that $1 \le r \le m$, $0 < \gamma_{rk} \le 1$ and there exists a function $v(r)$ such that

$$\lambda_{ij} a_i = c_{u(i-1)+j} = c_{v(r-1)+k} = \gamma_{rk} b_r \quad \text{and} \quad \sum_{k=1}^{q(r)} \gamma_{rk} = 1,$$

where $q(r) = v(r) - v(r-1)$. Hence if we define $f(i,j) = (r,k)$, f satisfies (1); this finishes the proof.

If $x^* = \sum_{i=1}^n \mu_i m_i$ with $m_i \in \Pi_{[0,N]}\mathcal{M}$, and $\mu_i > 0$, then the representation of x^* can be "improved" so that $\operatorname{sgn} m_i(\eta_t) = \operatorname{sgn} m_j(\eta_t)$ for all t with lev(t) = N, as is seen in the following proposition.

Proposition 3.d.14. Let $N \in \mathbb{N}$ and $x^* \in JT^*$, $x^* = \sum_{i=1}^n \mu_i m_i$ with $\mu_i > 0$ and $\sum_{i=1}^n \mu_i \le 1$, where $m_i = \sum_{r=1}^{s(i)} \lambda_{i,r} f_{S_{i,r}} \in \mathcal{M}$ and for $1 \le i \le n$ $\{S_{i,r}\}_{r=1}^{s(i)}$ is a set of pairwise disjoint segments contained in $\mathcal{T} \cap \{t : \operatorname{lev}(t) \le N\}$.
For every $t \in \mathcal{T}$ with lev(t) = N define

$$I_1 = \{i \in \{1,\ldots,n\} : m_i(\eta_t) \ge 0\}, \quad I_2 = \{i \in \{1,\ldots,n\} : m_i(\eta_t) < 0\},$$

$$d_i^t = |m_i(\eta_t)|, \quad 1 \le i \le n, \quad d^t = \min\Big(\sum_{i\in I_1} \mu_i d_i^t, \sum_{i\in I_2} \mu_i d_i^t\Big) \quad \text{if } I_j \neq \emptyset, \quad j=1,2,$$

$d^t = 0$ otherwise. Finally define $d = (\sum_{\{t:\operatorname{lev}(t)=N\}} (d^t)^2)^{1/2}$.

Then there exists a representation of x^*, $x^* = \sum_{i=1}^p \mu'_i m'_i$ with $m'_i \in \mathcal{M}$, $\mu'_i > 0$ and

$$\sum_{i=1}^p \mu'_i \le \sum_{i=1}^n \mu_i - d^2/2.$$

In particular it follows that

$$\|x^*\|_{JT^*} \le \sum_{i=1}^n \mu_i - d^2/2.$$

Proof: Let $t \in \mathcal{T}$ with lev(t) = N be such that $d^t > 0$; this implies that there exist i and j with $\operatorname{sgn} m_i(\eta_t) \neq \operatorname{sgn} m_j(\eta_t)$. Let

$$b^t = \max\Big(\sum_{i\in I_1} \mu_i d_i^t, \sum_{i\in I_2} \mu_i d_i^t\Big).$$

Then there exists $0<\lambda\le 1$ such that $\lambda b^t = d^t$; suppose that $b^t = \sum_{i\in I_1} \mu_i d_i^t$, the other case is similar. Thus

$$\sum_{i\in I_1} \lambda\mu_i m_i(\eta_t) = - \sum_{i\in I_2} \mu_i m_i(\eta_t).$$

Applying Lemma 3.d.13, there exist $\lambda_{ij} > 0$, for $i \in I_1$, $j = 1,\ldots,p(i)$, and $\gamma_{rk} > 0$, for $r \in I_2$, $k = 1,\ldots,q(r)$, and a bijective function f such that

(i) $\sum_{i\in I_1} p(i) = \sum_{r\in I_2} q(r) = s$,

(ii) $f : \{(i,j) : 1 \le j \le p(i), i\in I_1\} \longrightarrow \{(r,k) : 1 \le k \le q(r), r\in I_2\}$,

(iii) $\sum_{j=1}^{p(i)} \lambda_{ij} = \sum_{k=1}^{q(r)} \gamma_{rk} = 1$ for every $i \in I_1$ and $r \in I_2$,

(iv) $\lambda\lambda_{ij}\mu_i m_i(\eta_t) = -\gamma_{rk}\mu_r m_r(\eta_t)$, where $(r,k) = f(i,j)$
for $(i,j) \in \{(i,j):1 \le j \le p(i),\ i\in I_1\}$.

Since

$$x^*(\eta_t) = b^t - d^t = \lambda b^t - d^t + (1-\lambda)b^t$$

we obtain

$$x^* = \sum_{i\in I_1}\sum_{j=1}^{p(i)}\lambda\lambda_{ij}\mu_i m_i + \sum_{r\in I_2}\sum_{k=1}^{q(r)}\gamma_{rk}\mu_r m_r + \sum_{i\in I_1}(1-\lambda)\mu_i m_i.$$

We may write this as

(1) $$x^* = \sum_{j=1}^{s}\nu_j m_{j1} + \sum_{j=1}^{s}\rho_j m_{j2} + \sum_{i\in I_1}(1-\lambda)\mu_i m_i$$

with

(2) $$\nu_j m_{j1}(\eta_t) = -\rho_j m_{j2}(\eta_t) \quad \text{for } j = 1,\dots,s,$$

where every m_{j1} and m_{k2} is equal to m_i for some i and hence belongs to $\mathcal{M}$. Also

(3) $$\sum_{j=1}^{s}\nu_j + \sum_{j=1}^{s}\rho_j + \sum_{i\in I_1}(1-\lambda)\mu_i = \sum_{i=1}^{n}\mu_i.$$

Writing $d^t_{j1} = |m_{j1}(\eta_t)|$ and $d^t_{j2} = |m_{j2}(\eta_t)|$, by (2) we get

(4) $$\nu_j d^t_{j1} = \rho_j d^t_{j2},$$

and thus

(5) $$\sum_{j=1}^{s}\rho_j d^t_{j2} = \sum_{j=1}^{s}\nu_j d^t_{j1} = \sum_{i\in I_1}\sum_{j=1}^{p(i)}\lambda\lambda_{ij}\mu_i m_i(\eta_t) = \lambda b^t = d^t.$$

We will show that we may write

$$\sum_{j=1}^{s}\nu_j m_{j1} + \sum_{j=1}^{s}\rho_j m_{j2} = \sum_{j=1}^{s}\nu'_j m'_{j1} + \sum_{j=1}^{s}\rho'_j m'_{j2},$$

so that

$$\sum_{j=1}^{s}\nu'_j + \sum_{j=1}^{s}\rho'_j \le \sum_{j=1}^{s}\nu_j + \sum_{j=1}^{s}\rho_j - (d^t)^2/2.$$

Suppose $m_{j1}(\eta_t) \ne 0$ and $m_{j1} = \sum_{r=1}^{r(j)}\nu_{jr} f_{S^1_{jr}}$, $m_{j2} = \sum_{k=1}^{k(j)}\rho_{jk} f_{S^2_{jk}}$. Let S_{j1} be the segment $S^1_{jr_0}$ which contains t in the representation of m_{j1} and let S_{j2} be the segment containing t which appears in the representation of m_{j2}.

Let $S_j = S_{j1} \cap S_{j2}$; since lev(t) = N, both S_{j1} and S_{j2} end at node t; hence either $S_j = S_{j1}$ or $S_j = S_{j2}$. Now define

$$\mathcal{D}_j = \{s : \text{lev}(s) \le N \text{ and } s \notin S_j\}$$

and $P_{\mathcal{D}_j} : JT \longrightarrow JT$ as $P_{\mathcal{D}_j}(x) = \sum_{s\in\mathcal{D}_j}\langle\eta_s^*, x\rangle\eta_s$; then by (2) since m_{j1} and

m_{j2} are constant on S_j,

(6) $$\nu_j m_{j1} + \rho_j m_{j2} = \left(\nu_j m_{j1} + \rho_j m_{j2}\right) P_{\mathcal{D}_j} = \left(\nu_j m_{j1}\right) P_{\mathcal{D}_j} + \left(\rho_j m_{j2}\right) P_{\mathcal{D}_j}.$$

We have that $m_{j1} P_{\mathcal{D}_j}$, $m_{j2} P_{\mathcal{D}_j} \in \mathcal{M}$ because m_{j1}, $m_{j2} \in \mathcal{M}$ and $S^1_{jr} \cap \mathcal{D}_j$, $S^2_{jk} \cap \mathcal{D}_j$ are segments for $r = 1,\dots,r(j)$ and $k = 1,\dots,k(j)$.
Suppose that $S_{j1} = S_j$; then, since $S^1_{jr_0} = S_{j1} = S_j$, we get

$$m_{j1} P_{\mathcal{D}_j} = \sum_{r=1, r\neq r_0}^{r(j)} \nu_{jr} f_{S^1_{jr}} \cap \mathcal{D}_j,$$

and since $\nu_{jr_0} = d^t_{j1}$ and $r \le a \le 1$, the inequality $(a - r^2)^{1/2} \le 1 - r^2/2$ holds and we get

(7) $$\left(\sum_{r=1, r\neq r_0}^{r(j)} (\nu_{jr})^2\right)^{1/2} = \left(\sum_{r=1}^{r(j)} (\nu_{jr})^2 - \nu^2_{jr_0}\right)^{1/2} \le 1 - (d^t_{j1})^2/2.$$

Then, if we define

(8) $$\begin{cases} \nu'_j = \nu_j\left(1 - (d^t_{j1})^2/2\right), & m'_{j1} = \left(1 - (d^t_{j1})^2/2\right)^{-1} m_{j1} P_{\mathcal{D}_j}, \\ \rho'_j = \rho_j \text{ and } m'_{j2} = m_{j2} P_{\mathcal{D}_j}, & \end{cases}$$

by (7) and (8) m'_{j1}, $m'_{j2} \in \mathcal{M}$ and by (6),

(9) $$\nu_j m_{j1} + \rho_j m_{j2} = \nu'_j m'_{j1} + \rho'_j m'_{j2}.$$

Similarly, if $S_j = S_{j2}$ one obtains

(10) $$\begin{cases} \rho'_j = \rho_j\left(1 - (d^t_{j2})^2/2\right), & m'_{j2} = \left(1 - (d^t_{j2})^2/2\right)^{-1} m_{j2} P_{\mathcal{D}_j}, \\ \nu'_j = \nu_j \text{ and } m'_{j1} = m_{j1} P_{\mathcal{D}_j}. & \end{cases}$$

We proceed in the same manner for every j with $m_{j1}(\eta_t) \neq 0$. If $m_{j1}(\eta_t) = 0$, let $\nu'_j = \nu_j$ and $m'_{j1} = m_{j1}$, $\rho'_j = \rho_j$ and $m'_{j2} = m_{j2}$. Therefore using (1), (8), (9) and (10) we may represent x^* as

(11) $$x^* = \sum_{j=1}^{s} \nu'_j m'_{j1} + \sum_{j=1}^{s} \rho'_j m'_{j2} + \sum_{i\in I_1} (1 - \lambda)\mu_i m_i = \sum_i \mu'_i m'_i.$$

Let $A_1 = \{j : S_j = S_{j1}\}$, $A_2 = \{j : S_j = S_{j2}\}$, $A_3 = \{j : m_{j1}(\eta_t) = 0\}$.
By (4) and (5) we obtain

$$d^t = \sum_{j\in A_1}\nu_j d^t_{j1} + \sum_{j\in A_2}\rho_j d^t_{j2},$$

and using Lemma 3.d.12, (4) and (5) we get

$$\sum_{j=1}^s \nu'_j + \sum_{j=1}^s \rho'_j = \sum_{j\in A_1}(\nu'_j + \rho'_j) + \sum_{j\in A_2}(\nu'_j + \rho'_j) + \sum_{j\in A_3}(\nu'_j + \rho'_j) =$$

$$= \sum_{j\in A_1}(1 - (d^t_{j1})^2/2)\nu_j + \sum_{j\in A_2}(1 - (d^t_{j2})^2/2)\rho_j + \sum_{j\in A_1}\rho_j + \sum_{j\in A_2}\nu_j +$$

$$+ \sum_{j\in A_3}(\nu_j + \rho_j) \le \sum_{j\in A_1}\nu_j + \sum_{j\in A_2}\rho_j - \left(\sum_{j\in A_1}\nu_j d^t_{j1} + \sum_{j\in A_2}\rho_j d^t_{j2}\right)^2/2 +$$

$$+ \sum_{j\in A_1}\rho_j + \sum_{j\in A_2}\nu_j + \sum_{j\in A_3}(\nu_j + \rho_j) = \sum_{j=1}^s(\nu_j + \rho_j) - (d^t)^2/2.$$

By the above, by (3) and (11) we get

$$\sum_i \mu'_i \le \sum_{i=1}^n \mu_i - (d^t)^2/2.$$

Repeat this construction for another t with lev(t) = N, starting with the representation (11) for x^*. Applying the same procedure successively to every $t \in \mathcal{T}$ with lev(t) = N, we finally get $x^* = \sum_{j=1}^p \bar{\mu}_j \bar{m}_j$ with

$$\sum_{j=1}^p \bar{\mu}_j \le \sum_{i=1}^n \mu_i - \sum_{\{t:\mathrm{lev}(t)=N\}}(d^t)^2/2 = \sum_{i=1}^n \mu_i - d^2/2,$$

and this proves the proposition.

Observe that the above method to improve the norm can only be used once; if one finishes the above construction, in the new representation d = 0.

Proposition 3.d.15. Let $N \in \mathbb{N}$, $0 < \varepsilon < 1$ and $0 < \delta < 2^{-10}\varepsilon^3$. If $x^* \in JT^*$ with $\|x^*\| \le 1$ and $\|\Pi_{[0,N]}x^*\| > 1 - \delta$, then

$$\|\Pi_{[N,\infty)}x^*\| < \|\Pi_N x^*\| + \varepsilon.$$

Proof: By Proposition 3.d.7 it suffices to prove the result for functions of the type $x^* = \sum_{i=1}^n \mu_i m_i$ with $\mu_i > 0$ for $i = 1,\dots,n$ and $\sum_{i=1}^n \mu_i = 1$, where $m_i = \sum_{j=1}^{s(i)} \lambda_{i,j} f_{S_{i,j}} \in \mathcal{M}$.

For every i we divide the set $\{1,\dots,s(i)\}$ of indices into three parts:

$$I^i_1 = \{1 \le j \le s(i) : S_{i,j} \cap \{t \in \mathcal{T} : \mathrm{lev}(t) \le N\} = \emptyset\},$$

$$I^i_2 = \{1 \le j \le s(i) : \text{there exists } t \in S_{i,j} \text{ with } \mathrm{lev}(t) = N \text{ and}$$
$$\operatorname{sgn} x^*(\eta_t) \ne \operatorname{sgn} \lambda_{i,j}\},$$

$$I_3^i = \{1,\dots,s(i)\} \setminus (I_1^i \cup I_2^i),$$

where sgn a denotes the sign of a and by convention sgn $0 = +$.

For $r = 1, 2, 3$ let

$$m_{i,r} = \sum_{j\in I_r^i} \lambda_{i,j} f_{S_{i,j}} \quad \text{and} \quad x_r^* = \sum_{i=1}^n \mu_i m_{i,r}.$$

Observe that by the above, for every $t \in \mathcal{T}$ with lev$(t) = N$, sgn $m_{i2}(\eta_t)$ and sgn $m_{i3}(\eta_t)$ are constant for every $i = 1,\dots,n$.

Clearly x_3^* satisfies the hypotheses of Lemma 3.d.11; hence

(1) $$\|\Pi_{[N,\infty)} x_3^*\| = \|\Pi_N x_3^*\|.$$

Let $t \in \mathcal{T}$ with lev$(t) = N$. Then

$$x_2^*(\eta_t) = \sum_{\{i:\exists j\in I_2^i \text{ with } t\in S_{i,j}\}} \mu_i \lambda_{i,j}.$$

But sgn $x_2^*(\eta_t) =$ sgn $\lambda_{i,j}$ for every i for which there exists $j \in I_2^i$ with $t \in S_{i,j}$. Hence, again applying Lemma 3.d.11, we obtain

(2) $$\|\Pi_{[N,\infty)} x_2^*\| = \|\Pi_N x_2^*\|.$$

Since for every $t \in \mathcal{T}$ with lev$(t) = N$, $x^*(\eta_t) = x_2^*(\eta_t) + x_3^*(\eta_t)$ and sgn $x^*(\eta_t) =$ sgn $x_3^*(\eta_t) = -$sgn $x_2^*(\eta_t)$, we have that $|x_2^*(\eta_t)| \le |x_3^*(\eta_t)|$. For every t with lev$(t) = N$ and every $1 \le i \le n$, either $m_{i,2}(\eta_t) = 0$ or $m_{i,3}(\eta_t) = 0$. Therefore

(3) $$\|\Pi_N x_2^*\| = \Big(\sum_{\text{lev}(t)=N}\big(\sum_{i=1}^n \mu_i m_{i,2}(\eta_t)\big)^2\Big)^{1/2} = \Big(\sum_{\text{lev}(t)=N}(d^t)^2\Big)^{1/2} = d,$$

where d and d^t are the numbers associated to $\Pi_{[0,N]} x^*$ in Proposition 3.d.14. Thus using that proposition,

$$1 - d^2/2 = \sum_{i=1}^n \mu_i - d^2/2 \ge \|\Pi_{[0,N]} x^*\| > 1 - \delta > 1 - 2^{-10}\varepsilon^3 > 1 - \varepsilon^2/32,$$

and we obtain $d < \varepsilon/4$. Hence by (2) and (3)

(4) $$\|\Pi_{[N,\infty)} x_2^*\| < \varepsilon/4.$$

Now we will estimate $\|x_1^*\| = \|\Pi_{[N,\infty)} x_1^*\|$. Let

$$I_0 = \{i = 1,\dots,n : \|m_{i,2} + m_{i,3}\| < 1 - (\varepsilon^2/2^7)\}.$$

Since for every i, $\|m_{i,2} + m_{i,3}\| \le 1$, we get

$$1 - \delta < \|\Pi_{[0,N]}x^*\| = \|\Pi_{[0,N]}(x^*_2 + x^*_3)\| \le \sum_{i=1}^n \mu_i \|m_{i,2} + m_{i,3}\| \le$$

$$\le (1 - \sum_{i\in I_0} \mu_i) + (1 - (\varepsilon^2/2^7))\sum_{i\in I_0} \mu_i.$$

Hence by the hypothesis on δ,

$$\sum_{i\in I_0} \mu_i < 2^7\delta/\varepsilon^2 < \varepsilon/8.$$

As for every $i = 1,\dots,n$,

$$\|m_{i,1}\|^2 + \|m_{i,2} + m_{i,3}\|^2 \le \sum_{j\in I_1} \lambda^2_{ij} + \sum_{j\in I_2\cup I_3} \lambda^2_{ij} \le 1,$$

if $i \notin I_0$ we get

$$\|m_{i,1}\|^2 \le 1 - \|m_{i,2} + m_{i,3}\|^2 \le 1 - (1 - \varepsilon^2/2^7)^2 < (\varepsilon/8)^2.$$

Therefore $\|m_{i,1}\| < \varepsilon/8$ and

(5) $$\|\Pi_{[N,\infty)}x^*_1\| = \|x^*_1\| \le \sum_{i\in I_0} \mu_i \|m_{i,1}\| + \sum_{i\notin I_0} \mu_i \|m_{i,1}\| < \varepsilon/4.$$

Using (4) and $\|\Pi_N x^*_2\| \le \|\Pi_{[N,\infty)}x^*_2\|$,

$$\|\Pi_N x^*\| = \|\Pi_N x^*_2 + \Pi_N x^*_3\| \ge \|\Pi_N x^*_3\| - \|\Pi_N x^*_2\| \ge \|\Pi_N x^*_3\| - \varepsilon/4,$$

Finally by (1), (4) and (5)

$$\|\Pi_{[N,\infty)}x^*\| \le \|\Pi_{[N,\infty)}x^*_1\| + \|\Pi_{[N,\infty)}x^*_2\| + \|\Pi_{[N,\infty)}x^*_3\| \le$$

$$\le \|\Pi_N x^*_3\| + \varepsilon/2 \le \|\Pi_N x^*\| + \varepsilon.$$

From the above proposition we obtain the following interesting property of the norm in JT^*, and as a consequence of this, it follows that the predual $\mathfrak{B}$ of JT has a boundedly complete skipped blocking finite dimensional decomposition (see Definition 2.j.6).

Lemma 3.d.16. Let $\{x^*_n\}$ be a sequence with $x^*_n \in \Pi_n JT^*$ such that

(i) $\lim_{N\to\infty} \|\sum_{n=1}^N x^*_n\| < \infty$,

(ii) $\liminf_{n\to\infty} \|x^*_n\| = 0$.

Then $\{\sum_{n=1}^N x^*_n\}_N$ converges in norm.

Proof: If $\lim_{N\to\infty} \|\sum_{n=1}^N x^*_n\| = 0$, the result follows immediately; if $\lim_{N\to\infty} \|\sum_{n=1}^N x^*_n\| \ne 0$, then we may suppose that

(1) $$\lim_{N\to\infty} \|\sum_{n=1}^{N} x_n^*\| = 1.$$

Let $0 < \varepsilon < 1$, $0 < \delta < 2^{-10}\varepsilon^3$ and N be such that $\|x_N^*\| < \varepsilon(1 + \delta)$ and such that $1 + \delta > \|\sum_{n=1}^{M} x_n^*\| > 1 - \delta$ for every $M \geq N$; this may be done by (ii) and (1). Then by Proposition 3.d.15, for every $M \geq N$

$$\frac{1}{1+\delta}\|\Pi_{[N,\infty)}\sum_{n=1}^{M} x_n^*\| = \frac{1}{1+\delta}\|\sum_{n=N}^{M} x_n^*\| < \frac{1}{1+\delta}\|\Pi_N x_N^*\| + \varepsilon = \frac{1}{1+\delta}\|x_N^*\| + \varepsilon < 2\varepsilon.$$

This finishes the proof.

Corollary 3.d.17. The sequence of spaces $\{\Pi_n JT^*\}_{n=1}^{\infty}$ is a boundedly complete skipped blocking finite dimensional decomposition of the predual $\mathfrak{B}$ of JT.

Proof: First observe that $\Pi_n JT^* = \Pi_n \mathfrak{B}$ and thus, since $\mathfrak{B} = [\eta_t^*]_{t\in\mathcal{T}}$, obviously $\{\Pi_n JT^*\}_{n=1}^{\infty}$ is a monotone Schauder decomposition of $\mathfrak{B}$. Now let $\{n_k\}_{k=1}^{\infty}$ and $\{m_k\}_{k=1}^{\infty}$ be two sequences of natural numbers such that $m_k < n_k + 1 < m_{k+1}$. Let $x_k^* \in \sum_{i=m_k}^{n_k} \Pi_i JT^*$ be such that

$$A = \sup_n \|\sum_{k=1}^{n} x_k^*\| < \infty.$$

Thus $A = \lim_n \|\sum_{k=1}^{n} x_k^*\|$ and if $y_n^* = \Pi_n x_k^*$ for $m_k \leq n \leq n_k$ and $y_n^* = 0$ otherwise, then $y_{n_k+1}^* = 0$ and $\lim_N \|\sum_{n=1}^{N} y_n^*\| = A$. Hence by Lemma 3.d.16 $\sum_{k=1}^{\infty} x_k^* \in \mathfrak{B}$ and this finishes the proof.

A closely related notion to the Kadec-Klee property is that of points of weak star to norm continuity.

Definition 3.d.18. Let X be a Banach space, $x^* \in X^*$ with $\|x^*\| = 1$. Then x^* is a point of weak star to norm continuity of the unit ball of X^* if the identity function

$$I : B_{X^*} \to X^*$$

is continuous at x^*, where in B_{X^*} we consider the weak star topology and in X^* the norm topology.

After these preliminaries we are ready to prove the main results of this section.

First we will see that if $x^* \in \mathfrak{B}$ with $\|x^*\| = 1$, then x^* is a point of weak star to norm continuity of the unit ball of JT^*. This proves in particular that the relative norm neighborhoods in B_{JT^*} of x^* are relative weak star neighborhoods and therefore are also relative weak neighborhoods of x^* and also that the norm of $\mathfrak{B}$ has the Kadec-Klee property. Even more, the next two results prove that in fact the only points of weak star to norm continuity of the unit ball of JT^* are precisely those in $\mathfrak{B}$.

Proposition 3.d.19. Let $x^* \in \mathfrak{B}$ with $\|x^*\| = 1$. Then x^* is a point of weak star to norm continuity of the unit ball of JT^*. In fact, the sequence of relative weak star neighborhoods $\{\mathcal{V}_n(x^*)\}$ where

$$\mathcal{V}_n(x^*) = \{z^* \in JT^* : \|z^*\| \leq 1 \text{ and } \|\Pi_{[0,n]}(x^* - z^*)\| < 1/n\}$$

forms a relative neighborhood base in the norm topology in B_{JT^*}.

Proof: The proof that $\mathcal{V}_n(x^*)$ is indeed a relative w^*-neighborhood of x^* is similar to that given in Theorem 3.d.4.

Now let $0 < \varepsilon < 1$, $0 < \delta < 2^{-10}\varepsilon^3$ and $n \in \mathbb{N}$ with $n > 2/\delta$ be such that

$$(1) \qquad \|\Pi_{[0,n]}x^*\| > 1 - \delta/2$$

and

$$(2) \qquad \|\Pi_{[n,\infty)}x^*\| < \varepsilon.$$

This can be done by Lemma 3.d.8 and by the fact that for $x^* \in \mathfrak{B}$, $\Pi_\infty x^* = 0$. Let $z^* \in \mathcal{V}_n(x^*)$; by (1)

$$\|\Pi_{[0,n]}z^*\| = \|\Pi_{[0,n]}x^* - \Pi_{[0,n]}(z^* - x^*)\| > 1 - \delta.$$

Hence by Proposition 3.d.15

$$\|\Pi_{[n,\infty)}z^*\| \leq \|\Pi_n z^*\| + \varepsilon.$$

Therefore, since by (2)

$$\|\Pi_n z^*\| \leq \|\Pi_n x^*\| + \|\Pi_n(x^* - z^*)\| \leq \varepsilon + 1/n,$$

we get

$$\|x^* - z^*\| \leq \|\Pi_{[0,n-1]}(x^* - z^*)\| + \|\Pi_{[n,\infty)}x^*\| + \|\Pi_{[n,\infty)}z^*\| <$$
$$< \delta/2 + \varepsilon + 2\varepsilon + \delta/2 < 4\varepsilon,$$

and this finishes the proof.

Although for $x^* \in \mathfrak{B}$ with $\|x^*\| = 1$ the weak star neighborhoods are a relative base in the norm topology in B_{JT^*}, this is false for $x^* \in JT^* \setminus \mathfrak{B}$, as the following lemma shows.

Lemma 3.d.20. Let $x^* \in JT^* \setminus \mathfrak{B}$ with $\|x^*\| = 1$. Then, if S_{JT^*} denotes the unit sphere in JT^*,

$$\inf\{\operatorname{diam}(V \cap S_{JT^*}) : V \text{ is a } w^*\text{-neighborhood of } x^*\} \geq \|\Pi_\infty x^*\|.$$

Proof: Let $x^* \in JT^* \setminus \mathfrak{B}$ with $\|x^*\| = 1$ and

$$V = \{z^* \in S_{JT^*} : |\langle x^* - z^*, x_0 \rangle| < \varepsilon\}$$

for some $x_0 \in JT$. Let $n \in \mathbb{N}$ satisfy $1/n < \varepsilon/(2\|x_0\|)$ and $\|P_n x_0 - x_0\| < \varepsilon/4$, and let

(1) $$V_n(x^*) = \{z^* \in JT^* : \|z^*\| = 1 \text{ and } \|\Pi_{[0,n]}(x^* - z^*)\| < 1/n\}.$$

Then, if $z^* \in V_n(x^*)$, since $P_n^* = \Pi_{[0,n]}$, we get

$$|\langle x^* - z^*, x_0 \rangle| \leq |\langle x^* - z^*, P_n x_0 \rangle| + |\langle x^* - z^*, x_0 - P_n x_0 \rangle| \leq$$
$$\leq \|\Pi_{[0,n]}(x^* - z^*)\| \|x_0\| + \|x^* - z^*\| \|x_0 - P_n x_0\| \leq$$
$$\leq (1/n)\|x_0\| + 2\|x_0 - P_n x_0\| < \varepsilon.$$

Thus it is enough to prove the lemma for the neighborhoods of the form $V_n(x^*)$ as in (1).

Let $z^* = \Pi_{[0,\infty)} x^*$, then $\|\Pi_{[0,n]}(x^* - z^*)\| = 0$ and by Lemma 3.d.8

$$\|z^*\| = \|x^*\| = 1;$$

therefore $z^* \in V_n(x^*)$. But

$$\|x^* - z^*\| = \|\Pi_\infty x^*\|,$$

and the proof is complete.

As we saw above, the weak star neighborhoods of elements in $JT^* \setminus \mathfrak{B}$ have a large diameter, so in order to prove the Kadec-Klee property for JT^* we have to find enough weak neighborhoods with small diameter; for this the concept of slice will be very useful.

Definition 3.d.21. Let C be a closed, convex, bounded subset of a Banach space X. Then, if $f \in X^*$, $\|f\| = 1$, $\alpha > 0$ and $M_f = \sup\{\langle f, x \rangle : x \in C\}$, the set $\{x \in C : \langle f, x \rangle > M_f - \alpha\}$ is called a slice of C.

The next lemma gives a bound for the diameters of slices of balls in Hilbert spaces; since the subspace $\ell_2(\Gamma)$ of JT^* is a Hilbert space, this will enable us to find bounds for certain slices in JT^*.

Lemma 3.d.22. If H is a Hilbert space, $x \in H$, $\beta > 0$ and

$$S = S(x,\beta) = \{z \in H : \|z\| \le \|x\| \text{ and } \langle x, z\rangle_H > \|x\|^2 - \beta\},$$

where $\langle\ ,\ \rangle_H$ denotes the inner product in H, then diam $S < 2(2\beta)^{1/2}$.

Proof: Observe that S is a slice, taking C as the ball with center 0 and radius $\|x\|$, f as the function defined by $f(z) = \langle x/\|x\|, z\rangle$ and $\alpha = \beta/\|x\|$. Let $y \in S$, then

$$\|x - y\|^2 = \langle x - y, x - y\rangle = \langle x, x\rangle + \langle y, y\rangle - 2\langle x, y\rangle <$$
$$< 2\|x\|^2 - 2(\|x\|^2 - \beta) = 2\beta.$$

From here the result follows.

The following lemma describes a criterion to decide whether two elements in the unit ball of $\Pi_{[N,\infty)}JT^*$ are close to each other. Roughly speaking, it says that if two elements are close at level N and also at level ∞, they are close everywhere.

Lemma 3.d.23. Let $x^* \in JT^*$. If there exist $N \in \mathbb{N}$, $0 < \varepsilon < 2^{-6}$ and $0 < \delta < 2^{-30}\varepsilon^5$ such that

(i) $\|\Pi_{[N,\infty)}x^*\| \le 1$,

(ii) $\|\Pi_N x^*\| > 1 - \delta$ and $\|\Pi_\infty x^*\| > 1 - \delta$,

then, if $y^* \in JT^*$ is such that

(i') $\|\Pi_{[N,\infty)}y^*\| \le 1$,

(ii') $\|\Pi_N(x^* - y^*)\| < \delta$ and $\|\Pi_\infty(x^* - y^*)\| < \delta$,

we get

$$\|\Pi_{[N,\infty)}(x^* - y^*)\| \le \varepsilon.$$

Proof: We may suppose without loss of generality that x^*, $y^* \in \Pi_{[N,\infty)}JT^*$. It is not hard to show that by the hypotheses x^* and y^* belong to the subset of the unit ball,

(1) $$T(F,4\delta) = \{z^* \in JT^* : \|z^*\| \le 1 \text{ and } F(z^*) > 2 - 4\delta\},$$

with $F : JT^* \longrightarrow \mathbb{R}$ defined by

$$F(z^*) = \|\Pi_N x^*\|^{-1}\langle \Pi_N x^*, \Pi_N z^*\rangle_{H_1} + \|\Pi_\infty x^*\|^{-1}\langle \Pi_\infty x^*, \Pi_\infty z^*\rangle_{H_2},$$

where $H_1 = \mathbb{R}^{2^N}$, $H_2 = \ell_2(\Gamma)$, and for $i = 1, 2$, $\langle\ ,\ \rangle_{H_i}$ denotes the inner product in H_i. Note that $\|F\| \le 2$.

Using the density of the simple functions in B_{JT^*}, we will see that the diameter of $\Pi_{[N,\infty)}T(F,4\delta)$ is less than ε. In order to be able to do this, first we will prove that if m_0 and m_1 belong to $\mathcal{M}$ and to the subset of the unit ball $T(F,32\delta/\varepsilon)$, then

(2) $$\|\Pi_{[N,\infty)}(m_0 - m_1)\| < \varepsilon/2.$$

Now, since $x^* \in T(F,32\delta/\varepsilon)$, which is an open set in B_{JT^*}, by Proposition 3.d.7, $T(F,32\delta/\varepsilon) \cap \mathrm{co}\mathcal{M} \ne \emptyset$, and since $(B_{JT^*})\backslash T(F,32\delta/\varepsilon)$ is convex, it follows that $T(F,32\delta/\varepsilon) \cap \mathcal{M} \ne \emptyset$. Thus let

$$m_0 = \sum_{j=1}^{k}\lambda_j f_{S_j} \in T(F,32\delta/\varepsilon) \cap \mathcal{M}$$

with $\sum_{j=1}^{k}\lambda_j^2 \le 1$, where the set $\{S_j\}_{i=1}^{k}$ consists of pairwise disjoint segments in $\mathcal{T}$; then

(3) $$F(m_0) > 2 - 32\delta/\varepsilon.$$

To prove (2) we will use an auxiliary $m_0' \in \mathcal{M}$, which we obtain from m_0, by only considering the infinite segments appearing in the representation of m_0 which intersect level N . For this purpose define

$$I_0 = \{j = 1,\dots,k : S_j \text{ is infinite and } S_j \cap \{t \in \mathcal{T}: \mathrm{lev}(t) = N\} \ne \emptyset\},$$

$$I_1 = \{j = 1,\dots,k : S_j \cap \{t \in \mathcal{T}: \mathrm{lev}(t) = N\} = \emptyset\},$$

$$I_2 = \{j = 1,\dots,k : S_j \cap \{t \in \mathcal{T}: \mathrm{lev}(t) = N\} \ne \emptyset \text{ and } S_j \text{ is finite}\}.$$

We will show that $I_0 \ne \emptyset$. Indeed, suppose $I_0 = \emptyset$. Then, if for $j \in I_2$ t_j denotes the node with level N in S_j,

$$2 - 32\delta/\varepsilon < F(m_0) =$$

$$= \|\Pi_N x^*\|^{-1}\langle \Pi_N x^*, \Pi_N\textstyle\sum_{j\in I_2}\lambda_j f_{S_j}\rangle_{H_1} + \|\Pi_\infty x^*\|^{-1}\langle \Pi_\infty x^*, \Pi_\infty\sum_{j\in I_1}\lambda_j f_{S_j}\rangle_{H_2} =$$

$$= \|\Pi_N x^*\|^{-1}\textstyle\sum_{j\in I_2}\lambda_j\langle x^*, \eta_{t_j}\rangle + \|\Pi_\infty x^*\|^{-1}\sum_{j\in I_1}\lambda_j\lim_{t\in S_j}\langle x^*, \eta_t\rangle \le$$

$$\le \frac{1}{1-\delta}\left(\sum_{j=1}^{k}\lambda_j^2\right)^{1/2}\left(\sum_{j\in I_2}\langle x^*, \eta_{t_j}\rangle^2 + \sum_{j\in I_1}(\lim_{t\in S_j}\langle x^*, \eta_t\rangle)^2\right)^{1/2} =$$

$$= \frac{1}{1-\delta}\left(\sum_{j=1}^{k}\lambda_j^2\right)^{1/2}\left(\|\Pi_N x^*\|^2 + \|\Pi_\infty x^*\|^2\right)^{1/2} \le \frac{1}{1-\delta}\left(\sum_{j=1}^{k}\lambda_j^2\right)^{1/2} \le \frac{1}{1-\delta}.$$

Hence $\varepsilon < 32\delta\,\dfrac{1-\delta}{1-2\delta}$, and this is a contradiction by the hypothesis on δ.

Now define $m'_0 = \Pi_{[N,\infty)}(\sum_{j\in I_0}\lambda_j f_{S_j})$. Clearly $m'_0 \in \tilde{\mathcal{M}}$, and we will show that it is close to $\Pi_{[N,\infty)} m_0$. To this end, observe that using the inequality of Cauchy and Schwarz in $\Pi_N JT^*$ we obtain

$$(\sum_{j\notin I_1}\lambda_j^2)^{1/2} = \|\Pi_N(\sum_{j\notin I_1}\lambda_j f_{S_j})\| \ge$$

$$\ge \langle\|\Pi_N x^*\|^{-1}\Pi_N x^*, \Pi_N(\sum_{j\notin I_1}\lambda_j f_{S_j})\rangle_{H_1} = \langle\|\Pi_N x^*\|^{-1}\Pi_N x^*, \Pi_N\sum_{j=1}^{k}\lambda_j f_{S_j}\rangle_{H_1} =$$

$$= F(m_0) - \langle\|\Pi_\infty x^*\|^{-1}\Pi_\infty x^*, \Pi_\infty\sum_{j=1}^{k}\lambda_j f_{S_j}\rangle_{H_2} > 2 - 32\delta/\varepsilon - 1 = 1 - 32\delta/\varepsilon.$$

Therefore

(4) $$(\sum_{j\in I_1}\lambda_j^2)^{1/2} \le (1 - (1 - 32\delta/\varepsilon)^2)^{1/2} < (2^6\delta/\varepsilon)^{1/2} < 2^{-12}\varepsilon^2.$$

Similarly, using Π_∞ and I_2 instead of Π_N and I_1 we can prove

(5) $$(\sum_{j\in I_2}\lambda_j^2)^{1/2} < 2^{-12}\varepsilon^2.$$

Therefore we get

(6) $$\|\Pi_{[N,\infty)}(m_0 - m'_0)\| = \|\Pi_{[N,\infty)}(\sum_{j\in I_1}\lambda_j f_{S_j} + \sum_{j\in I_2}\lambda_j f_{S_j})\| \le$$
$$\le \|\sum_{j\in I_1}\lambda_j f_{S_j}\| + \|(\sum_{j\in I_2}\lambda_j f_{S_j}\| < 2^{-11}\varepsilon^2.$$

Since $\|F\| \le 2$ and $F(m_0 - m'_0) = F(\Pi_{[N,\infty)}(m_0 - m'_0))$,

(7) $$F(m_0 - m'_0) \le 2\|\Pi_{[N,\infty)}(m_0 - m'_0)\|.$$

From (3), (6) and (7)

(8) $$F(m'_0) > F(m_0) - 2\|\Pi_{[N,\infty)}(m_0 - m'_0)\| > 2 - 32\delta/\varepsilon - 2^{-10}\varepsilon^2 > 2 - 2^{-9}\varepsilon^2.$$

Since both $\langle\|\Pi_N x^*\|^{-1}\Pi_N x^*, \Pi_N m'_0\rangle_{H_1}$ and $\langle\|\Pi_\infty x^*\|^{-1}\Pi_\infty x^*, \Pi_\infty m'_0\rangle_{H_2}$ are smaller than or equal to one, from (8) and from the definition of F we obtain

$$\langle\|\Pi_N x^*\|^{-1}\Pi_N x^*, \Pi_N m'_0\rangle_{H_1} > 1 - 2^{-9}\varepsilon^2$$

and

$$\langle \|\Pi_\infty x^*\|^{-1}\Pi_\infty x^*, \Pi_\infty m_0'\rangle_{H_2} > 1 - 2^{-9}\varepsilon^2.$$

Applying Lemma 3.d.22 to $H_1 = \mathbb{R}^{2^N}$ and the slice $S_1(\|\Pi_N x^*\|^{-1}\Pi_N x^*, 2^{-9}\varepsilon^2)$ and afterwards to $H_2 = \ell_2(\Gamma)$ and the slice $S_2(\|\Pi_\infty x^*\|^{-1}\Pi_\infty x^*, 2^{-9}\varepsilon^2)$, we deduce

$$\|\Pi_N(x^* - m_0')\| = \|\Pi_N x^* - \Pi_N m_0'\| < 2(2^{-8}\varepsilon^2)^{1/2} = \varepsilon/8, \tag{9}$$

$$\|\Pi_\infty(x^* - m_0')\| < \varepsilon/8. \tag{10}$$

Now let $m_1 \in T(F,32\delta/\varepsilon) \cap \mathcal{M}$ and let $m_1' \in \mathcal{M}$ be the auxiliary simple function corresponding to m_1, constructed in the same manner as m_0' so that

$$\|\Pi_{[N,\infty)}(m_1 - m_1')\| < 2^{-11}\varepsilon^2, \tag{11}$$

$$\|\Pi_N(x^* - m_1')\| < \varepsilon/8, \tag{12}$$

$$\|\Pi_\infty(x^* - m_1')\| < \varepsilon/8. \tag{13}$$

Since the segments which appear in the representations of m_0' and m_1' all start at level N and are all infinite, by (10) and (13)

$$\|\Pi_{[N,\infty)}(m_0' - m_1')\| = \|\Pi_\infty(m_0' - m_1')\| < \varepsilon/4. \tag{14}$$

Hence by (6), (11) and (14) we obtain (2):

$$\|\Pi_{[N,\infty)}(m_0 - m_1)\| < \varepsilon/2.$$

Now we are able to prove that the diameter of $\Pi_{[N,\infty)}T(F,4\delta)$ is less than ε. Let $z^* \in T(F,4\delta)$, $z^* = \sum_{i=1}^k \mu_i m_i$ with $m_i \in \mathcal{M}$, $\mu_i > 0$, $\sum_{i=1}^k \mu_i = 1$ and let $I = \{i \in \{1,\dots,k\} : m_i \in T(F,32\delta/\varepsilon)\}$. Then

$$2 - 4\delta < \sum_{i=1}^k \mu_i F(m_i) = \sum_{i\in I} \mu_i F(m_i) + \sum_{i\notin I} \mu_i F(m_i) \le$$
$$\le 2(1 - \sum_{i\notin I} \mu_i) + (2 - 32\delta/\varepsilon)(\sum_{i\notin I} \mu_i),$$

and therefore

$$\sum_{i\notin I} \mu_i < \varepsilon/8. \tag{15}$$

Hence if $z' = \sum_{i\in I} \mu_i m_i$,

$$\|z^* - z'\| < \varepsilon/8. \tag{16}$$

Let $w^* \in T(F,4\delta)$ be another convex combination of simple functions n_j in $\mathcal{M}$, $w^* = \sum_{j=1}^r \nu_j n_j$ with $\nu_j > 0$, $\sum_{j=1}^r \nu_j = 1$, where we may suppose that the simple functions belonging to $T(F,32\delta/\varepsilon)$ appear in the expressions for z^* and w^* before the ones that don't belong. Then applying Lemma 3.d.13

to $\sum_{i=1}^{k}\mu_i = \sum_{j=1}^{r}\nu_j = 1$, there exists $\{\lambda_i\}_{i=1}^{s}$ with $\sum_{i=1}^{s}\lambda_i = 1$, $\lambda_i > 0$ and simple functions in $\mathcal{M}$, which by abuse of notation we will denote again by m_i and n_i (in fact the m_i's and n_i's are the same as before and appear in the same order, only some functions appear repeatedly), such that $z^* = \sum_{i=1}^{s}\lambda_i m_i$ and $w^* = \sum_{i=1}^{s}\lambda_i n_i$. Let

$$I_0 = \{i : m_i \in T(F,32\delta/\varepsilon)\} \subset \{1,\dots,s\}$$

and let

$$I'_0 = \{i : n_i \in T(F,32\delta/\varepsilon)\} \subset \{1,\dots,s\}.$$

Since $\sum_{i\in I_0}\lambda_i = \sum_{i\in I}\mu_i$, if $z' = \sum_{i\in I_0}\lambda_i m_i$, then as in (15) and in (16)

(17) $$\sum_{i\notin I_0}\lambda_i < \varepsilon/8 \quad\text{and}\quad \|z^* - z'\| < \varepsilon/8.$$

Similarly if $w' = \sum_{i\in I'_0}\lambda_i n_i$,

(18) $$\|w^* - w'\| < \varepsilon/8.$$

Therefore, assuming that $I_0 \supset I'_0$, by (17), (18) and (2) we have

$$\|\Pi_{[N,\infty)}(z^* - w^*)\| \le \|\Pi_{[N,\infty)}(z' - w')\| + \varepsilon/4 \le$$

$$\le \|\Pi_{[N,\infty)}(\textstyle\sum_{i\in I_0}\lambda_i(m_i - n_i))\| + \|\Pi_{[N,\infty)}(\sum_{i\notin I_0}\lambda_i(m_i - n_i))\| + \varepsilon/4 <$$

$$< \varepsilon/2 + 2\varepsilon/8 + \varepsilon/4 = \varepsilon.$$

Thus the lemma is true for $z^*, w^* \in \mathrm{co}\mathcal{M}$ and this together with Proposition 3.d.7 gives us diam $\Pi_{[N,\infty)}T(F,4\delta) \le \varepsilon$ and proves the desired result.

Now we are finally ready to prove the Kadec-Klee property for the norm of JT^*.

Theorem 3.d.24. The norm of JT^* has the Kadec-Klee property.

Proof: We will prove the theorem by showing that if for $x^* \in JT^*$ with $\|x^*\| = 1$, $n \in \mathbb{N}$ and $H_2 = \ell_2(\Gamma)$ we define

$$V_n = \{z^* \in JT^* : \|z^*\| \le 1 \text{ and } \|\Pi_{[0,n]}(x^* - z^*)\| < 1/n\},$$

$$S_n = \{z^* \in JT^* : \|z^*\| \le 1 \text{ and } \langle\Pi_\infty z^*, \Pi_\infty x^*\rangle_{H_2} > \|\Pi_\infty x^*\|^2 - 1/n\},$$

then $\{V_n \cap S_n\}_{n=1}^{\infty}$ forms a relative neighborhood base of x^* in the norm topology of the unit ball of JT^*.

Thus let $x^* \in JT^*$ with $\|x^*\| = 1$. First observe that $\mathcal{V}_n \cap S_n$ is a relative weak neighborhood of x^*, since as was mentioned in Proposition 3.d.19, $\mathcal{V}_n$ is a relative weak star neighborhood. Also S_n is a weak neighborhood of x^*, because the functional $\mathcal{g} : JT^* \longrightarrow \mathbb{R}$ given by

$$\mathcal{g}(z^*) = \langle \Pi_\infty z^*, \Pi_\infty x^* \rangle_{H_2}$$

is continuous.

If $\Pi_\infty x^* = 0$, then the result follows directly from the statement in Proposition 3.d.19; if $\Pi_\infty x^* \neq 0$, let $\mathfrak{b} = \|\Pi_\infty x^*\|$.

For every $0 < \delta < 1$, by Lemma 3.d.8 there exists $N_1 \in \mathbb{N}$ such that for $n \geq N_1$

$$\|\Pi_{[n,\infty)} x^*\| < \mathfrak{b}/(1 - \delta^2/16)$$

and

(1) $$\|\Pi_n x^*\| > \mathfrak{b}(1 - \delta/2).$$

Let $m \in \mathbb{N}$ be such that $\|\Pi_{[N_1,\infty)} x^*\| < (\mathfrak{b}/(1 - \delta^2/16)) - 1/m$. Then for $n \geq N_1$

(2) $$\|\Pi_{[n,\infty)} x^*\| \leq \|\Pi_{[N_1,\infty)} x^*\| < (\mathfrak{b}/(1 - \delta^2/16)) - 1/m.$$

Now let N_2 be such that for $n \geq N_2$

(3) $$\|\Pi_{[0,n]} x^*\| > 1 - 1/(2^{14} m^3).$$

Let $0 < \varepsilon < 2^{-3}$, $\varepsilon_1 < \varepsilon/8$, $0 < \delta < 2^{-31}\varepsilon_1^5$ and let n be fixed so that $n > N = \max(N_1, N_2, 2^{14}m^3, 16/(\mathfrak{b}\delta)^2)$. Fix $z^* \in \mathcal{V}_n \cap S_n$; since $z^* \in \mathcal{V}_n$, from $\|\Pi_{[0,n]}(x^* - z^*)\| < 1/n$, by (2)

(4) $$\|\Pi_n z^*\| < \|\Pi_n x^*\| + 1/n < (\mathfrak{b}/(1 - \delta^2/16)) + 1/n - 1/m$$

and by (3) we get that

$$\|\Pi_{[0,n]} z^*\| > 1 - 1/n - 1/(2^{14} m^3).$$

Since $1/n + (1/(2^{14}m^3)) < (1/2m)^3(1/2^{10})$, we conclude from Proposition 3.d.15 and (4) that

(5) $$\|\Pi_{[n,\infty)} z^*\| < \|\Pi_n z^*\| + 1/2m < (\mathfrak{b}/(1 - \delta^2/16)).$$

Now define

$$x_1^* = \mathfrak{b}^{-1}(1 - \delta^2/16)x^*, \quad z_1^* = \mathfrak{b}^{-1}(1 - \delta^2/16)z^*,$$

$$S = \{w^* \in JT^* : \|w^*\| \le 1 \text{ and } \langle \Pi_\infty w^*, \|\Pi_\infty x_1^*\|^{-1}\Pi_\infty x_1^*\rangle_{H_2} > 1 - \delta^2/8\}.$$

First we will show that $\Pi_{[n,\infty)} z_1^* \in S$.

By (5)

(6) $$\|\Pi_{[n,\infty)} z_1^*\| < 1.$$

On the other hand, since $z^* \in S_n$ and $n > 16/(b\delta)^2$, we have

$$\langle \Pi_\infty(\Pi_{[n,\infty)} z_1^*), \|\Pi_\infty x_1^*\|^{-1}\Pi_\infty x_1^*\rangle_{H_2} = \langle \Pi_\infty z^*, \Pi_\infty x^*\rangle_{H_2} (1-\delta^2/16)b^{-2} > (1-\delta^2/8).$$

Therefore

(7) $$\Pi_{[n,\infty)} z_1^* \in S.$$

Now, since $z^* \in V_n$, $\Pi_n(\Pi_{[n,\infty)} z_1^*) = \Pi_n z_1^*$ and $n > 16/(b\delta)^2 > (1-\delta^2/16)/b\delta$,

(8) $$\|\Pi_n(x_1^* - z_1^*)\| = b^{-1}(1 - \delta^2/16)\|\Pi_n(x^* - z^*)\| < \delta.$$

We will see that $\operatorname{diam}(\Pi_{[n,\infty)}(S)) < 2\varepsilon_1$; to this end observe that by (2)

(9) $$\|\Pi_{[n,\infty)}(x_1^*)\| \le 1,$$

and by (1)

(10) $$\|\Pi_n(x_1^*)\| > (1 - \delta^2/16)(1 - \delta/2) > 1 - \delta.$$

Also

(11) $$\|\Pi_\infty(x_1^*)\| = 1 - \delta^2/16 > 1 - \delta.$$

Define

$$S' = \{u \in H_2 : \|u\|_{H_2} \le 1 \text{ and } \langle u, \|\Pi_\infty x_1^*\|^{-1}\Pi_\infty x_1^*\rangle_{H_2} > 1 - \delta^2/8\}.$$

By Lemma 3.d.22 diam $S' < \delta$.

Let $w^* \in S$, then $\Pi_\infty w^* \in S'$ and since by (11), $\Pi_\infty x_1^* \in S'$, we have

$$\|\Pi_\infty(x_1^* - w^*)\| < \delta.$$

In particular by (7), since $\Pi_\infty \Pi_{[n,\infty)} z_1^* = \Pi_\infty z_1^*$,

(12) $$\|\Pi_\infty(x_1^* - z_1^*)\| < \delta.$$

Consequently by (6), (8), (9), (10), (11) and (12), x_1^* and z_1^* satisfy the hypotheses of Lemma 3.d.23 and therefore

$$\|\Pi_{[n,\infty)}(x_1^* - z_1^*)\| < \varepsilon_1.$$

It follows that

$$\|\Pi_{[n,\infty)}(x^* - z^*)\| < \varepsilon_1 b/(1 - \delta^2/16) < \varepsilon/4,$$

and

$$\operatorname{diam} \Pi_{[n+1,\infty)}(S_n \cap V_n) \le \operatorname{diam} \Pi_{[n,\infty)}(S_n \cap V_n) < \varepsilon/2.$$

Finally we get

$$\operatorname{diam} (S_n \cap V_n) \le \operatorname{diam} \Pi_{[0,n]}(S_n \cap V_n) + \operatorname{diam} \Pi_{[n+1,\infty)}(S_n \cap V_n) <$$

$$< 2/n + \varepsilon/2 < \varepsilon/2 + \varepsilon/2 = \varepsilon.$$

Hence $S_n \cap V_n$ is contained in the ball with center x^* and radius ε, and this finishes the proof.

Observe that the above result has as a consequence that the norm induced in the dual of $(J, \| \|_J)$ also has the Kadec-Klee property.

Now we will show that, although JT^* with the usual norm is not strictly convex (see Lemma 3.d.3), we can endow JT^* with a strictly convex equivalent norm which still possesses the Kadec-Klee property.

For notational convenience in what follows we will write $\| \|_*$ instead of $\| \|_{JT^*}$.

Theorem 3.d.25. JT has an equivalent norm such that the dual norm in JT^* is strictly convex and possesses the Kadec-Klee property.

Proof: We define in JT^* the norm $\nmid \nmid$ as follows: for $x^* \in JT^*$

$$\nmid x^* \nmid = \|x^*\|_* + (\textstyle\sum_{n=1}^{\infty} 2^{-2n} \langle x^*, x_n \rangle^2)^{1/2},$$

where $\{x_n\}$ is a dense sequence in the unit ball of $(JT, \| \|)$. Clearly

$$\|x^*\|_* \le \nmid x^* \nmid \le 2\|x^*\|_*.$$

We will see that there exists a norm in JT equivalent to $\| \|$ such that $\nmid \nmid$ is the corresponding dual norm. To this end (see e.g. Lemma 5.1, Singer [1]), it suffices to show that the set $A = \{x^* \in JT^* : \nmid x^* \nmid \le 1\}$ is weak star closed.

Let $x^* \in JT^*$ with $\nmid x^* \nmid > 1$; then by the density of $\{x_n\}$ in B_{JT} $\varepsilon = (\sum_{n=1}^{\infty} 2^{-2n} \langle x^*, x_n \rangle^2)^{1/2} > 0$ and there exists m such that

$$\langle x^*, x_m \rangle > \|x^*\|_* - \varepsilon/4.$$

Fix $N \in \mathbb{N}$, $N \ge m$ so that

(1) $$\begin{cases} (\sum_{n=1}^{N} 2^{-2n} \langle x^*, x_n\rangle^2)^{1/2} > \varepsilon^2/16, \\ \|x^*\|_* + (\sum_{n=1}^{N} 2^{-2n} \langle x^*, x_n\rangle^2)^{1/2} > 1 + 3\varepsilon/4. \end{cases}$$

Let
$$\varepsilon' = \min\{\varepsilon/4,\ \varepsilon^2/16(1 + \|x^*\|_*)\},$$
and let $\mathcal{V}$ be the weak star neighborhood of x^* given by
$$\mathcal{V} = \{y^* \in JT^* : |\langle y^* - x^*, x_i\rangle| < \varepsilon',\ i = 1,\dots,N\}.$$

Then for $y^* \in \mathcal{V}$ with $\|y^*\|_* \le 1$, $|\langle y^* + x^*, x_i\rangle| \le 1 + \|x^*\|_*$. Hence

(2) $$\langle y^*, x_i\rangle^2 > \langle x^*, x_i\rangle^2 - \varepsilon^2/16,$$

and

(3) $$\|y^*\|_* \ge \langle y^*, x_m\rangle > \langle x^*, x_m\rangle - \varepsilon' > \langle x^*, x_m\rangle - \varepsilon/4 > \|x^*\|_* - \varepsilon/2.$$

Hence by (1), (2), (3) and using the inequality for $a \ge b \ge 0$
$$(a - b) \le (a^2 - b^2)^{1/2},$$
we get for $y^* \in \mathcal{V}$ with $\|y^*\|_* \le 1$,
$$|\!|\!|y^*|\!|\!| > \|x^*\|_* - \varepsilon/2 + (\sum_{n=1}^{N} 2^{-2n} \langle x^*, x_n\rangle^2 - \sum_{n=1}^{N} 2^{-2n} \varepsilon^2/16)^{1/2} >$$
$$> \|x^*\|_* + (\sum_{n=1}^{N} 2^{-2n} \langle x^*, x_n\rangle^2)^{1/2} - 3\varepsilon/4 > 1.$$

On the other hand, if $y^* \in \mathcal{V}$ and $\|y^*\|_* > 1$, then clearly $|\!|\!|y^*|\!|\!| > 1$. Thus $\mathcal{V} \subset JT^* \setminus A$ and A is weak star closed.

Now we will show that the norm $|\!|\!|\ |\!|\!|$ is strictly convex.

Let x^*, $y^* \in JT^*$ with $|\!|\!|x^*|\!|\!| = |\!|\!|y^*|\!|\!| = 1$ and suppose $|\!|\!|x^* + y^*|\!|\!| = 2$. If $\|x^* + y^*\|_* < \|x^*\|_* + \|y^*\|_*$, then
$$2 = |\!|\!|x^* + y^*|\!|\!| < \|x^*\|_* + \|y^*\|_* + (\sum_{n=1}^{\infty} 2^{-2n} \langle x^* + y^*, x_n\rangle^2)^{1/2} \le$$
$$\le |\!|\!|x^*|\!|\!| + |\!|\!|y^*|\!|\!| = 2.$$

Thus we may assume that $\|x^* + y^*\|_* = \|x^*\|_* + \|y^*\|_*$, and this implies
$$(\sum_{n=1}^{\infty} 2^{-2n} \langle x^* + y^*, x_n\rangle^2)^{1/2} =$$
$$= (\sum_{n=1}^{\infty} 2^{-2n} \langle x^*, x_n\rangle^2)^{1/2} + (\sum_{n=1}^{\infty} 2^{-2n} \langle y^*, x_n\rangle^2)^{1/2}.$$

Since this is an equality in ℓ_2, either $\langle y^*, x_n\rangle = 0$ for every $n \in \mathbb{N}$, or otherwise there exists $\lambda \in \mathbb{R}$ such that $\langle y^*, x_n\rangle = \lambda\langle x^*, x_n\rangle$ for every $n \in \mathbb{N}$. In the first case, by the density of the set $\{x_n\}$ in B_{JT}, we get $y^* = 0$ and in the second $y^* = \lambda x^*$, and since $⦀x^* + y^*⦀ = 2$, this implies $x^* = y^*$. Hence the norm $⦀\ ⦀$ is strictly convex.

In particular the above also shows that every point of the unit sphere of $(JT^*, ⦀\ ⦀)$ is extreme.

Finally we will see that $⦀\ ⦀$ possesses the Kadec-Klee property.

Let $\{x_\alpha\}$ be a net in JT^* converging weakly to x_0^* with $⦀x_\alpha^*⦀ = ⦀x_0^*⦀ = 1$. Then, using the inequality

$$|a^{1/2} - b^{1/2}| \le |a - b|^{1/2} \text{ for } a, b \ge 0,$$

we get

$$\big|\|x_\alpha^*\|_* - \|x_0^*\|_*\big| = \big|(\textstyle\sum_{n=1}^{\infty} 2^{-2n}\langle x_\alpha^*, x_n\rangle^2)^{1/2} - (\sum_{n=1}^{\infty} 2^{-2n}\langle x_0^*, x_n\rangle^2)^{1/2}\big| \le$$

$$\le \big|\textstyle\sum_{n=1}^{\infty} 2^{-2n}\langle x_\alpha^*, x_n\rangle^2 - \sum_{n=1}^{\infty} 2^{-2n}\langle x_0^*, x_n\rangle^2\big|^{1/2} =$$

$$= \big|\textstyle\sum_{n=1}^{\infty} 2^{-2n}(\langle x_\alpha^*, x_n\rangle - \langle x_0^*, x_n\rangle)(\langle x_\alpha^*, x_n\rangle + \langle x_0^*, x_n\rangle)\big|^{1/2} \le$$

$$\le \sqrt{2}\,\big|\langle x_\alpha^* - x_0^*, \textstyle\sum_{n=1}^{\infty} 2^{-2n}(-1)^{m_n} x_n\rangle\big|^{1/2},$$

where $(-1)^{m_n}$ is the sign of $\langle x_\alpha^*, x_n\rangle - \langle x_0^*, x_n\rangle$.

Since $\sum_{n=1}^{\infty} 2^{-2n}(-1)^{m_n} x_n \in JT$, by the weak convergence of the net $\{x_\alpha^*\}$ the last term in the above inequality tends to zero. Therefore

$$\lim_\alpha \|x_\alpha^*\|_* = \|x_0^*\|_*, \tag{1}$$

and also $\|x_\alpha^*\|_*^{-1} x_\alpha^*$ converges weakly to $\|x_0^*\|_*^{-1} x_0^*$. Applying Theorem 3.d.24 we have

$$\big\|\|x_\alpha^*\|_*^{-1} x_\alpha^* - \|x_0^*\|_*^{-1} x_0^*\big\|_* \longrightarrow 0,$$

or equivalently

$$\|x_0^* - (\|x_\alpha^*\|_*^{-1}\|x_0^*\|_*)\, x_\alpha^*\|_* \longrightarrow 0. \tag{2}$$

Therefore by (1) and (2)

$$\|x_\alpha^* - x_0^*\|_* \le \|x_\alpha^* - (\|x_\alpha^*\|_*^{-1}\|x_0^*\|_*)x_\alpha^*\|_* + \|(\|x_\alpha^*\|_*^{-1}\|x_0^*\|_*)x_\alpha^* - x_0^*\|_* =$$

$$= \big|1 - \|x_\alpha^*\|_*^{-1}\|x_0^*\|_*\big|\ \|x_\alpha^*\|_* + \|(\|x_\alpha^*\|_*^{-1}\|x_0^*\|_*)x_\alpha^* - x_0^*\|_* \longrightarrow 0.$$

From $⦀x_\alpha^* - x_0^*⦀ \le 2\|x_\alpha^* - x_0^*\|_*$, we get

$$\|x_\alpha^* - x_0^*\| \longrightarrow 0,$$

and this finishes the proof.

In addition to the results described in this section, Schachermayer [1] also showed that the strongly exposing functionals for every w^*-compact convex subset of JT^* form a dense $\mathfrak{G}_\delta$-subset of JT^{**}, and from that he obtained an equivalent norm $|\ |$ on JT such that every point of the unit sphere of $(JT, |\ |)^*$ is strongly exposed.

CHAPTER 4. WHAT ELSE IS THERE ABOUT J AND JT?

The ways of all the woodland
Gleam with a soft and golden fire—
For whom does all the sunny woodland
Carry so brave attire?

James Joyce

The object of this chapter is to serve as a reference for further study of J and JT.

On the one hand we will state, with brief comments, several results about J and JT that go beyond the basic properties discussed in the previous chapters. However, since the proofs of most of them either are too complicated or require tools which have not been introduced in this book, either we will not prove them at all or, in some cases, we will just give a sketch of the proof.

On the other hand, we will also talk about several generalizations of James spaces which appear in the literature, such as the space JF of Lindenstrauss and Stegall and the long James space of Edgar.

4.a. More about J

4.a.1. J and J^* do not have local unconditional structure.

The concept of local unconditional structure (l.u.st.) provides a "localization" of the Banach lattice structure and is defined as follows:
X is said to have l.u.st. provided $X = \bigcup_\alpha E_\alpha$, where the E_α are finite dimensional subspaces forming an increasing net when directed by inclusion, and E_α has a basis $\{e_i^\alpha\}_{i=1}^{n(\alpha)}$ for which $\sup_\alpha U(e_i^\alpha)_{i=1}^{n(\alpha)} = K < \infty$, where $U(e_i)_{i=1}^n$ denotes the unconditional constant of $\{e_i\}_{i=1}^n$. This

definition is due to Dubinsky, Pelczynski and Rosenthal, but it is worthwhile to point out that there is another definition of local unconditional structure given by Gordon and Lewis [1], and it is not known if they are equivalent, although the first one implies the second.

The result for J and J^* is a corollary of the next two theorems, the first one is due to Johnson and Tzafriri and can be found in Lindenstrauss and Tzafriri [2] and the second one is by Figiel, Johnson and Tzafriri [1] and is valid for both definitions of l.u.st.

Theorem. Let X be a complemented subspace of a Banach lattice. Then X is reflexive if and only if it doesn't contain c_0 or ℓ_1.

Theorem. If X has local unconditional structure, then X^{**} is isomorphic to a complemented subspace of a Banach lattice.

Corollary. J and J^* do not have local unconditional structure.

4.a.2. J has the Gordon-Lewis property.

A Banach space X is said to have the Gordon-Lewis property, if every one-summing operator from X into an arbitrary Banach space Y factors through L_1.

This concept was studied by Gordon and Lewis in [1] to answer a question posed by Grothendieck. They showed that there exist Banach spaces without this property which now bears their names.
The notions of l.u.st. and the Gordon-Lewis property (GLP) are intimately related; in fact every space that has l.u.st has the GLP. However, Pisier in [1] proved that, even if J does not have a local unconditional structure, it has the GLP.

4.a.3. Other results on the extreme points in $(J, \| \|)$ and J^{**}.

In Section 2.e we gave Bellenot's characterization of the extreme points of $(J^{**}, \| \|)$. In the same paper, [4], he further showed that the set of extreme points in J as well as that in J^{**} is closed and nowhere dense and $x \in J$ is extreme if and only if x is exposed.

4.a.4. J as a Banach algebra.

Andrew and Green in [1] gave a detailed study of the James space J from the point of view of Banach algebras. They proved that under pointwise multiplication, if $x \in J$ and

$$\interleave x \interleave = \sup\{ \|\, \|xy\| \,\| : y \in J,\ \|\, \|y\| \,\| \le 1\},$$

then $\interleave\ \interleave$ is an equivalent norm which makes J and J^{**} into commutative semisimple Banach algebras, moreover J^{**} is just the algebra obtained from J by adjoining an identity.

We now list some results of Andrew and Green:

(i) The closed ideals of J are in one to one correspondence with the closed ideals of c_0.
(ii) The multiplier algebras of J and J^{**} can be identified isometrically and isomorphically with the Banach algebra J^{**}.
(iii) Every automorphism of J is bounded; indeed the group of automorphisms Aut(J) is a proper subgroup of the group of permutations of the natural numbers $\Pi(\mathbb{N})$ and the only automorphism with norm less than $\sqrt{2}$ is the identity.
(iv) There exists a metric d on $\Pi(\mathbb{N})$ such that $(\mathrm{Aut}(J), d)$ is a topological group and the topology induced by d coincides with the strong operator topology on norm bounded subsets of Aut(J).

4.a.5. The general linear group $GL(J^n)$.

Another study of the structure of the group of automorphisms of J and more generally J^n, considered only with its vector space structure, was given by Mityagin and Èdel'shtein [1].

Based on the fact that J^{**}/J is isomorphic to $\mathbb{K}$, where $\mathbb{K}$ is $\mathbb{R}$ or $\mathbb{C}$, depending on whether J is considered as a space over $\mathbb{R}$ or $\mathbb{C}$, the above authors proved that this group, denoted by $GL(J^n)$, is the direct product of the finite dimensional group $GL(n, \mathbb{K})$ and an infinite dimensional contractible group.

In particular, it follows that the homotopy type of $GL(J^n)$ is the same as that of the maximal compact subgroup of GL(n), either O(n) or U(n), according to the base field. For instance, in the special case n = 1 and $\mathbb{K} = \mathbb{R}$, it follows that GL(J) has two connected components, each of them homotopically trivial. This is to be contrasted with the typical situation for some reflexive spaces such as Hilbert spaces or ℓ_p for $1 < p < \infty$, but also for ℓ_1 and c_0, whose groups of automorphisms are contractible.

4.b. More about JT

4.b.1. Existence of a weakly measurable function $f : \Delta \longrightarrow JT^*$ which is not equivalent to any strongly measurable function.

This rather surprising property of JT^* can be found in Lindenstrauss and Stegall [1], where the authors showed that there is a weakly measurable function from the Cantor set Δ, endowed with the Haar measure μ, into JT^* which is not equivalent to any strongly measurable function. This is a counterexample to the question whether the existence of a weakly measurable function $\sigma : K \longrightarrow X$ (where K is a compact Hausdorff space and X is a separable Banach space) which is not equivalent to a strongly measurable function implies that X contains ℓ_∞.

4.b.2. $\mathfrak{B}$, JT and JT^* have the point of continuity property.

A Banach X space has the point of continuity property (PCP), if for every weakly closed bounded subset C of X, there is $x \in C$ such that the weak and norm topologies restricted to C coincide at x.

This notion was introduced by Bourgain and Rosenthal in [1], who also showed that this property is implied by the existence of a boundedly complete skipped blocking finite dimensional decomposition (BCSBD) (see Definition 2.j.6). Further, they proved that $\mathfrak{B}$ has a BCSBD and thus PCP. (In Corollary 3.d.17 we gave another proof of $\mathfrak{B}$ having a BCSBD). This was the first example of a space having PCP and yet failing the RNP (Proposition 3.c.10).

Edgar and Wheeler in [1] showed that JT^* also has the PCP and this was the first example of a dual space with PCP without the RNP.
That JT has the PCP is an immediate consequence from it having the RNP (Proposition 3.c.10) and the result of Edgar and Wheeler in [1], which says that Banach spaces with the RNP have the PCP.

4.b.3. $\mathfrak{B}$ $\mathfrak{G}_\delta$-embeds in ℓ_2 and JT^* nicely $\mathfrak{G}_\delta$-embeds in $\ell_2 \oplus \ell_2(\Gamma)$, where Γ is the set of 0-branches of $\mathcal{T}$.

If X is a Banach space it $\mathfrak{G}_\delta$-embeds (nicely $\mathfrak{G}_\delta$-embeds) in Y, if there is a one to one bounded linear operator $S : X \longrightarrow Y$ such that the image of every norm closed bounded separable subset of X under S is a norm $\mathfrak{G}_\delta$ (a weak $\mathfrak{G}_\delta$).
Ghoussoub and Maurey proved the first result in [1] and the second in [2], answering negatively Bourgain's and Rosenthal's question in [2] whether a separable Banach space which $\mathfrak{G}_\delta$-embeds into a RNP space has the RNP.

4.b.4. Some topological properties of $\mathfrak{B}$, JT and JT^*.

The James tree spaces have several topological properties which we will define next; all the concepts and results in this subsection can be found in Edgar and Wheeler [1].

Let B_X be the closed unit ball of a Banach space X.
(a) B_X is Čech complete if it is a $\mathfrak{G}_\delta$-set in some compactification of B_X.
(b) B_X is a Polish space, if considering the weak topology in B_X, it is a complete, separable metrizable space, and it can be shown that this definition is equivalent to X being separable plus B_X being Čech complete.

(c) X is an Asplund space if every separable subspace of X has a separable dual space or equivalently, if X^* has the Radon-Nykodim property.

(d) X is a Godefroy space if X^* is separable, X^* does not contain ℓ_1 and $X^\perp$ is weak star separable in X^{***}.

From the equivalence in definition (c) and Proposition 3.c.10, it follows immediately that both $\mathfrak{B}$ and JT^* are Asplund spaces.

$\mathfrak{B}$ is not a Godefroy space, since it was seen in Corollary 3.c.5 that if B is a branch $F_B \in JT^*$ is defined by $F_B(\eta_t) = 0$ for all $t \in \mathcal{T}$, $F_B(f_B) = 1$ and $F_B(f_{B'}) = 0$ for every branch $B' \neq B$, then $F_B \in (j_0\mathfrak{B})^{\perp}$, where j_0 is the canonical embedding of $\mathfrak{B}$ in $\mathfrak{B}^{**}$. This proves that $\mathfrak{B}^{\perp}$ is not weak star separable.

Neither B_{JT} nor B_{JT^*} is Čech complete and this follows from Corollary 3.c.7 and a theorem which says that if X is a space with Čech complete ball, then X and X^* are both WCG. Thus by the equivalence in (b), B_{JT} is not a Polish space. Finally, it can be shown that $B_{\mathfrak{B}}$ is Čech complete and thus, by (b), a Polish space

From the previous results and an another theorem also found in Edgar and Wheeler [1] which says that if X is an Asplund space with PCP, then X is somewhat reflexive, we obtain that JT^* is somewhat reflexive.

4.b.5. JT-type and JM-type decompositions.

If $\ell_2(J)$ is as in Section 2.i and I is the predual of J, then $\ell_2(J) = (\ell_2(I))^*$ and $\ell_2(J^*) = (\ell_2(J))^*$ (see e.g. Diestel and Uhl [1]); therefore $\ell_2(I)$ is a subspace of a separable conjugate space and thus it has the RNP. In 4.b.2 we also mentioned that the predual $\mathfrak{B}$ of JT has the PCP. In [2] Ghoussoub, Maurey and Schachermayer observed that all separable spaces with the PCP have a similar structure to $\mathfrak{B}$, while the separable spaces with the RNP have a structure similar to $\ell_2(I)$. More precisely they defined the concepts of JT-type and JM-type decompositions of a Banach space X as follows:

(i) $\{X_n\}_{n=1}^{\infty}$ is a JT-type decomposition of X, if $\{X_n\}_{n=1}^{\infty}$ is a boundedly complete skipped blocking decomposition of X with associated projections $p_n : X \longrightarrow X_n$, and there exists a sequence $\{\{y_n^k\}_{k=1}^{m_n}\}_{n=1}^{\infty}$ in X^* with $\|y_n^k\| = 1$, $y_n^k \in [X_m]_{m\neq n}^{\perp}$, such that for every $x^{**} \in X^{**}$ with $p_n(x^{**}) = x_n$ the condition

(1) $$\liminf_{n\to\infty} \max_{1\le k\le m_n} \langle x_n, y_n^k\rangle \le 0$$

implies that there is $x \in X$ with $\|x\| \le \|x^{**}\|$ such that $p_n(x) = x_n$ for every $n \in \mathbb{N}$.

(ii) $\{X_n\}_{n=1}^{\infty}$ is a JM-type decomposition of X, if it is a JT-type decomposition and in addition (1) may be replaced by

(2) $$\liminf_{n\to\infty} \langle x_n, y_n^k\rangle \le 0 \text{ for each } k \in \mathbb{N}.$$

In the particular case of $\mathfrak{B}$, taking $X_n = [\eta_t^*]_{lev(t)=n}$, $m_n = 2^{n+1}$ and

$$\{y_n^k\}_{k=1}^{2^{n+1}} = \{\eta_t\}_{lev(t)=n} \cup \{-\eta_t\}_{lev(t)=n},$$

it can be shown that $\{X_n\}$ is a JT-type decomposition of $\mathfrak{B}$.

In the case of $\ell_2(I)$, if $\{e_{nk}\}$ is its canonical basis, taking

$$X_n = [e_{nk}]_{k=1}^{n},\ m_n = 2n,\ y_n^{2k-1} = e_{nk}^* \text{ and } y_n^{2k} = -e_{nk}^*,$$

then $\{X_n\}$ is a JM-type decomposition of $\ell_2(I)$.

The main result of Ghoussoub, Maurey and Schachermayer in [2] can be stated as follows:

Let X be a separable Banach space. Then X has the PCP if and only if it has a JT-type decomposition and X has the RNP if and only if it has a JM-type decomposition.

4.c. Generalizations of J

4.c.1. The space JF.

This space is one of the first generalizations of James spaces. It was defined in the paper [1] by Lindenstrauss and Stegall as the closed linear span of the characteristic functions of subintervals of [0,1] with respect to the norm

$$\|f\| = \sup\Big(\sum_{i=0}^{k-1}\Big(\int_{t_i}^{t_{i+1}} f(t)dt\Big)^2\Big)^{1/2},$$

where the sup is taken over all partitions $0 = t_0 < t_1 <\ldots< t_k = 1$ of [0,1], $k = 1,2,\ldots$.

The following properties of JF were proved in Lindenstrauss and Stegall [1]:

(a) JF is separable but its dual is not.

(b) JF is contained in $L_1(0,1)$ and contains a subspace isomorphic to c_0.

(c) JF does not contain any subspace isomorphic to ℓ_1.

Also regarding JF, we can mention Bellenot's paper [2], where he defines the transfinite dual of a Banach space X and proves that J^{ω^2} is isometric to a complemented subspace of JF and vice versa, JF is isomorphic to a subspace of J^{ω^2}.

The following are generalizations of J considering norms determined by Banach spaces X with bases other than the unit vector basis in ℓ_2. The properties of these spaces depend on X as well as on the basis $\{x_i\}$ chosen in X. Several of the papers dedicated to this topic study the conditions under which J's properties are preserved, as for example: when is the generalized James space quasi-reflexive, when is it isomorphic to its double dual and so forth.

4.c.2. Casazza and Lohman's space $J(x_i)$.

Casazza and Lohman in [1] constructed the following space. Let X be a Banach space with basis $\{x_i\}$ and let $J(x_i)$ be the space of sequences in c_0 such that if $\alpha \in J(x_i)$, then

$$\|\alpha\| = \sup\left\|\sum_{i=1}^{n}(\alpha_{p_{2i-1}} - \alpha_{p_{2i}})x_i + \alpha_{p_{2n+1}} x_{n+1}\right\|_X < \infty,$$

where the sup is taken over all finite increasing sequences of natural numbers $p_1 < p_2 <\ldots< p_{2n+1}$, $n = 1,2,\ldots$.

They proved among other things that this construction yields a Banach space which has a basis when $\{x_i\}$ is normalized, symmetric and boundedly complete. Also if X is a reflexive Banach space with a symmetric, block p-Hilbertian basis, then $J(x_i)$ is isomorphic to its second dual, has a shrinking basis and is quasi-reflexive of order one. This gives a general method for constructing quasi-reflexive spaces of order one.

In [2] Casazza, Lin and Lohman showed that there is a reflexive Banach space X with symmetric basis $\{x_i\}$ such that $J(x_i)$ is quasi-reflexive of order one and contains no copy of c_0 or ℓ_p, $1 \le p < \infty$.

Continuing the study of the space $J(x_i)$ when $\{x_i\}$ is normalized, symmetric and boundedly complete and the norm in X is symmetric, Lin and Lohman in [1] proved that the summing basis $\{u_n\}$ in this space is nearly perfectly homogeneous, that is, $\{u_n\}$ is equivalent to all of its block basic sequences $\{y_n\}$ of the form

$$y_n = \sum_{i=p_n+1}^{p_{n+1}} u_i / \|\sum_{i=p_n+1}^{p_{n+1}} u_i\|.$$

Further, if $J(x_i)$ in addition is quasi-reflexive of order one, then it has an equivalent norm making it isometric to its second dual. They also showed that if X has a uniformly convex norm, then $J(x_i)$ has a complemented reflexive subspace which is not superreflexive.

Using the method of Casazza and Lohman [1], but considering different equivalent norms, Yao and Su in [1] constructed a class of quasi-reflexive spaces of order one, from Banach spaces with a spreading basis and block $\| \|_E$-control for suitable E. A basis $\{x_n\}$ is block $\| \|_E$-control if there is a constant K such that for each bounded block basis $\{z_k\}$ of $\{x_n\}$ and for each real sequence $\{a_n\}$ one has for m = 1, 2,...

$$\|\sum_{k=1}^{m} a_k z_k\| \le K(\sup_k \|z_k\|)\|\sum_{k=1}^{m} a_k e_k\|_E,$$

where $\{e_n\}$ is a basis of the Banach space E.

4.c.3. The Tsirelson-James space.

Beauzamy in [3] defined the Tsirelson-James space $T\mathcal{J}$ as follows:
Let T be the Tsirelson space (see Chapter 5 (iv)) and $\{t_i\}$ the canonical basis of T. $T\mathcal{J}$ is the completion with respect to $\| \|_{T\mathcal{J}}$ of the space of real sequences $\alpha = \{\alpha_i\}$ with finite support and

$$\|\alpha\|_{T\mathcal{J}} = \sup\|\sum_{i=1}^{n}(\alpha_{p_{2i-1}} - \alpha_{p_{2i}})t_i\|_T ,$$

where the sup is taken over all finite increasing sequences of positive integers $p_1 < p_2 < \ldots < p_{2n}$, n = 1,2....

$T\mathcal{J}$ has the following properties which can be found in Beauzamy and Lapresté [1]: $T\mathcal{J}$ has a shrinking basis, is not reflexive, does not contain ℓ_1, is quasi-reflexive of order one and has an equivalent norm which makes $T\mathcal{J}$ isometric to its double dual. Further ℓ_1 is its only spreading model and c_0 is the only spreading model of $T\mathcal{J}^*$.

4.c.4. Spaces of type ESA.

Brunel and Sucheston in [2] defined the "equal signs additive" (ESA) space which arises from spaces X with a monotone, unconditional, invariant under spreading (IS) basis $\{x_i\}$ (see Definition 2.d.10). The ESA space $G(x_i)$ is then defined as the completion of the space generated by the elements of the form $a = \sum_{i=1}^{k} a_i x_i$ with respect to the norm

$$M(a) = \sup_{\pi} \|(\textstyle\sum_{j\in I_1} a_j)x_1 + \ldots + (\sum_{j\in I_r} a_j)x_r\|_X,$$

where π denotes the set of partitions of $\{1,\ldots,k\}$ in consecutive disjoint intervals $I_1,\ldots,I_r$.

If $G(x_i)$ is regular, that is it does not contain c_0 or a complemented subspace isomorphic to ℓ_1, then $G(x_i)$ is quasi-reflexive of order one and admits an equivalent ESA norm which makes $G(x_i)$ isometrically isomorphic to its second dual. They also showed that there is an ESA space which is B-convex.

Finet in [1] proved that if $\{x_n\}$ is a subsymmetric, monotone, normalized basic sequence, then the spaces $J(x_n)$ in 4.c.2, and $G(x_n)$ defined above, are isomorphic.

4.c.5. The James-Orlicz space JO(X,M).

In Semenov's generalization of J in [1], $\mathbb{R}$ is replaced by a Banach space X and in the definition of the norm, $M(t) = t^2$ is replaced by any convex continuous even function M(t) with

$$\lim_{t\to\infty} (M(t)/t) = \infty \quad \text{and} \quad \lim_{t\to 0} (M(t)/t) = 0.$$

Such a function is called an Orlicz function. The James-Orlicz space JO(X,M) is the vector space of all sequences $x = \{x_i\} \subset X$ with $\lim_{i\to\infty} x_i = 0$ such that there is a t > 0 with

$$S_x(t) = \sup\sum_{i=1}^{k-1} M(\|x_{p_{i+1}} - x_{p_i}\|/t) < \infty,$$

where the sup is taken over all sequences of positive integers $1 \le p_1 < p_2 <\ldots< p_k$, endowed with the norm

$$\|x\| = \inf\{t > 0 : S_x(t) \le 1\}.$$

Under different restrictions on the Orlicz function M and on the space X, JO(X,M) recovers the main properties of J; for instance if X is reflexive and M satisfies certain technical conditions,

$$JO(X,M)^{**} \approx j_{JO}(JO) \oplus \Pi(X),$$

where j_{JO} denotes the canonical embedding of JO in JO^{**} and

$$\Pi(x) = (x, x, \ldots) \in JO^{**}.$$

4.c.6. Bellenot, Haydon and Odell's generalization of J.

Bellenot, Haydon and Odell in [1] generalized the James space J using its summing basis $\{\xi_i\}$ as follows: Let X be a Banach space with basis $\{x_i\}$. For a sequence $\alpha = \{a_i\}$ with finite support let

$$\|\alpha\| = \sup\|\sum_{i=1}^{k}(\sum_{j=p_i}^{q_i} a_j)x_{p_i}\|_X,$$

where the sup is taken over all sequences of positive integers $1 \le p_1 \le q_1 < p_2 \le q_2 <\ldots< p_k \le q_k$. $\mathcal{J}(x_i)$ is then defined as the Banach space generated by such α's.

They proved that the sequence of unit vectors $\{e_i\}$ in this space is a normalized, monotone basis and that if X is reflexive, then $\mathcal{J}(x_i)$ is quasi-reflexive of order one.

The following generalizations of J deal with an uncountable index set instead of using sequences, but maintaining the ℓ_2 norm.

4.c.7. The long James space J(η).

Edgar in [1] defined the long James space J(η) with an uncountable ordinal η as index set, which he studied from the point of view of its measurability properties. It is defined as follows: Let η be an ordinal and let $f : [0,\eta] \longrightarrow \mathbb{R}$ be a function. Then J(η) is the space of continuous functions f on $[0,\eta]$ with $f(0) = 0$ such that

$$\|f\| = \sup(\textstyle\sum_{i=1}^{n}|f(\alpha_i) - f(\alpha_{i-1})|^2)^{1/2} < \infty,$$

where the sup is taken over all finite sequences $\alpha_0 < \alpha_1 <\ldots< \alpha_n$ in $[0,\eta]$.

He proved that J(η) is a second dual space with a transfinite basis so that J(η) and J(η)* both have the RNP. For $\eta = \omega_1$, the least uncountable ordinal, the space J(ω_1) admits no equivalent weakly locally uniformly convex dual norm. He also showed that J(ω_1) is not real compact, not measure-compact, not Lindelöf, not WCG, not isomorphic to a subspace of a WCG space and fails the Pettis integral property; finally he proved that in this space the weak and weak star Borel sets are not the same, answering with this several questions about measurability properties of dual RNP spaces.

Edgar and Wheeler in [1] observed that if η is a countable ordinal, then its unitary ball is a regularly embedded Čech complete ball. Further, J(ω_1) has PCP, is an Asplund space and hence J(ω_1) is somewhat reflexive.

Zhao in [1] showed that the bidual of J(ω_1) can be identified with the set of all real valued functions g on the ordinal interval $[0,\omega_1]$ with bounded square variation and $g(0) = 0$ and that the transfinite sequence $\{\chi_{[\alpha,\omega_1]} : \alpha \in (0,\omega_1]\}$ of characteristic functions is a transfinite basis for this space.

4.c.8. Zhao's long James sum J(η,X).

Let η be as above and X a Banach space. Zhao in [2] defined a long James sum of η copies of X as a collection of long sequences $\{x_\alpha\}_{\alpha<\eta}$ with finite square variation. He proved that if X has a basis, then J(η,X) has a transfinite basis; in particular if $X = \ell_p$, $1 < p < \infty$, then the basis is shrinking.

4.c.9. Representability of a Banach space X in the form $Y^{**}/j(Y)$ for some Banach space Y.

The James space J is the first example of a space such that $(J^{**}/j(J)) \approx \mathbb{R}$. The generalizations that follow are focused on this point of view.

(i) In [6] James proved that if X is a Banach space with a monotone boundedly complete basis, there exists a Banach space B with a shrinking basis such that $(B^{**}/j(B)) \approx X$. As an application he proved the existence of a Banach space B_n whose n-th conjugate space is the first non-separable conjugate space of B_n.

(ii) Lindenstrauss in [1] generalized the result to any separable Banach space X.

(iii) Davis, Figiel, Johnson and Pelczynski in [1] generalized the result to WCG spaces X.

(iv) In [1] Bellenot defined the J-sum of a sequence (X_n, ϕ_n), where $(X_n, \| \ \|_n)$ is a Banach space, $\phi_n : X_n \longrightarrow X_{n+1}$ is a linear continuous map with $\|\phi_n\| \le 1$ and $X_0 = \{0\}$ as follows:

For $x_n \in X_n$ let $x_n^m = \phi_{m-1} \circ \phi_{m-2} \circ \ldots \circ \phi_n(x_n) \in X_m$. For a sequence $\{x_n\}$ with $x_n \in X_n$ and such that there is N with $x_n = 0$ for $n \ge N$, define

$$\|\{x_n\}\|_J^2 = \sup\left(\frac{1}{2} \sum_{i=1}^{k-1} \| x_{p(i)}^{p(i+1)} - x_{p(i+1)} \|_{p(i+1)}^2 + \|x_{p(k)}\|_{p(k)}^2 \right),$$

where the sup is taken over all increasing sequences of integers $0 \le p(1) < p(2) < \ldots < p(k)$. The completion of this space is called the J-sum $J(X_n, \phi_n)$. If $X_n = \mathbb{R}$ and ϕ_n = identity for $n = 1, 2, \ldots$, then $J(X_n, \phi_n) \approx J$.

As an application he proved the previous results **(i)-(iii)**, obtaining a concrete space B from a concrete space X. He also constructed a space Z isomorphic to both Z^* and $Z^{**}/j(Z)$.

(v) Valdivia in [1] proved the existence of a Banach space Z with Z^{**} separable such that for every separable Banach space X there exists a closed subspace Y of Z such that Z/Y has a shrinking basis and $(Z/Y)^{**}/j(Z/Y)$ is isometric to X.

4.d. Generalizations of JT

The James tree space JT is the first example of a Banach space whose members are sequences with index on a binary tree. Since then several generalizations have appeared in the literature and we now survey some of them. We can divide them into two categories: those generalizing the structure of the tree and "preserving" the definition of the norm in J, using it as a main ingredient, and those using different Banach spaces instead of J to define the norm.

4.d.1. Hagler and Odell's space.

In [1] Hagler and Odell constructed a non-separable analogue of JT, which is somewhat reflexive and thus does not contain any subspace isomorphic to ℓ_1. They used this space to construct the first example of a Banach space X whose dual ball is not w^*-sequentially compact but such that X does not contain any subspace isomorphic to ℓ_1.

4.d.2. Amemiya and Ito's spaces.

Amemiya and Ito in [1] constructed tree spaces branching an arbitrary finite number of times at each node of the tree. They proved that each normalized weakly null sequence in the space has a subsequence $(1 + \varepsilon)$-isomorphic to the unit vector basis of ℓ_2.

4.d.3. The space JT_∞.

Ghoussoub and Maurey in [2] used a tree where each node has an infinite countable set of offspring, to construct the space JT_∞ whose predual $\mathfrak{B}_\infty$ $\mathfrak{G}_\delta$-embeds in ℓ_2 but fails PCP.

In [1] Ghoussoub, Maurey and Schachermayer proved that $\mathfrak{B}_\infty$ has the convex point of continuity property (CPCP) and that hence PCP and CPCP are not equivalent, answering with this a question posed by Edgar and Wheeler in [1].

4.d.4. The class $\mathcal{J}$.

Brackebusch in [1] studied a class $\mathcal{J}$ of spaces similar to JT but modeled on an arbitrary tree which include the spaces mentioned in 4.d.1, 4.d.2 and 4.d.3. She established several results for these spaces, for instance every space in $\mathcal{J}$ is a dual space and has the RNP. Also if $X \in \mathcal{J}$, then X^{**} is also in $\mathcal{J}$ but based on a different tree.

4.d.5. The space JH.

Hagler in [2] constructed the space JH based on a binary tree whose norm is defined in terms of "admissible" segments. He proved that its dual JH^* is not separable, that each subspace of JH has a subspace isomorphic to c_0, that each subspace of JH^* has a subspace isomorphic to ℓ_1. Further JH has a subspace Y such that Y^* is separable and Y does not embed in c_0, JH^* has a separable subspace Z such that JH^*/Z is isometrically isomorphic to $c_0(\Gamma)$, where Γ has cardinality c, and weak sequential convergence and norm convergence in JH^* coincide, that is JH^* is a Schur space.

Odell and Schumacher in [1] proved that if X is JT or JH, then X^* contains a separable subspace which has the PCP but does not have the RNP. It is also shown that JH^* has the PCP.

Leung in [1] proved that ℓ_1 is not isomorphic to a subspace of $JH \tilde{\otimes}_\varepsilon JH$ while it was shown by Ruess [1] that ℓ_1 embeds in $JT \tilde{\otimes}_\varepsilon JT$. Here $X \tilde{\otimes}_\varepsilon Y$ is the closure of the finite rank operators in $K_{w^*}(X^*,Y)$, the space of compact weak star to weak continuous operators from X^* into Y with the operator norm.

Edgar and Wheeler in [1] observed that JH^* does not have Čech complete ball.

4.d.6. Hagler's space based on a binary transfinite tree.

Hagler in [1] defined for every infinite cardinal number $\mathfrak{m}$ satisfying $\mathfrak{m}^{\aleph_0} < 2^{\mathfrak{m}}$ a space X using a binary tree $\mathcal{T}_\beta$, where β is the initial ordinal corresponding to $\mathfrak{m}$ such that,
(i) the dimension of X is $\mathfrak{m}$,
(ii) X^* contains a subspace isometrically isomorphic to $\ell_1(\Gamma)$, where Γ has cardinality $2^{\mathfrak{m}}$,
(iii) X does not contain a subspace isomorphic to $\ell_1(\Lambda)$ for any uncountable set Λ,
(iv) X^* does not contain a subspace isomorphic to $L_1\{0,1\}^{\mathfrak{n}}$ for any uncountable cardinal $\mathfrak{n}$,
(v) X is not WCG.

4.d.7. The tree-like Tsirelson space ST_p.

Schechtman in [1] constructed for each $1 \le p \le \infty$ a "tree-like Tsirelson" space ST_p which is reflexive, has a 1-unconditional basis, does not contain ℓ_p and yet is uniformly isomorphic to $(\sum_{i=1}^n ST_p)_{\ell_p^n}$ for $n = 1, 2, \ldots$.

Andrew in [4] showed that if X is either one of the spaces ST_p (4.d.7) and JH (4.d.5), then X has the following property: if $U : X \longrightarrow X$ is a bounded linear operator, then there is a subspace $W \subset X$ such that $U|_W$ (or $(I - U)|_W$) is an isomorphism and UW (or (I - U)W) is complemented in X.

4.d.8. Bellenot, Haydon and Odell's generalization of JT.

Bellenot, Haydon and Odell in [1] constructed the space $JT(x_i)$, where $\{x_i\}$ is a normalized basis of a Banach space X. $JT(x_i)$ is constructed similarly to JT but taking the norm in X instead of the ℓ_2 norm. They showed that if X is reflexive, then $JT(x_i)$ is a dual space not containing ℓ_1, and if $\{x_i\}$ is boundedly complete, then the unit vector basis in $JT(x_i)$ is also boundedly complete.

Odell in [1] had already defined in 1985 the space $JT(e_n)$, where $\{e_n\}$ is the unit vector basis of the modified Tsirelson space T_M. The dual of this space is an example of a non-separable Banach space which does not contain a subsymmetric basic sequence. He further showed that $JT(e_n)^{**}$ is isomorphic to $JT(e_n) \oplus \ell_1(\Delta)$, where Δ is the Cantor set, and that all spreading models in $JT(e_n)$ are equivalent to the unit vector basis of ℓ_1 and the spreading models in $JT(e_n)^*$ are equivalent to either the summing basis or the unit vector basis of c_0.

CHAPTER 5. OTHER PATHOLOGICAL SPACES

Unmöglich ist's, drum eben glaubenswert

Johann Wolfgang von Goethe

This final chapter is devoted to a summary description of other noteworthy examples of pathological Banach spaces which are not directly related to the James spaces.

Undoubtedly, in the history of the geometry of Banach spaces a major role has been played by conjectures and questions about their properties, and the material in the previous chapter gives a good idea of the interplay between these conjectures and the appearance of many of today's well known spaces, which in turn gave rise to new questions and further conjectures.... Indeed, James' spaces J and JT are only two examples, albeit very important ones, of a vast class of spaces with unusual properties that have prompted the use of the adjective "pathological" when referring to them. Furthermore, the field is still very active, as testified by the recent constructions of Schlumprecht and of Gowers and Maurey. Thus it seems appropriate to conclude our exposition with a brief survey of some of these spaces, so that the readers get an idea of how things stand today.

(i) Schreier's space S.

One of the earliest of these spaces with unusual properties was constructed in 1930 by Schreier [1], which provided a counterexample to the question of Banach and Saks whether each weakly convergent sequence in C[0,1] contains a subsequence whose arithmetic means converge in norm. In fact, what is nowadays called Schreier's space S is a slightly different version of Schreier's original construction which can be found for instance in Beauzamy and Lapresté [1] and Casazza and Shura [1].

To define it we need the notion of admissible set:

A finite subset $E = \{n_1 < n_2 <...< n_k\}$ of natural numbers is called admissible if $k \le n_1$. Let N be the class of all admissible sets. S is then defined as the $\| \|_S$-completion of the space of real sequences $x = \{x\}$ of finite support, where

$$\|x\|_S = \sup_{E \in N} \sum_{k \in E} |x_k|.$$

S has the following properties: it is a non-reflexive Banach space with a 1-unconditional basis which does not contain ℓ_1, but it contains c_0 isometrically; furthermore, c_0 embeds in every infinite dimensional subspace of S. The spreading model based on the canonical basis in S is isometric to ℓ_1. The canonical injection from ℓ_1 into S is weakly compact and has norm 1.

Beauzamy and Maurey in [1] constructed a generalization of S, the Schreier-Orlicz space S_ϕ, and studied its spreading models.

(ii) James uniformly non-octahedral space.

In 1964 James in [7] proved that a uniformly non-square Banach space is reflexive and conjectured that a Banach space is reflexive if its subspaces are uniformly non-ℓ_1^n for some n. He himself in [8] constructed in 1974 a space $\tilde{X}$ which is a counterexample to this and also to the conjecture that a Banach space is superreflexive if and only if it is isomorphic to a uniformly non-octahedral space, where uniformly non-octahedral means uniformly non-ℓ_1^3. $\tilde{X}$ has a monotone shrinking basis and is quasi-reflexive of order one. Thus $\tilde{X}$ is of type p for some $p > 1$; in fact this was the first non-reflexive space known to have this property.

Later in [1] James and Lindenstrauss improved the definition of non-octahedral spaces and defined a non-octahedral space X_p for every p, $p \ge 1$, and in [9] James proved that X_p is of type 2 if $p > 2$. This was the first example of a non-reflexive space of type 2.

(iii) Baernstein's space B.

This space was constructed by Baernstein in [1] in 1972 answering the question raised by S. Sakai whether there was an example of a reflexive Banach space not satisfying the Banach-Saks property.

If for a set $A \in N$ and a sequence $x = \{x_n\}$ we define

$$P_A x(j) = \begin{cases} x_j & \text{if } j \in A, \\ 0 & \text{otherwise.} \end{cases}$$

and for a sequence x with finite support we let

$$\|x\|_B = \sup\left\{ \left(\sum_{k=1}^n \|P_{A_k} x\|_{\ell_1}^2\right)^{1/2} : A_1 < A_2 < \ldots < A_n \right\},$$

where the sup is taken over all finite strings of consecutive admissible sets, then B is the $\| \|_B$-completion of the space of the real sequences with finite support. It can be seen in Beauzamy and Lapresté [1] and Casazza and Shura [1] that B is a reflexive Banach space with a monotone unconditional basis, so that neither ℓ_1 nor c_0 embeds in B; however, B does contain an isomorphic copy of ℓ_2, and B has ℓ_1 and ℓ_2 as spreading models.

More generally, if $1 < p < \infty$ and in the definition of $\| \|_B$ the number 2 is replaced by p, the resulting Banach space B_p has the following properties as is shown by Seifert in [1]: B_p is reflexive and neither ℓ_1 nor c_0 embeds in B_p, B_p^* has the Banach-Saks property, but B_p has not. If $\{b_n\}$ is the canonical unit vector basis of B_p, $\{b_{2^n}\}$ is isometrically equivalent to the unit vector basis of ℓ_p. If X is an infinite dimensional subspace of B_p, then X contains a subspace Y such that Y is isomorphic to ℓ_p and Y is complemented in B_p. In addition B_p and B_q are totally incomparable if $p \neq q$.

Beauzamy in [3] defined the Baernstein-Orlicz space B_ϕ. He proved that it has an unconditional monotone basis and does not contain c_0. Also he showed that its spreading models are either isometric to ℓ_2 or isomorphic to some Orlicz space ℓ_ψ.

(iv) Tsirelson's space.

In 1973 in [1] Tsirelson constructed his space as a counterexample to the classical conjecture that every Banach space contains a copy of either c_0 or ℓ_p for some p, $1 \leq p < \infty$. Tsirelson's space turned out to be one of the most important sources of counterexamples in Banach space theory, and a nice and thorough monograph on this space has been written by Casazza and Shura [1]; the results we list can be found there. The definition they use in their monograph is due to Figiel and Johnson and describes the norm in the dual space of the original space of Tsirelson, which is the space T that is known as Tsirelson's space today. It is defined as follows:

Let $x = \{a_n\}$ be a real sequence of finite support. Let

$$\begin{cases} \|x\|_0 = \max_n |a_n|, \\ \|x\|_{m+1} = \max\left\{\|x\|_m, \frac{1}{2}\max\left\{\sum_{j=1}^k \|P_{A_j} x\|_m\right\}\right\}. \end{cases}$$

where the inner max is taken over all finite consecutive subsets $\{A_j\}_{j=1}^k$ of $\mathbb{N}$ such that $k \le \min A_1$ and P_A is as in **(iii)**. Then $\lim_m \|x\|_m$ exists; denote this limit by $\|x\|$. It is easily shown that this is indeed a norm. Tsirelson's space T is the $\|\ \|$-completion of the space of real sequences with finite support.

(a) T is a reflexive Banach space with a shrinking 1-unconditional basis $\{t_n\}$.

(b) T and T^* have no embedded copies of c_0 or any ℓ_p for $p \ge 1$ and don't contain any subsymmetric basic sequence.

(c) Neither T nor T^* is primary.

(d) T is the first Banach lattice to fail the Banach-Saks property; however, T^* has it.

(e) T contains no minimal subspaces but T^* is minimal. (An infinite dimensional Banach space is minimal if it embeds into each of its infinite dimensional subspaces.)

(f) T does not contain any infinite dimensional superreflexive subspaces.

(g) All isometries of T are surjective and the norm of T has the fixed point property.

(h) c_0 is finitely representable in T^*.

As is the case with James spaces, Tsirelson's space has many variations and generalizations.

(v) Schlumprecht's space.

In [1] Schlumprecht answered affirmatively the question of whether there exists a Banach space that is λ-distortable for every $\lambda > 1$. A Banach space $(X, \| \ \|)$ is λ-distortable if there is an equivalent norm $|\!|\!| \ |\!|\!|$ on X such that for every infinite dimensional subspace Y of X

$$\sup_{x\in Y}(|\!|\!|x|\!|\!|/\|x\|) \geq \lambda \inf_{x\in Y}(|\!|\!|x|\!|\!|/\|x\|).$$

Schlumprecht's space, constructed in the spirit of Tsirelson's space, is defined as follows:

For a finite support sequence $x = \{a_n\}$ let

$$\begin{cases} \|x\|_0 = \max_n |a_n|, \\ \|x\|_m = \max \left\{ \log_2(k+1)^{-1}(\sum_{j=1}^{k} \|P_{A_j} x\|_{m-1}) \right\} \quad m \geq 1. \end{cases}$$

where the max is taken over all finite consecutive subsets $\{A_j\}_{j=1}^{k}$ of $\mathbb{N}$ and P_{A_j} is as in **(iii)**. Set $\|x\| = \lim_n \|x\|_n$. Schlumprecht's space is the $\| \ \|$-completion of the space of sequences with finite support.

He shows that the usual basis $\{e_n\}$ is a 1-subsymmetric, 1-unconditional basis of this space and that the space is arbitrarily distortable, that is λ-distortable for every $\lambda > 1$.

(vi) Gowers-Maurey space.

The latest addition to this forest of strange spaces appeared in 1991, when Gowers and Maurey independently constructed a space X related to Schlumprecht's space, which solved the long standing unconditional basic sequence problem, that is X does not contain an infinite unconditional basic sequence. They published their results in a joint paper (Gowers and Maurey [1]).

The space X is reflexive and has a monotone basis, but most important, X is hereditarily indecomposable (h.i.) which means that no closed infinite dimensional subspace is the topological sum of two infinite dimensional closed subspaces, or equivalently if Y and Z are two infinite dimensional subspaces of X and $\varepsilon > 0$ then there exist $y \in Y$ and

$z \in Z$ such that $\|y\| = \|z\| = 1$ and $\|y - z\| < \varepsilon$. From this they derive that X does not contain any unconditional basic sequence, that X is not isomorphic to any of its proper subspaces and that every bounded operator on X is of the form $\lambda I_X + S$ where S is a strictly singular operator, i.e., S restricted to an infinite dimensional subspace Z cannot be an isomorphism.

In a second paper, [2], Gowers and Maurey, generalizing their construction, find a new prime Banach space, a space isomorphic to its subspaces of codimension two but not to its hyperplanes and a space isomorphic to its cube but not to its square.

Donc, c'en est fait. Ce livre est clos. Chères Idées
Qui rayiez mon ciel gris de vos ailes de feu
Dont le vent caressait mes tempes obsédées,
Vous pouvez revoler devers l'Infini bleu!

Paul Verlaine

REFERENCES

Alspach D. [1] A fixed point free non-expansive map. *Proc. A.M.S.* **82** (1981), 423-424.

Amemiya I., Ito T. [1] Weakly null sequences in James' spaces on trees. *Kodai Math. J.* **4** (1981), 418-425.

Andrew A. [1] James' quasi-reflexive space is not isomorphic to any subspace of its dual. *Israel J. Math.* **38** (1981), 276-282.

Andrew A. [2] Spreading basic sequences and subspaces of James' quasi-reflexive space. *Math. Scand.* **48** (1981), 109-118.

Andrew A. [3] The Banach space JT is primary. *Pac. J. Math.* **108** (1983), 9-17.

Andrew A. [4] Projections on tree-like Banach spaces. *Can. J. Math.* **37** (1985), 908-920.

Andrew A., Green W. [1] On James' quasi-reflexive Banach space as a Banach algebra. *Can. J. Math.* **32** (1980), 1080-1101.

Baernstein A.H. [1] On reflexivity and summability. *Studia Math.* **42** (1972), 91-94.

Banach S. [1] *Théorie des opérations lineaires*. Warszawa 1932, reprinted by Chelsea, New York 1972.

Banach S., Saks S. [1] Sur la convergence forte dans les champs L^p. *Studia Math* **2** (1930), 51-57.

Beauzamy B. [1] *Introduction to Banach spaces and their geometry*. North-Holland Mathematics Studies 68, Amsterdam 1985.

Beauzamy B. [2] Banach-Saks properties and spreading models. *Math Scand.* **44** (1979), 357-384.

Beauzamy B. [3] Deux espaces de Banach et leurs modèles étalés. Publ. du Dép. de Math. Univ. Lyon I (1980), t. 17/2.

Beauzamy B., Lapresté J.T. [1] *Modèles étalés des espaces de Banach*. Travaux en Cours. Hermann, Paris 1984.

Beauzamy B., Maurey B. [1] Iteration of spreading models. *Arkiv för Mat.* **17/2** (1979), 193-198.

Bellenot S. [1] The J-sum of Banach spaces. *J. Func. Analysis* **48** (1982), 95-106.

Bellenot S. [2] Transfinite duals of quasi-reflexive Banach spaces. *Trans. A.M.S.* **273** (1982), 551-577.

Bellenot S. [3] Isometries of James spaces. Contemp. Math. 85. A.M.S., Providence, R.I. 1989, 1-18.

Bellenot S. [4] The maximum path theorem and extreme points of James' space. *Studia Math.* **94** (1989), 1-15.

Bellenot S. [5] Banach spaces with trivial isometries. *Israel J. Math.* **56** (1986), 89-96.

Bellenot S., Haydon R., Odell E. [1] Quasi-reflexive and tree spaces constructed in the spirit of James. Contemp. Math. 85. A.M.S., Providence, R.I. 1989, 19-43.

Bessaga C., Pelczynski A. [1] Banach spaces non-isomorphic to their Cartesian squares I. *Bull. Acad. Sci. Pol.* **8** (1960), 77-80.

Bessaga C., Pelczynski A. [2] Some aspects of the present theory of Banach spaces. Oeuvres Stefan Banach. Vol. II. P.W.N., Warszawa 1979, 221-302.

Bombal F. [1] Propiedades locales y ultraproductos de espacios de Banach Publ. del Depto. de Análisis Mat. 7. Fac. de Mat. Universidad Complutense, Madrid 1982.

Bourgain J., Rosenthal H. [1] Geometrical implications of certain finite dimensional decompositions. *Bull. Soc. Math. Belg.* **32** (1980), 57-82.

Bourgain J., Rosenthal H. [2] Applications of the theory of semi-embeddings to Banach space theory. *J. Func. Analysis* **52** (1983), 149-188.

Brackebusch R. [1] James' space in general trees. *J. Func. Analysis* **79** (1988), 446-475.

Brodskii M.S., Milman D.P. [1] On the center of a convex set. *Dokl. Akad. Nauk. SSSR* **59** (1948), 837-840.

Brown L., Ito T. [1] Isometric preduals of James spaces. *Can. J. Math.* **32** (1980), 59-69.

Brunel A., Sucheston L. [1] On B-convex Banach spaces. *Math. Systems Theory* **7** (1973), 294-299.

Brunel A., Sucheston L. [2] Equal signs additive sequences in Banach spaces. *J. Func. Analysis* **21** (1976), 286-304.

Casazza P.G. [1] James' quasi-reflexive space is primary. *Israel J. Math.* **26** (1977), 294-305.

Casazza P.G., Kottman C.A., Lin B.L. [1] On some classes of primary Banach spaces. *Can. J. Math.* **29** (1977), 856-873.

Casazza P.G., Lin B.L., Lohman R.H. [1] On James' quasi-reflexive Banach space. *Proc. A.M.S.* **67** (1977), 265-271.

Casazza P.G., Lin B.L., Lohman R.H. [2] On non-reflexive Banach spaces which contain no c_0 or ℓ_p. *Can. J. Math.* **32** (1980), 1382-1389.

Casazza P.G., Lohman R.H. [1] A general construction of spaces of the type of R.C. James. *Can. J. Math.* **27** (1975), 1263-1270.

Casazza P.G., Shura T.J. [1] *Tsirelson's Space.* Lecture Notes in Mathematics 1363, Springer Verlag, Berlin 1989.

Ciesielski Z. [1] On Haar functions and on the Schauder basis of the space C[0,1]. *Bull. Acad. Sci. Pol.* **7** (1959), 227-232.

Ciesielski Z. [2] On the isomorphisms of the spaces H^α and *m*. *Bull. Acad. Sci. Pol.* **8** (1960), 217-222.

Civin P., Yood B. [1] Quasi-reflexive spaces. *Proc. A.M.S.* **8** (1957), 906-911.

Davis W.J. [1] Separable Banach spaces with only trivial isometries. *Rev. Roum. Math. Pures et Appl.* **16** (1971), 1051-1054.

Davis W.J, Figiel T., Johnson W.B., Pelczynski A. [1] Factoring weakly compact operators. *J. Func. Analysis* **17** (1974), 311-327.

Davis W.J., Singer I. [1] Boundedly complete M-basis and complemented subspaces in Banach spaces. *Trans. A.M.S.* **175** (1973), 187-194.

Day M.M. [1] *Normed Linear Spaces.* Springer Verlag, Berlin 1973.

Day M.M., James R.C., Swaminathan S. [1] Normed linear spaces that are uniformly convex in every direction. *Can. J. Math.* **23** (1971), 1051-1059.

Diestel J. [1] *Geometry of Banach Spaces - Selected Topics.* Lecture Notes in Mathematics 485, Springer Verlag, Berlin 1975.

Diestel J. [2] *Sequences and Series in Banach Spaces.* Springer Verlag, Berlin 1984.

Diestel J., Uhl J.J. [1] *Vector Measures.* Math. Surveys 15, A.M.S., Providence, R.I. 1977.

Dieudonné J. [1] Complex structures on real Banach spaces. *Proc. A.M.S.* **3** (1952), 162-164.

Edgar G.A. [1] A long James space. Measure Theory, Oberwolfach 1979, Lecture Notes in Math. 794, Springer Verlag, Berlin 1980, 31-37.

Edgar G.A., Wheeler R.F. [1] Topological properties of Banach spaces. *Pac. J. Math.* **115** (1984), 317-350.

Enflo P. [1] A counterexample to the approximation property in Banach spaces. *Acta Math.* **130** (1973), 309-317.

Figiel T., Johnson W.B., Tzafriri L. [1] On Banach spaces having local unconditional structure, with applications to Lorentz function spaces. *J. Approximation Theory* **13** (1975), 395-412.

Finet C. [1] Espaces de James généralisés et espaces de type ESA. Séminaire de Géometrie des Espaces de Banach. Paris VII (1982), 139-155.

Ghoussoub N., Maurey B. [1] Counterexamples to several problems concerning G_δ-embeddings. *Proc. A.M.S.* **92** (1984), 409-412.

Ghoussoub N., Maurey B. [2] G_δ-embeddings in Hilbert space. *J. Func. Analysis* **61** (1985), 72-97.

Ghoussoub N., Maurey B., Schachermayer W. [1] A counterexample to a problem on points of continuity in Banach spaces. *Proc. A.M.S.* **99** (1987), 278-282.

Ghoussoub N., Maurey B., Schachermayer W. [2] Geometric implications of certain infinite dimensional decompositions. *Trans. A.M.S.* **317** (1990), 541-584.

Giesy D.P., James R.C. [1] Uniformly non-$\ell^{(1)}$ and B-convex Banach spaces. *Studia Math.* **48** (1973), 61-69.

Godefroy G. [1] Espaces de Banach: existence et unicité de certains préduaux. *Ann. Ins. Fourier, Grenoble* **28** (1978), 87-105.

Gordon Y., Lewis D.R. [1] Absolutely summing operators and local unconditional structures. *Acta Math.* **133** (1974), 27-48.

Gowers W.T., Maurey B. [1] The unconditional basic sequence problem *J. A.M.S.* **6** (1993), 851-874.

Gowers W.T., Maurey B. [2] Banach spaces with small spaces of operators (available on the Banach space Bulletin Board).

Guerre S., Lapresté J.T. [1] Quelques propriétés des modèles étalés sur les espaces de Banach. *Ann. Inst. H. Poincaré, Sect. B (N.S.)* **16** (1980), 339-347.

Hagler J. [1] Non-separable "James tree" analogues of the continuous functions on the Cantor set. *Studia Math.* **61** (1977), 41-53.

Hagler J. [2] A counterexample to several questions about Banach spaces. *Studia Math.* **60** (1977), 289-308.

Hagler J., Odell E. [1] A Banach space not containing ℓ_1 whose dual ball is not weak* sequentially compact. *Illinois J. Math.* **22** (1978), 290-294.

Herman R., Whitley R. [1] An example concerning reflexivity. *Studia Math.* **28** (1967), 289-294.

Hoffmann-Jørgensen J. [1] Sums of independent Banach space valued random variables. *Studia Math.* **52** (1974), 159-186.

James R.C. [1] Bases and reflexivity of Banach spaces. *Ann. Math.* **52** (1950), 518-527.

James R.C. [2] A non-reflexive Banach space isometric with its second conjugate space. *Proc. Nat. Acad. Sci. U.S.A.* **37** (1951), 174-177.

James R.C. [3] A separable somewhat reflexive Banach space with non-separable dual. *Bull. A.M.S.* **80** (1974), 738-743.

James R.C. [4] Banach spaces quasi-reflexive of order one. *Studia Math.* **60** (1977), 157-177.

James R.C. [5] Bases in Banach spaces. *Amer. Math. Monthly* **89** (1982), 625-640.

James R.C. [6] Separable conjugate spaces. *Pac. J. Math.* **10** (1960), 563-571.

James R.C. [7] Uniformly non-square Banach spaces. *Ann. Math.* **80** (1964), 542-550.

James R.C. [8] A non-reflexive Banach space that is uniformly non-octahedral. *Israel J. Math.* **18** (1974), 145-175.

James R.C. [9] Non-reflexive spaces of type 2. *Israel J. Math.* **30** (1978), 1-13.

James R.C. [10] Super-reflexive spaces with bases *Pac. J. Math.* **41** (1972), 409-419.

James R.C., Lindenstrauss J. [1] The octahedral problem for Banach spaces. Proc. of the Seminar on Random Series, Convex Sets and Geometry of Banach Spaces. Various Publ. Ser. 24 Mat. Inst. Aarhus, Univ. Aarhus, Denmark 1975, 100-120.

Johnson W.B. [1] On quotients of L_p which are quotients of ℓ_p. *Compositio Math.* **34** (1977), 69-89.

Johnson W.B., Rosenthal H.P., Zippin M. [1] On bases, finite dimensional decompositions and weaker structures in Banach spaces. *Israel J. Math.* **9** (1971), 488-506.

Karlovitz L.A. [1] Existence of fixed points for non-expansive mappings in a space without normal structure. *Pac. J. Math.* **66** (1976) 153-159.

Kelley J.L., Namioka I. [1] *Linear Topological Spaces.* Springer Verlag, Berlin 1963.

Khamsi M.A. [1] James quasireflexive space has the fixed point property. *Bull. Austral. Math. Soc.* **39** (1989), 25-30.

Khamsi M.A. [2] Normal structure for Banach spaces with Schauder decomposition. *Can. Math. Bull.* **32** (1989), 344-351.

Kirk W.A. [1] A fixed point theorem for mappings which do not increase distance. *Amer. Math. Monthly* **72** (1965) 1004-1006.

Lacey H.E. [1] A unified approach to the principle of local reflexivity. Notes in Banach Spaces. Bernau S.J., University of Texas Press, Austin and London 1980, 427-439.

Leung D. [1] Embedding ℓ^1 into tensor products of Banach spaces. Functional Analysis. Lecture Notes in Math. 1470, Springer Verlag, Berlin 1991, 171-176.

Lin B.L., Lohman R.H. [1] On generalized James quasi-reflexive Banach spaces. *Bull. Inst. Math. Academia Sinica* **8** Part II (1980), 389-399.

Lindenstrauss J. [1] On James' paper "Separable conjugate spaces". *Israel J. Math.* **9** (1971), 279-284.

Lindenstrauss J., Stegall C. [1] Examples of separable spaces which do not contain ℓ^1 and whose duals are not separable. *Studia Math.* **54** (1975), 81-105.

Lindenstrauss J., Tzafriri L. [1] *Classical Banach spaces I.* Springer Verlag, Berlin 1977.

Lindenstrauss J., Tzafriri L. [2] *Classical Banach spaces II.* Springer Verlag, Berlin 1979.

Lindenstrauss J., Tzafriri L. [3] *Classical Banach spaces.* Lecture Notes in Mathematics 338, Springer Verlag, Berlin 1973.

Maurey B., Pisier G. [1] Séries de variables aléatoires vectorielles indépendantes et propriétés géométriques des espaces de Banach. *Studia Math.* **58** (1976), 45-90.

Mc.Williams R.D. [1] On certain Banach spaces which are w^*-sequentially dense in their second duals. *Duke Math. J.* **37** (1970), 121-126.

Mityagin B.S., Èdel'shtein I.S. [1] Homotopy type of linear groups of two classes of Banach spaces. *Func. Analysis and its Appl.* **4** (1970), 221-231.

Odell E. [1] A nonseparable Banach space not containing a subsymmetric basic sequence. *Israel J. Math.* **52** (1985), 97-109.

Odell E., Schumacher C.S. [1] JH^* has the P.C.P. Contemp. Math. 85. A.M.S., Providence, R.I. 1989, 387-403.

Pelczynski A. [1] Structural theory of Banach spaces and its interplay with analysis and probability. Proc. of the I.C.M.(Warsaw, 1983), P.W.N., Warsaw 1984, 237-269.

Pisier G. [1] The dual J^* of the James space has cotype 2 and the Gordon Lewis Property. *Math. Proc. Camb. Phil. Soc.* **103** (1988), 323-331.

Pisier G. [2] *Factorization of linear operators and Geometry of Banach spaces.* CBMS Regional Conference Series no.60 A.M.S. 1986.

Pisier G. [3] Probabilistic methods in the geometry of Banach spaces. Department of Mathematics. Texas A&M University, 1-75.

Rosenthal H.P. [1] A characterization of Banach spaces containing ℓ_1. *Proc. Nat. Acad. Sci. U.S.A.* **71** (1974), 2411-2413.

Rudin W. [1] *Functional Analysis.* McGraw-Hill, New York 1973.

Rudin W. [2] *Real and Complex Analysis.* McGraw-Hill, New York 1974.

Ruess W. [1] Duality and geometry of spaces of compact operators. Functional Analysis: Surveys and Recent Results III. Math. Studies 90, North-Holland, Amsterdam 1984, 59-78.

Schachermayer W. [1] Some more remarkable properties of James-tree space. Contemp. Math. 85 A.M.S., Providence, R.I. 1989, 465-496.

Schachermayer W., Sersouri A., Werner E. Moduli of non-dentability and the Radon-Nikodym property in Banach spaces. *Israel J. Math.* **65** (1989), 225-257.

Schaefer H.H. [1] *Topological Vector Spaces.* Springer Verlag, Berlin 1971.

Schechtman G. [1] A tree-like Tsirelson space. *Pac. J. Math.* **83** (1979), 523-530.

Schlumprecht T. [1] An arbitrarily distortable Banach space. *Israel J. Math.* **76** (1991), 81-95.

Schreier J. [1] Ein Gegenbeispiel zur Theorie der schwachen Konvergenz. *Studia Math.* **2** (1930), 58-62.

Seifert C.J. [1] Averaging in Banach spaces. (Dissertation), Kent State University Ohio 1977.

Semenov P.V. [1] James-Orlicz spaces. *Russian Math. Surveys* **34** (1979), 185-186.

Semenov P.V., Skorik A.I. [1] Isometries of James spaces. *Math. Notes* **38** (1985), 804-808. (Translated from Russian: *Mat. Zametki* **38** (1985), 537-544.)

Sersouri A. [1] On James' type spaces. *Trans. A.M.S.* **310** (1988), 715-745.

Singer I. [1] *Bases in Banach spaces I*. Springer Verlag, Berlin 1970.

Singer I. [2] Bases and quasi-reflexivity of Banach spaces. *Math. Ann.* **153** (1964), 199-209.

Tsirelson B.S. [1] Not every Banach space contains an embedding of ℓ_p or c_0. *Func. Analysis and its Appl.* **8** (English Ed.) (1974), 138-141.

Valdivia M. [1] Shrinking bases and Banach spaces Z^{**}/Z. *Israel J. Math.* **62** (1988), 347-354.

Van Dulst D. [1] Equivalent norms and the fixed point property for non-expansive mappings. *J. Lond. Math. Soc.* **25** (1982), 139-144.

Varopoulos N. [1] Sous-espaces de $\mathbb{C}(G)$ invariants par translations et de type $\mathcal{L}_1$. Exposé no. 12, Séminaire Maurey-Schwartz, 1975-1976. École Polytechnique, Paris.

Wojtaszczyk P. [1] *Banach spaces for Analysts*. Cambridge Studies in Advanced Mathematics 25, Cambridge University Press 1991.

Yao Z.A., Su L.N. [1] A generalization of James' space. *J. Math. Res.* **6** (1986), 47-50. (Chinese).

Zhao J.F. [1] The transfinite basis of the bidual space of the long James space. *Acta Math. Sci.* **5** (English Ed.) (1985), 295-301.

Zhao J.F. [2] The vector-valued long James type Banach spaces $J(\eta,\ell_p)$, $1 \le p < \infty$. J. *Systems Sci. Math. Sci.* **5** (1985), 52-62.

INDEX OF CITATIONS

Alspach D. 165

Amemiya I., Ito T. 227

Andrew A. 52,59,60,83,100,135,138,147,149,229

Andrew A., Green W. 216

Baernstein A.H. 232

Banach S. 11,15,134

Banach S., Saks S. 92

Beauzamy B. 4,8,10,28,35,48,60,81,90,92,100,103,113,118,176,222,233

Beauzamy B., Lapresté J.T. 94,100,223,231,233

Beauzamy B., Maurey B. 232

Bellenot S. 61,72,215,221,226

Bellenot S., Haydon R., Odell E. 224,229

Bessaga C., Pelczynski A. 15,18,27

Bombal F. 18,27

Bornemann W. v

Bourgain J., Rosenthal H. 217,218

Brackebusch R. 228

von Brandenburg G.W. 4

Brodskii M.S., Milman D.P. 161,163

Brown L., Ito T. 81

Brunel A., Sucheston L. 94,98,223

Casazza P.G. 37,41,60,124

Casazza P.G., Kottman C.A., Lin B.L. 32

Casazza P.G., Lin B.L., Lohman R.H. 29,30,44,46,59,222

Casazza P.G., Lohman R.H. 221,222

Casazza P.G., Shura T.J. 231,233,234

Ciesielski Z. 126

Civin P., Yood B. 10,73

Davis W.J. 61
Davis W.J, Figiel T., Johnson W.B., Pelczynski A. 226
Davis W.J., Singer I. 135
Day M.M. 27,178,182
Day M.M., James R.C., Swaminathan S. 167
Diestel J. 35,82,164,178,181,182,188
Diestel J., Uhl J.J. 129,219
Dieudonné J. 16
Dubinski E., Pelczynski A., Rosenthal H.P. 215

Edgar G.A. 225
Edgar G.A., Wheeler R.F. 218,219,225,228
Eluard P. 11
Enflo P. 35,119

Figiel T., Johnson W.B., Tzafriri L. 215
Finet C. 223

Ghoussoub N., Maurey B. 218,227
Ghoussoub N., Maurey B., Schachermayer W. 219,220,228
Giesy D.P., James R.C. 17,18
Godefroy G. 81
von Goethe J.W. 133,231
Gordon Y., Lewis D.R. 215
Gowers W.T., Maurey B. 235,236
Gray T. x
Guerre S., Lapresté J.T. 100,101

Hagler J. 228,229
Hagler J., Odell E. 227
Herman R., Whitley R. 46,48
Hoffmann-Jørgensen J. 102

James R.C. 6,11,12,17,18,35,72,75,82,104,134,139,226,232
James R.C., Lindenstrauss J. 232
Johnson W.B. 124
Johnson W.B., Rosenthal H.P., Zippin M. 116,121

Johnson W.B., Tzafriri J. 215
Joyce J. 214

Karlovitz L.A. 165
Kelley J.L., Namioka I. 121
Khamsi M.A. 161,167
Kirk W.A. 161
Krein M.G., Milman D.P., Rutman M.A. 7

Lacey H.E. 121
Leung D., 228
Lin B.L., Lohman R.H. 222
Lindenstrauss J. 182,226
Lindenstrauss J., Rosenthal H.P. 121
Lindenstrauss J., Stegall C. 168,217,220,221
Lindenstrauss J., Tzafriri L. 4,37,48,59,82,101,124,125,139,176,179,215

Maurey B., Pisier G. 103,104
Mazur S. 139
Mc Williams R.D. 176
Mityagin B.S., Èdel'shtein I.S. 216

Odell E. 230
Odell E., Schumacher C.S. 228

Paz O. 134
Pelczynski A. xi
Pisier G. 104,105,131,215

Rosenthal H.P. 168
Rudin W. 4,81,192
Ruess W. 228

Sakai S 232
Schachermayer W. 183,213
Schaefer H.H. 16
Schechtman G. 229
Schlumprecht T. 235

Schreier J. 231
Seifert C.J. 233
Semenov P.V. 223
Semenov P.V., Skorik A.I. 62
Sersouri A. 62
Singer I. 4,8,10,14,210

Tsirelson B.S. 233

Valdivia M. 227
Van Dulst D. 167
Varopoulos N. xi
Verlaine P. 236

Wojtaszczyk P. xi

Yao Z.A., Su L.N. 222

Zhao J.F. 225

LIST OF SPECIAL SYMBOLS

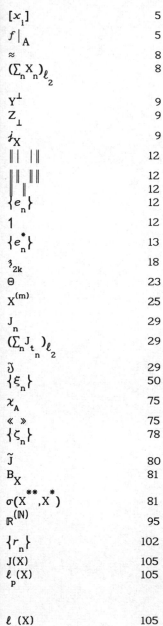

$[x_i]$	5
$f\|_A$	5
$\approx$	8
$(\sum_n X_n)_{\ell_2}$	8
$Y^{\perp}$	9
$Z_{\perp}$	9
$\jmath_X$	9
$\|\!\|\ \ \|\!\|$	12
$\|\!\|\!\|\ \ \|\!\|\!\|$	12
$\|\ \|$	12
$\{e_n\}$	12
$\mathbb{1}$	12
$\{e_n^*\}$	13
$\mathfrak{z}_{2k}$	18
Θ	23
$X^{(m)}$	25
J_n	29
$(\sum_n J_{t_n})_{\ell_2}$	29
$\mathfrak{J}$	29
$\{\xi_n\}$	50
χ_A	75
$\langle\!\langle\ \rangle\!\rangle$	75
$\{\zeta_n\}$	78
$\tilde{J}$	80
B_X	81
$\sigma(X^{**}, X^*)$	81
$\mathbb{R}^{(\mathbb{N})}$	95
$\{r_n\}$	102
$J(X)$	105
$\ell_p(X)$	105
$\ell_\infty(X)$	105

$\mathrm{Lip}_\alpha([0,1],\ X)$	106
$\mathrm{Lip}_\alpha(X)$	107
I_X	117
$\{\hbar_n\}$	127
$L_p(\Omega,\Sigma,\mu,X)$	130
$L_\infty(\Omega,\Sigma,\mu,X)$	130
$L_p(X)$	130
$L_p^N(X)$	131
$\mathcal{R}_n(X^*)$	132
$\\|\ \\|_J$	135
$\mathcal{T}$	135
$\mathrm{lev}(t)$	136
$\mathrm{lev}_\varphi(s)$	136
$\mathcal{T}_\nu$	136
$\eta_t, \eta_{(i,j)}, \eta_{ij}, \eta_{i,j}$	136
$\eta_t^*, \eta_{(i,j)}^*, \eta_{ij}^*, \eta_{i,j}^*$	136
$\mathrm{supp}\,x$	136
f_S	137
f_B	137
P_N	137
P_S	137
P_B	137
Q_t	137
$[\ ]_k$	140
$S_1 \succ S_2$	152
$\mathrm{co}A$	162
$d(x,\ A)$	162
$\mathrm{diam}\ K$	162
Γ	170
e_B	171
$[\![\ ]\!]$	174
$\dashv\ \vdash$	176
F_B	177
Δ	180
$C(\Delta)$	180
S_X	183

$\mathcal{M}$	186
$\mathcal{MI}$	186
$\overline{\mathtt{A}}$	186
$\overline{\mathtt{A}}^{w*}$	186
$\overline{\mathtt{co}}\mathtt{A}$	186
$\overline{\mathtt{co}}^{w*}\mathtt{A}$	186

SUBJECT INDEX

admissible set 231,233
approximation property
—— ——, bounded (B.A.P.) 37,39,119
—— ——, μ (μ-A.P.) 119

B-convex 28
Banach-Saks (B.S.) property 92
—— ——, alternate 93
—— ——, weak 93
basic sequence, basis 4
——, block 5,44,84,86
——, block $\| \|_E$-control 222
——, boundedly complete 5,50
——, invariant under spreading (IS) 50,51
——, k-shrinking 14,15
——, monotone 4,50
——, nearly perfectly homogeneous 222
——, seminormalized 7
——, shrinking 5,12,78
——, spreading 44,50,52,58,59,83
——, subsymmetric 50,59
——, summing 50,51,83
——, unconditional 4,44,47,48,49
basis constant 4
biorthogonal functionals 5
Bochner integrable function 129,130,131
Bochner measurable function 131,138
branch 136
——, n- 136

Cantor set 180
Čech completeness 218
complexification 16
convexity
——, locally uniform 183
——, strict 183
——, uniform 35
cotype 103,104,115

decomposition (Schauder) 119
——, blocking of a 120,125
——, boundedly complete 120
——, boundedly complete skipped blocking (BCSBD) 120
——, finite dimensional (F.D.D.) 37,120,123,124,125

——, shrinking 37,39,120,123,125
decomposition constant 120
dentable set 181
descendant 135
distortable space 235
——, λ- 235
Dvoretzky-Rogers theorem 27

finite representability 17,18,27,28
fixed point property 161,164,166

gap 138
$\mathfrak{G}_\delta$-embedding 218
general linear group of J^n ($GL(J^n)$) 216
Gordon-Lewis property 215

Haar functions 108,127
Haar system 127
hereditarily indecomposable (h.i.) space 235

isometry 61,70

James space J 12
James tree space JT 136
JM-type decomposition 219,220
JT-type decomposition 219,220

Kadec Klee property 183,184,207,210
Kahane's inequality 103
Khintchine's inequality 113
Kirk's theorem 164
Krein-Milman property (KMP) 182

local unconditional structure 214,215

minimal space 234
μ-measurable function 130

node 135
——, level of a 136
normal structure 163,164,166
——, weak 163,164,166

Odell-Rosenthal-Haydon theorem 188
offspring 135
Orlicz function 223,224

π_λ-space 37,116,117,118,122,123
point
——, diametral 162
——, extreme 62,63,64

—— of w^* to norm continuity 200,201
point of continuity property (PCP) 217,218,219
predual, isometric 73,74,81,82
——, isomorphic 73
primary space 44,50,74,160
principle of local reflexivity 121

quasi-reflexivity 10,11,14,15,73,79

Rademacher functions 102
Radon-Nikodym property (RNP) 181,219
Ramsey's theorem 96
real underlying space 15,16,17
Rosenthal's dichotomy theorem 101

Schauder basis 4
segment 135
sequence
——, complementary 30
——, diametral 162
——, good 95,98
——, normalized 7
——, proper 30,31,32
——, seminormalized 7
simple function 185
slice 202
somewhat reflexivity 44,48,91,139
space,
——, Amemiya-Ito 227
——, Asplund 218,219
——, $\mathfrak{B}$ 171
——, $\mathfrak{B}_\infty$ 227
——, Baernstein - B 232
——, Baernstein-Orlicz - B_ϕ 233
——, equal signs additive (ESA) 223
——, $G(x_i)$ 223
——, Godefroy 218,219
——, Gowers-Maurey 235
——, Hagler-Odell 227
——, I 72,75
——, $\mathfrak{J}$ 29
——, J_n 29
——, J-sum - $J(X_n,\phi_n)$ 259
——, James - J 12
——, James-Orlicz - JO(X,M) 226
——, James tree - JT 136
——, James uniformly non-octahedral 232
——, $J(x_n)$ 221
——, $\mathcal{J}(x_n)$ 224

——, JF 220
——, JH 228
——, JT_∞ 227
——, $JT(x_i)$ 229
——, long James - $J(\eta)$ 225
——, long James sum - $J(\eta,X)$ 225
——, Polish 218,219
——, Schlumprecht 235
——, Schreier - S 231
——, Schreier-Orlicz - S_ϕ 232
——, Schur 228
——, tree-like Tsirelson - ST_p 229
——, Tsirelson - T 222,233
——, Tsirelson-James - $T\mathcal{J}$ 222
spreading model 94,100,101
strongly disjoint intervals 75
strongly disjoint (S.D.) step-function 75
subtree 136
superreflexivity 35,36

tree 136
——, binary 135
type 103,104

uniform convexifiability 35

vertex 135

weakly compactly generated (WCG) 178,179